Geometry in the Mathematics Curriculum

Yearbooks
published by the
National Council of Teachers of Mathematics

First Yearbook: *A General Survey of Progress in the Last Twenty-five Years*
Second Yearbook: *Curriculum Problems in Teaching Mathematics*
Third Yearbook: *Selected Topics in the Teaching of Mathematics*
Fourth Yearbook: *Significant Changes and Trends in the Teaching of Mathematics throughout the World since 1910*
Fifth Yearbook: *The Teaching of Geometry*
Sixth Yearbook: *Mathematics in Modern Life*
Seventh Yearbook: *The Teaching of Algebra*
Eighth Yearbook: *The Teaching of Mathematics in the Secondary School*
Ninth Yearbook: *Relational and Functional Thinking in Mathematics*
Tenth Yearbook: *The Teaching of Arithmetic*
Eleventh Yearbook: *The Place of Mathematics in Modern Education*
Twelfth Yearbook: *Approximate Computation*
Thirteenth Yearbook: *The Nature of Proof*
Fourteenth Yearbook: *The Training of Mathematics Teachers*
Fifteenth Yearbook: *The Place of Mathematics in Secondary Education*
Sixteenth Yearbook: *Arithmetic in General Education*
Seventeenth Yearbook: *A Source Book of Mathematical Applications*
Eighteenth Yearbook: *Multi-Sensory Aids in the Teaching of Mathematics*
Nineteenth Yearbook: *Surveying Instruments: Their History and Classroom Use*
Twentieth Yearbook: *The Metric System of Weights and Measures*
Twenty-first Yearbook: *The Learning of Mathematics: Its Theory and Practice*
Twenty-second Yearbook: *Emerging Practices in Mathematics Education*
Twenty-third Yearbook: *Insights into Modern Mathematics*
Twenty-fourth Yearbook: *The Growth of Mathematical Ideas, Grades K–12*
Twenty-fifth Yearbook: *Instruction in Arithmetic*
Twenty-sixth Yearbook: *Evaluation in Mathematics*
Twenty-seventh Yearbook: *Enrichment Mathematics for the Grades*
Twenty-eighth Yearbook: *Enrichment Mathematics for High School*
Twenty-ninth Yearbook: *Topics in Mathematics for Elementary School Teachers*
Thirtieth Yearbook: *More Topics in Mathematics for Elementary School Teachers*
Thirty-first Yearbook: *Historical Topics for the Mathematics Classroom*
Thirty-second Yearbook: *A History of Mathematics Education in the United States and Canada*
Thirty-third Yearbook: *The Teaching of Secondary School Mathematics*
Thirty-fourth Yearbook: *Instructional Aids in Mathematics*
Thirty-fifth Yearbook: *The Slow Learner in Mathematics*
Thirty-sixth Yearbook: *Geometry in the Mathematics Curriculum*

Geometry in the Mathematics Curriculum

Thirty-sixth Yearbook

**National Council of
Teachers of Mathematics
1973**

Copyright © 1973 by
THE NATIONAL COUNCIL OF TEACHERS OF MATHEMATICS, INC.
1906 Association Drive, Reston, Virginia 22091

Library of Congress Cataloging in Publication Data:

National Council of Teachers of Mathematics.
  Geometry in the mathematics curriculum.

  (Its Yearbook, 36th)
  Includes bibliographies.
  1. Geometry—Study and teaching. I. Title.
II. Series.
QA1.N3 36th [QA461]   375'.51'6   73-16458

Printed in the United States of America

Contents

Preface vii

Introduction

1. Disparities in Viewing Geometry 3
   James F. Ulrich, Arlington High School, Arlington Heights,
   Illinois

Part One: Informal Geometry

2. Informal Geometry in Grades K–6 11
   Paul R. Trafton, Indiana University, Bloomington, Indiana
   John F. LeBlanc, Indiana University, Bloomington, Indiana

3. Informal Geometry in Grades 7–14 52
   John C. Peterson, Eastern Illinois University, Charleston, Illinois

Part Two: Formal Geometry in the Senior High School

4. Conventional Approaches Using Synthetic Euclidean
   Geometry 95
   Charles Brumfiel, The University of Michigan, Ann Arbor,
   Michigan

5. Approaches Using Coordinates 116
   Lawrence A. Ringenberg, Eastern Illinois University, Charleston,
   Illinois

v

6. A Transformation Approach to Euclidean Geometry 136
    Arthur F. Coxford, Jr., The University of Michigan, Ann Arbor,
    Michigan

7. An Affine Approach to Euclidean Geometry 201
    C. Ray Wylie, Furman University, Greenville, South Carolina

8. A Vector Approach to Euclidean Geometry 232
    Steven Szabo, University of Illinois at Urbana-Champaign,
    Urbana, Illinois

9. Geometry in an Integrated Program 303
    Harry Sitomer, C. W. Post College, Greenvale, New York

10. An Eclectic Program in Geometry 334
    Jack E. Forbes, Purdue University, Calumet Campus, Hammond,
    Indiana

Part Three: Contemporary Views of Geometry

11. Geometry as a Secondary School Subject 369
    Howard F. Fehr, Teachers College, Columbia University, New
    York, New York

12. An Evolutionary View 381
    Seymour Schuster, Carleton College, Northfield, Minnesota

13. A Quick Trip through Modern Geometry, with Implications
    for School Curricula 397
    Ross L. Finney, University of Illinois at Urbana-Champaign,
    Urbana, Illinois

Part Four: The Education of Teachers

14. The Education of Elementary School Teachers in Geometry.... 435
    Henry Van Engen, University of Wisconsin, Madison, Wisconsin

15. The Education of Secondary School Teachers in Geometry ... 446
    F. Joe Crosswhite, The Ohio State University, Columbus, Ohio

Preface

The suggestion that a yearbook on the teaching of geometry be considered was advanced by the Yearbook Planning Committee of the NCTM. This seemed quite appropriate in light of the continued interest in curriculum development in mathematics. That there was general interest in the particular case of geometry might be inferred from the fact that the Mathematical Association of America at its Houston meeting in 1969 conducted a conference on the teaching of this subject. In addition, many articles on the teaching of geometry were appearing. These articles reflected the variety of opinions concerning how geometry might best be taught. One conclusion from the discussions and writings was that there was no consensus.

The Yearbook Planning Committee felt that a contribution could be made by preparing a yearbook in which the various points of view were presented as clearly as possible with illustrations of how geometry might be taught under each point of view. It then secured authorization and financial support from the Board of Directors for such a yearbook. An editor and editorial committee were appointed in 1969 and directed to plan the scope and content of the yearbook.

At its first meeting early in 1970, the Editorial Committee developed a prospectus, including the rationale of the yearbook and the scope and sequence of possible chapters. This prospectus was sent to some thirty-six individuals—classroom teachers, mathematicians, and mathematics educators—for criticism and suggestions. At its second meeting, the prospectus was revised on the basis of the suggestions received. A list of possible authors was developed with the expectation that the authors would prepare the first draft of their chapters by February 1971.

As the first drafts were received, they were duplicated by the Headquarters Office of the NCTM, and a copy of every chapter was sent to each author. This was done because a conference of the authors was planned for April 1971 to consider the suggestions provided by each author and to coordinate the chapters. Following this conference, each author, or pair of authors, submitted the final draft of his chapter.

The editor would like to express his appreciation for the advice and help so generously provided by the other members of the Editorial Committee: Arthur F. Coxford, Lawrence A. Ringenberg, and James F. Ulrich. He would also like to thank Charles Hucka, associate executive secretary of the Council, and his competent editorial assistants, Charles Clements, David Roach, Dorothy Broderick, Wanda Stepanek, and Deborah Mullins. They took the burden of details off the shoulders of the editor. Finally, to the writers of the chapters—the ones who supplied the ideas—both the editor and those who profit from this yearbook are indebted.

Grateful acknowledgment is made for permission to reprint extracts and reproduce figures from the copyrighted sources named below. Individual articles are listed under the name of the journal or book in which each appears. (More complete bibliographical information appears in the reference lists at the ends of the chapters.)

Annals of Mathematics. "Some Wild Cells and Spheres in Three-Dimensional State," by Ralph H. Fox and Emil Artin.

Geometry with Coordinates, by the School Mathematics Study Group.

Guidelines and Standards for the Education of Secondary School Teachers of Science and Mathematics, by the American Association for the Advancement of Science.

Mathematics Education. "Teacher Education," by Roy Dubisch. Reprinted by permission of the National Society for the Study of Education.

The Reorganization of Mathematics in Secondary Education, by the National Committee on Mathematical Requirements of the Mathematical Association of America. Reprinted by permission of the MAA.

Kenneth B. Henderson, *Editor*

Introduction

1

Disparities
in Viewing Geometry

JAMES ULRICH

CURRICULUM reform in mathematics has made generally good progress; however, about geometry in particular, there is less than complete satisfaction. In a backhanded way, geometry itself is at fault because it is so all-embracing, so ubiquitous, so important, and so capable of being interpreted in many different ways. Geometry can be taught intuitively or rigorously or by methods that lie somewhere between these extremes; to the learner, the study of geometry can be dull or exciting or somewhere in between. The mathematician whose specialty is algebra sees geometry as a vehicle for algebraic expression; if the mathematician's specialty is vectors, he believes that vectors should be used to interpret geometry. The theory-oriented mathematician views geometry as an excellent example of a group. The teacher who believes mathematics should be taught as it was developed historically by man will prefer Euclidean geometry; those who believe that mathematics education should exploit the most recent and modern advances in mathematics will favor an approach that features coordinates, vectors, or transformations.

In view of these obvious disparities in points of view that relate to both content and method, it is not surprising that there is a lack of consensus concerning the nature of geometry and the manner in which it should be taught. In grades 1 through 14, algebra and geometry essentially provide two different viewpoints of the same subject matter. There are occasions at

3

the secondary level when they are presented together because they support and reinforce each other so beautifully. For example, the solution of a system of two independent equations in two variables corresponds to the geometric figure in which two intersecting lines have a unique common point. Each equation is the algebraic counterpart of one of the given lines, and the ordered pair of numbers that is the solution is an algebraic way of specifying the point. The study of the conic sections serves as another example of the way in which algebra and geometry can work together to elucidate the topic under scrutiny. In these examples the algebra and the geometry of the given concept can be dealt with simultaneously by the student. Owing to the nature of each subject and to the level of sophistication and maturity of the student, this cannot always be done. A third-grade student has no difficulty recognizing the geometric figure called a circle, yet this student ought not to be exposed to the equation $x^2 + y^2 = r^2$, which describes the circle algebraically. A student can grasp the meaning of the concept of the distance from a given line to a point not on the line long before he can understand and use the algebraic formula for this concept. For some concepts this order is reversed: at an early age a student knows that there are many, many pairs of numbers he can add to get the sum of 20, but a few years pass before he knows that when these number pairs are plotted, the line with equation $x + y = 20$ results.

The foregoing remarks are not intended to imply that by some arbitrary age of the student each concept ought to be developed fully *both* algebraically and geometrically. There are times when the algebraic approach to a problem is far superior to the geometric approach, and vice versa. When this occurs, there is usually no question that the better, perhaps the more efficient in terms of time and space, approach should be used. For example, suppose the question is this: For what real numbers x is the sentence $x^2 - x - 1 = x^2 - 2x + 2$ true? The typical algebraic solution is as follows:

(1) $$x^2 - x - 1 = x^2 - 2x + 2$$
(2) $$-x - 1 = -2x + 2 \qquad \text{subtracting } x^2$$
(3) $$x - 1 = 2 \qquad \text{adding } 2x$$
(4) $$x = 3 \qquad \text{adding } 1$$

The four steps in the solution are not difficult and are straightforward. The solution by the methods of coordinate geometry is quite another story. The left side of (1) is the function $x^2 - x - 1$, whose graph is a parabola; the right side, $x^2 - 2x + 2$, is also a parabola. The graphs of these two parabolas intersect at the point whose abscissa is 3. Each side of (2) is also a function, and the graphs of these lines intersect at the point whose abscissa is 3. Similar statements can be made about (3) and (4). Thus the geometric solution that parallels the algebraic solution consists of four pairs of graphs, and for each pair the intersection of the curves occurs at the point whose abscissa is 3. Although it is probably worthwhile for students to be made aware of this geometric approach during the study of secondary mathematics, it can hardly be recommended as the most efficient method for general use. The important point here is that one approach is arbitrarily emphasized to the exclusion of the other as a matter of expediency. Teachers apparently do not question the legitimacy of these arbitrary decisions, for they appear to be decisions that are pedagogically sensible.

What has been said thus far can be condensed to something like the following: A given mathematical concept may be treated algebraically or geometrically; for some concepts the algebra precedes the geometry, and for others the order is reversed. Certain concepts lend themselves so nicely to the one approach that the other approach is virtually ignored. Years of experience (or is it tradition?) have sorted and sifted the various concepts and linked them with one approach or the other. The results are found in the textbook materials in use today.

If two additional approaches, that of vectors and that of transformations, are added to the algebraic and geometric approaches, the situation becomes somewhat more complex. That these four approaches—algebraic, geometric, vector, and transformational—are mutually exclusive is an oversimplification, of course, but they are part of today's mathematics, nevertheless. Other approaches could also be mentioned, but they only complicate the picture. It is reasonable to expect that the earlier remarks concerning the tug-of-war between an algebraic and a geometric approach to the teaching of a concept may be escalated to a four-way contest whose participants are the proponents of algebra, geometry, vectors, and transformations. It is of some interest to note that in the historical development of mathematics, vectors and transformations are relatively recent creations. This may be part of the reason why they do not have "equal time" in comparison to algebra and geometry, which have been around much longer.

When a mathematician attacks a problem, he will use the approach that suits his purposes best. His preparation and prior experience equip him with many techniques and methods, and from these he may choose those that will lead to the desired solution. Do we want students to act like amateur mathematicians? If we do, we should see to it that they are exposed to several approaches so that they will have options from which to choose the most suitable method.

Many of the foregoing paragraphs raise important questions for the curriculum specialist, the textbook writer, the teacher, and the person who is to train teachers. The reader probably has thought of some of these questions already. The authors of this book are aware of these questions and consider them in the chapters that follow. As the reader deliberates on these questions, he should perhaps be reminded initially that he is not likely to find one single set of clearly defined answers that will tell him exactly what to do and how to do it. What he will find here is a careful exposition of the major issues surrounding the teaching of geometry together with some suggested alternatives that he may weigh and employ as he sees fit. The inspired and creative teacher prefers, not to be told exactly what to do, but to be given sufficient background information and knowledge so that he may be selective in determining the direction that to him seems best. It is the purpose of this yearbook to describe alternative approaches to geometry, to inform the reader about some of the newer programs and approaches that have been tried to date, to indicate the directions that future developments may include, and to provide helpful suggestions for the education of teachers of geometry.

Organization of the Yearbook

The yearbook is divided into the following four parts:
  Part I: Informal Geometry
  Part II: Formal Geometry in the Senior High School
  Part III: Contemporary Views of Geometry
  Part IV: The Education of Teachers
Chapters 2 and 3 constitute Part I. Some students may not have the opportunity or the ability to study formal geometry. What kind of geometry should they study? What approaches should be used for them? It is recognized that the geometry of grades K through 9 should be intuitive and informal, but what content should be included? When should it be taught? And how should it be taught? Chapters 2 and 3 outline the objectives of informal geometry, describe its current status, and recommend some of the most promising ways to be used in teaching it.

Part II consists of chapters 4 through 10. They address themselves to the various approaches that may be used to present formal geometry in the secondary school. Each of these chapters describes the rationale of the approach, its philosophy and objectives, the kinds of students for whom the approach is most appropriate, the possible scope and content of a course that uses the particular approach, and special considerations and variations related to the approach.

In Part III, chapters 11 through 13, each of three mathematicians explains his conception of contemporary geometry and includes the underlying theory, the kinds of problems being investigated, and the implications for school geometry that may result.

The last two chapters, 14 and 15, make up Part IV of the yearbook. The current status of the education of teachers is reviewed, comments concerning experimental or innovative programs are given, and recommendations for the improvement of the training of teachers of geometry are included.

Uses of the Yearbook

This book may be used both by preservice students who are preparing to teach and by persons already engaged in teaching. If the persons studying this book are members of a class, the instructor may want to collect as many of the references listed in the bibliographies as he can. As his students proceed through this book, they will profit much by turning to the references as often as necessary.

For persons teaching, or preparing to teach, in grades K through 8, Part I of the book is of major importance; yet these persons should not be unaware of the material in Part II. Although the main concentration may be on Part I, Part II has much to offer those who teach the elementary and junior high grades.

Preservice and in-service education courses for secondary school teachers should include a thorough coverage of Parts I and II. It is important that secondary school teachers be knowledgeable about the geometry their students have been taught in grades K through 8, for it is on this foundation that the geometry of the senior high school must be built.

Finally, every mathematics teacher who wishes to keep abreast of the important developments in his field will find the yearbook of considerable merit as a resource book.

PART

Informal Geometry

2

Informal Geometry in Grades K-6

PAUL R. TRAFTON

JOHN F. LeBLANC

THE INCLUSION of a substantial amount of informal geometry has been identified as one major characteristic of the "modern mathematics" movement at the elementary school level. The geometry of the elementary school is frequently characterized as *informal* geometry to distinguish it from the familiar secondary school *formal* geometry with its emphasis on proof. Among the terms that have been associated with elementary school geometry are *point set* geometry and *nonmetric* geometry. The former term is suggested by the fact that geometry involves the study of sets of points in space. The term nonmetric has been used to highlight the focus on geometric ideas, figures, and relations apart from measurement aspects, which have traditionally received some emphasis in the curriculum.

Within a decade the area of geometry has become accepted by curriculum planners and, increasingly, by classroom teachers as a necessary and integral dimension of the elementary school mathematics curriculum. Although the idea of including geometry is well accepted, a wide divergence in thinking exists regarding the particulars of what constitutes an appropriate program for boys and girls. There is general agreement that elementary school pupils find an exposure to geometry a pleasurable experience. Children's natural affinity for geometry has stimulated interest in how children formulate geometric ideas. The development of geometric

11

thinking has its roots in the very early contact of children with an inherently geometric environment. The study of children's drawings through the early childhood years is one approach that has been used to provide clues about how the child's perception of his environment develops.

This chapter presents the development and status of geometry as a necessary element of an effective mathematics curriculum in the elementary school. A discussion of the past and present influences that have resulted in the current position of geometry is followed by a description of the content that in 1970 was commonly included in the school curriculum through grade 6. The chapter continues with a discussion of possible contributions of geometry to pupils' learning, together with guidelines for maximizing the effectiveness of a study of geometry. Finally, the issue of creating an adequate framework for guiding the selection of learning experiences in the continuing development of the geometry program is presented.

The Development of Geometry
in the Elementary School Curriculum

Historical background

It was about 1800 when Lacroix said:

> Geometry is possibly of all the branches of mathematics that which should be understood first. It seems to me a subject well adapted to interest children, provided it is presented to them chiefly with respect to its applications. . . . The operations of drawing and measuring cannot fail to be pleasant, leading them, as by the hand, to the science of reasoning. [13, p. 240]

In 1890 a committee of the National Education Association concluded:

> Geometry should be taught in the elementary school, beginning with lessons in form, and progressing through a course of instruction which will furnish the graduate of the common school with such geometrical concepts and facts as are needed by all. The course should be one of observation, construction, and representation, rather than of demonstration; yet the processes of training in logical thought and expression through form, will lead to practical demonstration. [7, p. 351]

As the quotations indicate, informal geometry has long been recognized as an appropriate area of study for young children. Until the past decade, however, little geometry was included in the American elementary school curriculum, apart from traditional aspects of shape recognition and the arithmetic elements of measurement. Nevertheless, a rich and extensive body of literature exists on this topic. Dating back over 150 years, it documents the many attempts to build programs of informal geometry for elementary school pupils. Not only was the learning of geometry prominent in the thinking of those primarily concerned with the mathematics curriculum; it was also considered important by those concerned with the nature of elementary education. (For a comprehensive survey of the development of informal geometry as well as further documentation of the historical background in this section, the interested reader is referred to the works of Coleman, Betz, Reeve, and Shibli cited in the references at the end of this chapter.)

The case for informal geometry in the elementary school has been justified from three different, yet at times related, positions. The first of these is the mathematical viewpoint. During much of the past 150 years, formal geometry has held an accepted position in the high school curriculum. As a result, informal geometry, consisting of the presentation of geometric facts and relationships in an exploratory and intuitive manner, came to be viewed as a means for providing a more effective foundation or preparation for demonstrative geometry. Although this position more directly influenced the junior high school program in the first half of the twentieth century, the elementary school program was at times influenced also.

A second justification was presented in terms of the contributions of geometry in meeting the general educational goals for all children, apart from any future mathematical value. Reforms in elementary education in the 1800s emphasized psychological principles of learning in the development of educational goals and programs for children. Learning came to be viewed through the eyes of the child, and the importance of drawing learning experiences from, and relating them to, the physical world was recognized. The learning of geometry fit well into the general methodologies being recommended. One emphasis growing out of this position stressed the close relationship between geometry and the natural world. Some advocated that informal geometry should be the "geometry of everyday life" and that geometry must "flow from nature and return to it."

A third justification for informal geometry was promoted by those who stressed the practical application of mathematical ideas and techniques in science and technology. The emphasis on shape, size, and position in informal geometry was seen to provide a foundation for later experiences in industry. Thus, early geometry instruction in the elementary school as well as in the later grades frequently became interwoven with training in industrial arts and instruction in mechanical drawing.

Mathematics education has frequently been concerned with maintaining a balance among the needs of the child, the needs of society, and the needs of the discipline. Such is the case in the development of informal geometry. The literature reveals the struggle over the relative importance of the psychological, practical, and mathematical aspects of learning geometry and the corresponding difficulty of developing programs that reflected a balance. Issues debated have included the content of informal geometry, initial approaches to geometry, the sequencing of instruction, and the appropriate emphasis of instruction. The long list of adjectives applied at various times to initial work with geometry reflects the lack of consensus. *Intuitive* is but one of the descriptions applied to early work. Others include *concrete, observational, inventional, mensurational, experimental, constructive, propædeutic, preparatory, empiric,* and *informal* (1, p. 55). Although similarity exists among these terms, many of them reflect different emphases in the early work.

Geometry in European elementary education

A major emphasis on the general educational values of geometry is found in the pioneering works of Pestalozzi, Herbart, and Froebel in the early part of the nineteenth century. Their contributions in revolutionizing elementary education through emphasizing the role of the learner and psychological principles of learning are well recognized. Thus it is of particular interest that geometric ideas were integrally involved in their approaches to learning. For Pestalozzi, geometric figures were viewed as a principal means of developing the mental faculty of sense perception, which he considered to be the basis of learning. Herbart accepted much of Pestalozzi's work with the initial stages of learning but was also concerned with extending the work to older children, recognizing the importance of geometric learning throughout a child's education.

The approach used by both Pestalozzi and Herbart was a synthetic one. They started with basic elements of plane geometry and combined them in elaborate and often pedantic detail to build more complex notions. Although the geometry work of Pestalozzi and Herbart contained many defects, it nonetheless had significance for the future development of informal geometry in the elementary school. Not only did they recognize the contribution of geometry to general educational goals, but their work served as a strong stimulus to their followers. Throughout the nineteenth century and early part of the twentieth century, there was great interest in developing geometry programs for elementary schools. Detailed geometry programs of widely diverse natures were written and implemented in the schools. The impact of this early work is revealed by surveys of mathematics curricula in foreign countries in the first part of the present century. These surveys indicated that a substantial amount of geometry was included in the elementary schools of several countries. (See 9.)

Like Pestalozzi, Froebel was primarily concerned with the initial levels of learning, thereby becoming identified with the kindergarten movement. His major contribution lay in recognizing the importance of play and self-activity in children's learning. Geometry had a central role in his programs also; however, his approach to it differed greatly from that of Pestalozzi and Herbart. In contrast to their simple-to-complex, or synthetic, approach, Froebel began with solid shapes, such as the cube and the sphere, moved to the point, and finally returned to solids again. His ideas were worked out in a highly imaginative set of structured materials presented as a series of thirteen "gifts" and ten "occupations." Like the work of Pestalozzi and Herbart, Froebel's ideas had great influence on the work of others both in Europe and in the United States.

Geometry in American elementary education

Interest in systematic instruction in informal geometry in American elementary education can be traced to the mid-1800s and was an outgrowth of the work done in German elementary schools. An early text was that of Thomas Hill, whose *First Lessons in Geometry,* first published in 1854, presented a thorough development of informal geometry for children of from six to twelve years of age. The following remarks from the preface of the book provide insight into Hill's purpose and remain of interest to the present day:

> I have addressed the child's imagination, rather than his reason, because I wished to teach him to conceive of forms. The child's powers of sensation are developed, before his powers of conception, and these before his reasoning powers. . . . I have, therefore, avoided reasoning, and simply given interesting geometric facts, fitted, I hope, to arouse a child to the observation of phenomena, and to the perception of forms as real entities. [10, p. 10]

Several other texts also appeared in the latter half of the nineteenth century and stimulated discussion about teaching geometry to young children. Despite the interest, however, geometry as such was not found on a widespread basis in elementary school programs, and the geometry that was included was frequently interwoven with other instruction. Coleman reports four sources of instruction in informal geometry during the latter half of the nineteenth century. They were in (1) form study related to object-teaching based on Pestalozzian ideas; (2) drawing instruction, which also had its roots in Pestalozzian ideas; (3) workshop or industrial arts instruction; and (4) kindergarten, where the ideas of Froebel were influential (4, p. 105).

The period of years beginning in 1890 and extending into the 1920s was one in which the necessity for reform in mathematics education was recognized. Instruction in informal geometry was viewed as an important element in this reform, as it has been in the reforms of our present era.

The Report of the National Education Association Committee on Elementary Education, published in 1890 and cited earlier (see 7), was one of the first reports to recognize the importance of geometry instruction based on the general educational goals for all pupils. The Report of the Committee of Ten on Secondary School Studies, published in 1893, not only represents an important document in the development of mathematics education but also represents a most significant and comprehensive statement on instruction in informal geometry. Because of the completeness and foresightedness of this report, the entire section on concrete geometry is given here:

The Conference recommends that the child's geometrical education should begin as early as possible; in the kindergarten, if he attends a kindergarten, or if not, in the primary school. He should at first gain familiarity through the senses with simple geometrical figures and forms, plane and solid; should handle, draw, measure, and model them; and should gradually learn some of their simpler properties and relations. It is the opinion of the Conference that in the early years of the primary school this work could be done in connection with the regular courses in drawing and modelling without requiring any important modification of the school curriculum.

At about the age of ten for the average child, systematic instruction in concrete or experimental geometry should begin, and should occupy about one school hour per week for at least three years. During this period the main facts of plane and solid geometry should be taught, not as an exercise in logical deduction and exact demonstration, but in as concrete and objective a form as possible. For example, the simple properties of similar plane figures and similar solids should not be proved, but should be illustrated and confirmed by cutting up and re-arranging drawings or models.

This course should include among other things the careful construction of plane figures, both by the unaided eye and by the aid of ruler, compasses and protractor; the indirect measurement of heights and distances by the aid of figures carefully drawn to scale; and elementary mensuration, plane and solid.

The child should learn to estimate by the eye and to measure with some degree of accuracy the lengths of lines, the magnitudes of angles, and the areas of simple plane figures; to make accurate plans and maps from his own actual measurements and estimates; and to make models of simple geometrical solids in pasteboard and in clay.

Of course, while no attempt should be made to build up a complete logical system of geometry, the child should be thoroughly convinced of the correctness of his constructions and the truth of his propositions by abundant concrete illustrations and by frequent experimental tests; and from the beginning of the systematic work he should be encouraged to draw easy inferences, and to follow short chains of reasoning.

From the outset the pupil should be required to express himself verbally as well as by drawing and modelling, and the language employed should be, as far as possible, the language of the science, and not a temporary phraseology to be unlearned later.

It is the belief of the Conference that the course here suggested, if skilfully taught, will not only be of great educational value to all children, but will also be a most desirable preparation for later mathematical work.

Then, too, while it will on one side supplement and aid the work in arithmetic, it will on the other side fit in with and help the elementary instruction in physics, if such instruction is to be given. [8, pp. 110–11]

In the report, both the general educational goals of informal geometry and the importance of continuity in geometry instruction as a preparation for later mathematical study are supported. The recognition by the Committee of dual purposes of early instruction was important and helped establish both aspects of instruction, a consideration that is recognized today as well.

The role of informal geometry was recognized in many of the individual and group reports that followed these early documents. The address of E. H. Moore to the American Mathematical Society in 1902 has long held an important place in the history of mathematics education. Moore, an eminent research mathematician, addressed a portion of his remarks on reform to the elementary school, calling for an enriched and vitalized approach to the methods and materials of mathematics in primary education. He states:

> Would it not be possible for the children in the grades to be trained in power of observation and experiment and reflection and deduction so that always their mathematics should be directly connected with matters of thoroughly concrete character? [6, p. 45]

He later adds:

> In particular, the grade teachers must make wiser use of the foundations furnished by the kindergarten. The drawing and the paper folding must lead on directly to systematic study of intuitional geometry, . . . including the construction of models and the elements of mechanical drawing, with simple exercises in geometrical reasoning. The geometry must be closely connected with the numerical and literal arithmetic. The cross-grooved tables of the kindergarten furnish an especially important type of connection, viz., a conventional graphical depiction of any phenomenon in which one magnitude depends upon another. These tables and the similar cross-section blackboards and paper must enter largely into all the mathematics of the grades. [6, pp. 45–46]

One notable emphasis in Moore's remarks is the relationship between geometry and arithmetic as represented through graphing.

Strong support for work in informal geometry in the elementary school is found in the Final Report of the National Education Association Committee of Fifteen on Geometry. In a section of the report devoted to this topic, they state:

> It is of the utmost importance that some work in geometry be done in the graded schools. For this there are at least two very strong reasons. In the first place, geometric forms certainly enter into the life of every child in the grades. The subject matter of geometry is therefore particularly suitable for instruction in such schools.
>
> Moreover, the motive for such teaching is direct. The ability to control geometric forms is unquestionably a real need in the life of every individual even as early as the graded school. For those who cannot proceed further this need is pressing. . . . For those who are going on to the high school, the development of the appreciation of geometric forms is almost an absolute prerequisite for any future work in geometry. [5, p. 89]

In the report it is evident that dual purposes of instruction are again stressed. The Committee observed that the initial work would likely be so informal that a separate course in geometry would not be required until the final grades of grammar school. In discussing the organization of content, the Committee recommended that solid shapes be emphasized before plane shapes, since the former were more real to pupils and could be presented more concretely.

Although the reform movement resulted in an awareness of the need to have a geometric emphasis in the mathematics program of the elementary school and stimulated thinking on this topic, evidence from curriculum surveys and other sources indicated no widespread implementation of the many proposals. In 1933 Betz wrote:

> For more than seventy years the attempt has been made by American teachers and authors to secure a more adequate provision in the elementary mathematical curriculum for systematic instruction in geometry. And yet, aside from the emphasis put on the fundamental rules of mensuration in the traditional arithmetic course, very little progress was noticeable until quite recently. [1, p. 57]

The progress to which Betz referred was the growth of the junior high school movement. It was at this level of instruction that some informal geometry gained a foothold in the curriculum, and little further discussion appears on informal geometry below grade 7. Among reasons cited for the lack of acceptance of geometry in the elementary school curriculum were (1) the overcrowding of the curriculum, together with resistance to replacing arithmetic topics with instruction in geometry; (2) the inadequate preparation of teachers to teach geometry; (3) the inclusion of some geometric work in art and shop work; and (4) the assertion of some that informal geometry would lessen pupils' interest in "real" geometry in high school.

Thus, despite the realization of the importance of informal geometry in the learning of children and the efforts of many individuals and organizations over a long span of years, it was not until the present age of reform that the recognized goal of systematic, informal geometry programs in the elementary school curriculum became a reality in American schools.

Recent developments

The years 1955 to 1970 are recognized as a major reform period in school mathematics. These were exciting years for mathematics educators, for not only was the attention of mathematics educators and educators in general focused on curriculum revision, but even the interest and concern of the public was centered on these efforts for reform. New topics and different emphases were brought into the curriculum at this time. Informal geometry formed an integral element of this reform. It seems as if the cumulative efforts and beliefs of mathematicians and educators over the previous 150 years to have geometry play an important role in the elementary school program would be realized.

Indeed, by 1970 geometry seemed to have gained a permanent place in the elementary school mathematics program. Though topics such as numeration in bases other than ten, which received heavy emphasis in the early days of the school mathematics reform period, had lost some of their support by 1970, the emphasis on informal geometry has grown and developed into a securely held position in the elementary school mathematics program. Even today, although informal geometry is presently firmly established in the curriculum, there is no general agreement among mathematics educators on what content should be taught. However, active and continuing dialogue continues on this issue as well as the issue of how the content should be developed.

Several forces shaped the direction of the program in informal geometry during these years. It is difficult to cite any one source as having greater influence than another. Among the forces that made an impact on elementary school geometry were curriculum development projects, professional meetings, journal articles, in-service education, and national conferences. Undoubtedly another force at work operating during the reform period was the historical weight of influence on the importance of informal geometry in the elementary school. The reform atmosphere and attitude provided an opportunity to implement a geometry program in the elementary school.

The efforts of experimental and developmental projects concerned with informal geometry at the elementary school level took many directions. Prominent among the groups that influenced the direction of elementary school geometry was the School Mathematics Study Group (SMSG). The first significant impact of SMSG curriculum related to geometry was the amount and scope of the geometry. A second note of significance in SMSG's curriculum in geometry was the structure developed in the program. This structure was essentially an informal development of Euclidean geometry. The elements of Euclidean geometry were presented in an intuitive but logically ordered development. The axioms and theorems were not explicitly stated, nor was the rigor of the traditional "statement-proof" of a typical program in geometry contained. Many familiar theorems in Euclidean geometry were presented with informal statements of justification.

In 1970, the directors of SMSG felt that there was a need and a demand for more specific attention to the geometry that they developed. Thus, they made the decision to edit and reorganize the geometry contained in Books 1–6 into a series of eleven units, called *Geometry Units for Elementary School*. The eleven units were these:

Unit 1:	Some Sets of Points
Unit 2:	Congruence
Unit 3:	Congruence and Familiar Geometric Figures
Unit 4:	Measurement of Curves (Length)
Unit 5:	Measurement of Plane Regions (Area)
Unit 6:	Measurement of Space Regions (Volume)
Unit 7:	Measurement of Angles
Unit 8:	Side and Angle Relationships for Triangles
Unit 9:	Circles and Constructions
Unit 10:	Whole Numbers as Coordinates of Points
Unit 11:	Integers as Coordinates of Points

No particular grade level was associated with a given unit. All units were considered appropriate for the intermediate grades, and at least portions of the first several units in the series were appropriate for the primary grades.

Efforts to develop a program in elementary school geometry were also made by other individuals and groups. Not all these efforts centered on an informal or intuitive development of Euclidean geometry. Some efforts were specifically directed at primary-grade children; others were aimed at intermediate-grade pupils.

One project in geometry for the primary grades made heavy use of ruler and compass. It was evident to the developers that young children could learn some concepts related to the Euclidean plane through construction activities. In another project some elements of motion geometry were introduced. Again, the developers reported that intermediate-grade pupils could indeed capture some intuitive notions of transformations. Research and experimental projects rarely reported that children could not understand the concepts of geometry that were developed in the project. The clamor of voices from these projects claiming that children could learn this idea and that idea left a residual question on the minds of responsible mathematics educators: What *should* be taught in elementary school geometry?

Another force of significance for the inclusion of geometry was the training of preservice teachers and the retraining of teachers through in-service institutes. These programs reflected the conviction that geometry ought to be part of the elementary school program. At first the preservice training of teachers began to include a unit or two related to geometry. Now a full course, generally titled "Geometry for Elementary School Teachers," is offered and even required for elementary education majors in several teacher-training institutions. Even the minimal preparation of elementary school teachers includes some geometry.

The emphasis on in-service training that occurred during the period of reform was, in itself, unprecedented. The motivating force for in-service work was often simple. Quite often programs adopted by local school districts contained content that teachers did not understand. A substantially increased amount of geometry had been included in these new programs. The in-service training sessions facilitated the implementation of the geometry program in the elementary school. The contribution of conventional textbook programs as an influencing force on the growth of elementary school geometry must also be recognized. Although curriculum projects influenced the thinking of educators, textbooks had an immediate impact on the program in geometry for a wide segment of the population.

During the initial stages of the changing mathematics curriculum, notable differences existed in the amount and scope of geometry. Some textbooks contained only isolated ideas and topics in geometry; others made significant attempts to organize the child's learning. The success of the program in geometry as measured by pupils' and teachers' enthusiasm for the subject led to an increased acceptance of geometry in the elementary school. Textbook authors, on the whole, reflected this acceptance by giving more careful attention to the developmental sequence of the program in geometry.

National conferences were another force that influenced the development and retention of geometry in the elementary school. One such conference was the Cambridge Conference on School Mathematics held in 1963. This report called for geometry to be integrated into the elementary school program beginning in kindergarten. As stated in the report, aims for the study of geometry were "to develop the planar and spatial intuition of the pupil, to afford a source of visualization for arithmetic and algebra, and to serve as a model for that branch of natural science which investigates physical space by mathematical methods" (2, p. 33).

A sequel to the conference discussed above was the Cambridge Conference on Teacher Training. This conference took place in 1966 and the report is contained in *Goals for Mathematics Education of Elementary School Teachers*. There is considerable speculation over the immediate and future impact of these conferences. One fact from the report is evident—namely, that geometry should play an important role in an elementary school program.

Finally, the number of talks on elementary school geometry at professional conferences increased during this period. This increased interest in geometry can be interpreted in different ways. The fact that more speakers chose the topic is a testimony to the greater attention given to geometry. Again, this fact can be considered a force influencing the spread of geometry in the elementary school.

Summary

In this section the development and status of geometry as an aspect of the elementary school mathematics curriculum has been presented. The current geometry program has been viewed as the consequence of several forces of historical and recent origins. How the past decade will be assessed by future historians of mathematics education can only be speculated on. However, it is likely that the establishment of a geometry program for boys and girls will be seen as a major accomplishment of this period, for in less than a decade the geometry program of the elementary school has been transformed from one of minimal consideration to a program of broad scope and significance.

Content of Elementary School Geometry in 1970

By 1970, informal geometry was established in the curriculum of elementary school mathematics. Guidelines for elementary school mathematics programs published by state departments of public education contained specific sections related to geometry. Checklists for the selection of curriculum material published by various professional organizations listed important objectives related to geometry. Local curriculum guides reflected the belief that informal geometry had a place in the curriculum of the elementary school child.

Thus, the evidence from several sources indicated that geometry was firmly established in the curriculum, and yet questions remain even today: How much geometry was learned by the elementary school pupil in 1970? What was the typical program in elementary school geometry in 1970?

These questions are raised for several reasons. There seems to be a general opinion among mathematics educators that some elementary school teachers either omit geometry or introduce it only if there is extra time. The number of teachers who comprise this group is a matter of speculation. On the one hand it seems true that to some classroom teachers the teaching and learning of geometry is not an essential part of the "real" mathematics program. For this group of teachers, the teaching of mathematics centers on the acquisition of computational skills. On the other hand, other elementary classroom teachers are equally convinced of the importance of experiences in informal geometry.

What was the typical program in elementary school geometry in 1970? In some ways, this is difficult to assess, for many teachers always provide more learning experiences than those included in any planned curriculum. Since textbooks frequently do define the basic, or core, mathematics curriculum in the elementary school, a survey of textbooks used in 1970 provides a reasonable means of describing a typical geometry program. Even so, a summary of the geometry programs existing in 1970 still presents some difficulty.

If one attempted to summarize the *arithmetic* program by surveying various elementary school mathematics textbooks, the task would be relatively simple. Not only was there general agreement on the content of arithmetic, but there was even considerable agreement on the grade placement of several topics. Most textbooks presented, for example, the concept of adding whole numbers with renaming as a late second- or early third-grade topic. The multiplication algorithm for whole numbers was usually introduced in grade 3 and more extensively developed in grade 4.

The development of *geometry* as represented in texts, however, was an entirely different matter. Neither was there agreement on the scope of the content, nor was there agreement on grade placement of topics. The program in geometry depended on the text used in the school. This fact, coupled with the differing attitudes a teacher may have had toward the importance of geometry, makes the description of a typical program quite tenuous. The summary, then, of the geometry in the elementary school presented in this chapter should be clearly understood as an "averaging" approximation of what existed in the 1970 textbooks.

A general observation evident in looking at each series of textbooks was that the cumulative review of geometry from grade to grade was extensive. In some series the same set of terms and concepts was reviewed and redeveloped at several grade levels. It is tempting to speculate whether curriculum developers and textbook writers felt geometry might have been omitted for the pupils in previous years.

In consulting the table of contents, topics in geometry accounted for approximately 15 percent of the total number of pages in the textbook. In most of the textbooks, geometry was presented in units or chapters. The programs contained in different textbook series represented different emphases in approach to geometry. Some programs made serious attempts to sequence carefully the development of the concepts; some placed the emphasis on the metric aspects of geometry. Other programs stressed the special terminology and symbolism associated with geometry, whereas still others presented informal geometry as a series of nearly unrelated experiences.

For purposes of reference, a summary of the major topics in the geometry programs of 1970 is presented. Although the programs are found within textbooks for the individual grades, the summary will be presented in two sections: kindergarten through grade 2, and grade 3 through grade 6. The presentation is organized in this fashion to give some general perspective to the program, which would be missed in a grade-by-grade summary.

Kindergarten through grade 2

The development of geometry in the kindergarten did not vary greatly from program to program. The content at this level generally included identification and discrimination among shapes and forms such as circles, triangles, rectangles, and squares. Some programs placed emphasis on the terminology associated with the shapes; other programs developed lessons in which children identified objects about them that resembled a square or a rectangle or a circle. Geometric forms and shapes were also used in activities that were not specifically related to geometry content. For example, activities that involved finding a pattern in a sequence of drawings might use geometric shapes in the patterns; or a child might be asked to color the *triangles* red, the *circles* blue, and the *squares* yellow.

The geometry programs in grades 1 and 2 had wide variance. For example, one program had no pages in the pupils' text devoted to geometry—only a few suggestions in the teacher's edition of this program formed the basis of the geometry. Another program began grade 1 with lessons on geometry and devoted a substantial number of pages to it, developing some intuitive concepts of interior and exterior points of a region as well as betweenness of points on a curve.

Since most mathematics programs in the primary grades were in workbook form where few, if any, instructions to children were given, the child was quite dependent on his teacher for his mathematical growth. The decision to include or exclude geometry at this level was totally up to the teacher. In the upper grades a lesson was often developed in the text sufficiently enough for an interested student to study geometry on his own even if his teacher decided to omit the unit or chapter.

Although there was a great amount of variation in the programs in the primary grades, there were several common elements. First, geometry was approached from an informal point of view, providing the child with intuitive experiences. At this level there was generally little emphasis on precise terminology or, in some programs, on refinements of a concept. Second, the geometry program in grade 2 represented for the most part a review of the program in grade 1. This feature of review of the previous grade's content was present in all programs and was repeated at each grade level. Third, there was an observable trend in several of the programs at this level to introduce concepts related to point, line, and plane. These lessons provided a readiness for the more formal efforts in geometry reflected in grades 3 through 6.

The geometry program in the primary grades reflected an effort on the part of curriculum builders to introduce an awareness of terms and concepts. This effort served at least two purposes: first, to help the child articulate some of the observations he had made about the space around him; and second, to serve as a readiness for more formal developmental lessons in the upper grades. There was consensus among all the authors that the primary experiences in geometry ought to take place in a pleasant and informal atmosphere.

Grade 3 through grade 6

Several differences in the mathematics program for the upper grades were found. Whereas the programs in the primary grades were printed on consumable materials so that children could write and draw in the text itself, the program in the upper grades (beginning in grade 3) was contained in a hardback text. Another difference was the dramatic increase in the verbal level of the text. Instruction and directions for each lesson were addressed directly to the children in the pupils' text rather than being only in the teacher's edition. A third difference was the increased formality in the development of the geometry program. Terms and concepts that had been introduced in an intuitive manner in earlier grades were now in the upper grades more precisely approached. The use of mathematical symbolism was also substantially (if not needlessly) increased at this level. Finally, the scope and quantity of the geometry program typically was greater in the upper grades than in the primary grades.

Many of the programs started the upper-grade geometry program through a point-line-plane approach. After a brief discussion of the notion of a point, the concepts of path, line, line segment, and curve were introduced as connected sets of points. The distinctions among these concepts, together with descriptive definitions, were developed in varying degrees of rigor. Polygons were often introduced and defined using the concepts previously discussed. The notion of a plane, the regions of the plane, and the exterior and interior points in the plane were presented in conjunction with the concept of open and closed curves.

After this general introduction to point, line, and plane, the geometry programs were developed in several directions. Some presented lines and line segments, developing ideas about angles, perpendicularity, parallelism, and intersection, and then used these learning concepts to study the same concepts for lines and planes that intersected, were parallel, and so on. Other programs studied the circle and its properties, thereby introducing intersecting radii, diameters, and the angles formed. Still other programs seemed to lack inherent cohesiveness, developing without any apparent direction a lesson or two on angles and a lesson or two on closed figures.

By the end of grade 6, however, each program had included a point-line-plane approach to geometry. Several programs, in fact, reviewed or retaught the initial concepts of point, line, and plane at *each* grade level (3 through 6) and then proceeded to develop a new area of exploration. No matter what program was used, the child studied something about intersecting, perpendicular, and parallel lines; open and closed curves; circles and angles; congruent, similar, and nonsimilar polygons; and about any of the usual topics commonly associated with a point-line-plane development.

Solid geometry was developed to some extent in the upper grades in most programs. Often this development was stimulated by pictures of familiar objects and the association of the name with the geometric solid. Generally, the lessons on solid geometry were not well integrated into the other lessons on geometry. Often the pages on solid geometry were even separate from the other pages on geometry. The lack of direction in this phase was more evident than in the point-line-plane phase of geometry.

In the development of solid geometry, all programs included somewhere in the upper grades a discussion of the relationship of the number of edges, vertices, and faces. Most programs included lessons on the construction of various geometric solids from paper or cardboard. Some programs developed ideas relating to conic sections. Lessons as well as suggestions to teachers included some discussion of predicting what surface would result from cutting through a solid. Throughout the lessons and activities related to solid geometry was an attempt to use real-world situations.

Metric geometry was included in all the geometry programs. Some programs discussed the various measures of lines, regions, and solids as these figures were being discussed. The measurement of the area of a square or a rectangle, for instance, was developed as the quadrilateral was taught. In other programs, the metric aspects of figures were developed in a separate section devoted to measurement. But in whatever way metric geometry was developed, by the end of the sixth grade a pupil would have had lessons on the measurement of the perimeter, area, and volume of the standard, regular geometric figures. In several programs the pupil would even have explored ways of measuring irregular figures.

Coordinate geometry, like solid geometry, was generally developed as a separate topic. The locating and plotting of points in the first quadrant on a coordinate grid was the only common feature of all programs. Some programs in the later lessons of upper-grade geometry introduced the graphing of a linear function. In general, however, the coordinate geometry in the elementary school seemed to serve the purpose of familiarizing the pupil with the coordinate grid as a way of picturing points as locations.

In the upper-grade geometry programs of the late sixties, an attempt was made to approach the study of some phases of geometry in a different way. All the programs in the early sixties developed the concept of congruence from a traditional point of view (side-angle-side comparison of triangles, for example). In the early development, the figures remained static. In some of the later programs, congruence was approached using some ideas from transformational geometry. Geometric figures were not static and could be moved by rotation, translation, or reflection. Thus some programs claimed to present elements of motion, or transformational, geometry. (This idea will be discussed again in the last section of this chapter.)

Summary

Elementary school geometry in 1970 was a program in a state of flux, a program that was generally agreed to be evolving. The fact that geometry was, and should be, an integral part of elementary school mathematics was established; the purpose and direction of the program was not. Although a pupil would have had many common experiences in completing the sixth grade (point-line-plane geometry, solid geometry, metric geometry, coordinate geometry), the nature of these experiences varied greatly. The goals of the programs often accounted for the differences in the nature of the experiences.

Some programs indicated that geometry in the elementary school should prepare the pupil for later work in junior and senior high school, or even for college; some developed geometry simply as good mathematics, appropriate for any pupil. Other programs reflected the thinking that geometry offered an opportunity for the "process" goals of mathematics, emphasizing discovery and finding relations among geometric shapes; still others emphasized the relationship of geometry and the real world. Finally, some programs reflected the belief that geometry was to be enjoyed, providing interesting but unrelated experiences for the pupils. In general, however, most programs reflected several of the purposes listed. This mixture of purposes left the effect, in many cases, of a program without structure and without direction.

Contributions of Geometry to the Elementary School Curriculum

The acceptance of geometry as an appropriate area of study in the mathematics curriculum of the elementary school implies that geometry contributes to the learning of children. In recent years, as well as in the past, many arguments have been advanced to justify its inclusion. In this section three broad areas of contribution to children's learning are presented.

First, the area of geometry seems well suited to the psychology of the learner. Geometric ideas are an integral part of nature, design, and the physical world. Educators have repeatedly stressed the importance of relating the learning of mathematics to the physical world, particularly with young learners. In the study of geometry the interplay between mathematics and the physical world can be effectively and naturally emphasized. It further appears evident that children have a natural curiosity and interest about geometric ideas.

The child's development of geometric ideas is a natural outgrowth of his early experiences. He lives in a world of geometric objects that he encounters visually and tactually. These objects have shape, size, and location. Thus, a geometry can be constructed to describe what the child has already informally encountered. In a natural and intuitive way, the child, for example, sees similarities between a window pane, a section of sidewalk, the front of a cereal box, an attaché case, a wooden block, a piece of paper, and the cover of a book. From these experiences, the mathematical concept of rectangle and rectangular region are later abstracted. From experience the child also possesses an intuitive awareness of many properties of a rectangle: that opposite sides "match," that opposite sides remain the same distance apart, that the corners are "square," and that a paper model can be folded so that one part exactly fits on the other part. Many similar illustrations demonstrate the type of informal experience that a child has with geometric ideas. Within a school classroom alone are many examples suggesting such important geometric concepts and relations as point, line, plane, angle, parallelism, perpendicularity, congruence, and symmetry. The teaching of geometry draws on these experiences and helps a child sharpen his perception of space and organize his previously haphazard notions.

Although a precise formulation of geometric notions may require a degree of sophistication on the part of the learner, the young learner likely possesses an intuitive awareness of important ideas that can be used in a more formal development. For example, children may realize that certain things are the same in some sense, such as the faces of a cube or the piece of a puzzle that exactly fits a space in the puzzle. In such awareness is embedded the notion of congruence, that is, figures having the same size and shape. The child has also seen designs that are the same on both sides when folded, such as the napkin shown in figure 2.1, or has made a paper doll by cutting a folded piece of paper, as shown in figure 2.2. Such situations are simple instances of the ideas of symmetry in the plane.

Fig. 2.1

Thus a child's constant contact with an implicitly geometric environment, together with his interest in the geometric elements of his world, provides a natural and fertile foundation that is psychologically appropriate for development in a more explicit manner with children.

Fig. 2.2

A second justification for its inclusion in the elementary school curriculum is that the study of geometry contributes to the "process" goals of learning mathematics. The qualitative aspects of children's reasoning have long been recognized as an integral part of mathematics education. Therefore it is important that pupils be involved in active inquiry, in discovering relationships, in formulating and testing conjectures, and in critical and analytical thinking. One aspect of developing these characteristics is the involvement of learners in appropriate activities. Geometry is recognized as an area particularly suited to these aspects of learning.

Many of the geometric concepts, facts, and relationships about which educators desire that children become knowledgeable can be presented through the active involvement of learners with materials. For example, children in the primary grades are well aware of many physical examples of space figures or closed surfaces, such as the rectangular prism, cube, pyramid, cylinder, cone, and sphere (see fig. 2.3). A section of tree trunk, a can of soup, a drum, and a water pipe are physical-world examples that suggest the cylinder. A child has seen sugar cubes, ice-cream cones, balls, shoeboxes, and popcorn boxes, which are examples of the other closed surfaces.

Through handling these objects as well as carefully constructed models of these shapes, pupils discover that some objects have only flat surfaces, some have only curved surfaces, and some have both. The child can determine the edges of these solids—that is, where the surfaces intersect— as well as the corners, or vertices, of the figures. This information, found through direct experience with models, can be organized in a table such as

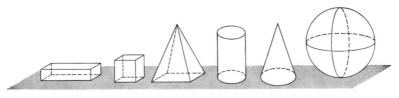

Fig. 2.3

the one shown in figure 2.4. As a summary experience, pupils will enjoy trying to identify a closed surface through a description of the number and kinds of surfaces, edges, and corners.

	Rectangular prism	Cube	Pyramid	Cylinder	Cone	Sphere
Flat surfaces	6					
Curved surfaces	0					
Edges	12					
Corners	8					

Fig. 2.4

Through paper folding, pupils form "square corners," from which the idea of a right angle is developed (see fig. 2.5), and can discover that the angle bisectors of a triangle meet in a single point (fig. 2.6), which can later be found by constructing the angle bisectors using a compass (fig. 2.7).

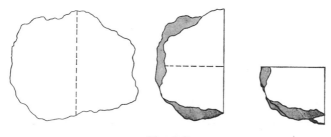

Fig. 2.5

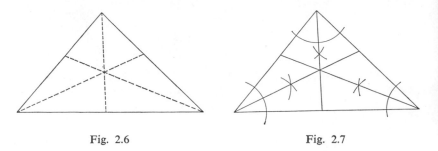

Fig. 2.6   Fig. 2.7

Using a note card and a pencil as models, pupils can explore the possibilities for the intersection of a line and a plane. (See fig. 2.8.) Then examples in the physical world that suggest this relationship can be sought.

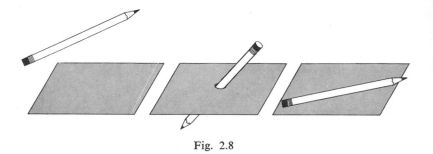

Fig. 2.8

Through straightedge and compass constructions, pupils can learn many properties of figures and have enjoyable experiences in making designs, such as the one shown in figure 2.9.

Fig. 2.9

The geoboard, a flat surface on which rubber bands can be stretched around equally spaced pegs, is an aid for exploring many ideas related to plane figures (see fig. 2.10). Elementary kinds of experiences can be provided by the geoboard, such as finding the different kinds of four-sided figures that can be formed or determining how many different squares can be found in a 4-by-4-unit section. Later, work with the geoboard can be extended to ideas of greater sophistication, such as area measure and coordinate geometry.

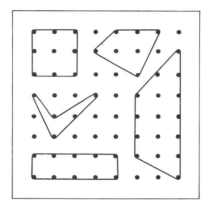

Fig. 2.10

Geometry provides many opportunities for experiences in analytical thinking. In figure 2.11, the labeled dots can be classified as being inside, outside, or on each figure. Dots that fit two classifications, such as *on the triangle and on the circle,* can also be identified.

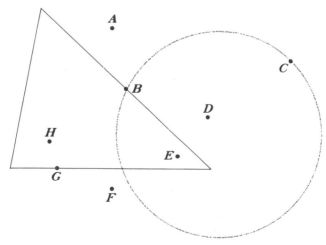

Fig. 2.11

The theorem stating that *the line segment connecting the midpoints of two sides of a triangle is one-half the length of the third side and parallel to it* is one of the many relationships that can be a vehicle for pupil exploration and discovery in the elementary school. Pupils can be asked to connect the midpoints of two sides of a triangle with a line segment and compare its length to that of the third side (see fig. 2.12). After several examples with a variety of triangles, pupils will discover the generalization for themselves. The relationship depicted in figure 2.12 can also be extended. See figure 2.13. The segments connecting the three midpoints form a second triangle. Pupils can be asked what relationship exists between the perimeters of the two triangles. Some will be challenged to explain why the relationship exists.

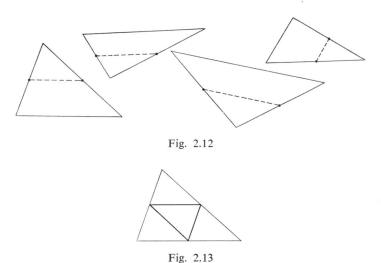

Fig. 2.12

Fig. 2.13

Geometry, then, is a most valuable vehicle for stimulating imaginative thinking by pupils—a process that frequently involves spatial perception—as well as for engaging pupils in productive activities.

A third argument for its inclusion in the elementary school is that geometry represents a unifying theme of mathematics. Geometry is a major branch of mathematics that has its roots in the early history of mankind and has many significant applications today. Thus it is appropriate that pupils gain an early awareness of basic concepts and also develop an appreciation of the breadth of the discipline of mathematics. A study of geometry in the elementary school means that pupils will have continual exposure to important ideas of geometry and learn that geometric ideas can be used to develop and clarify ideas in other areas of the mathematics curriculum.

Geometric concepts also contribute effectively to the learning of number and measurement ideas. Perhaps the most familiar example of relating geometry and number is the number line (fig. 2.14). On a line, a beginning point, A, and a unit segment, $\overline{AB}$, are selected. A series of units conguent to segment $\overline{AB}$ is then marked off along the line. The whole numbers, if made to correspond to the points on the line marking the congruent units, show the number of segments from the starting point, or 0.

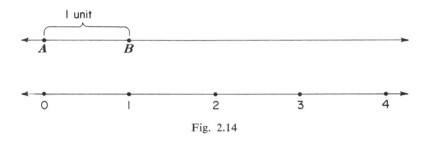

Fig. 2.14

The number line provides a way of "picturing" the set of whole numbers, fractional numbers, and integers as they are studied in the elementary school. Carefully used, the number line can be an effective vehicle for developing important number concepts, such as order relations and operations on numbers.

Regions are effectively used in initial developmental work with fractional numbers. In figure 2.15, a rectangular region has been partitioned into two congruent regions. One of the two congruent regions, or one-half of the figure, is shaded.

Fig. 2.15

In figure 2.16, two of the three congruent regions into which the triangle has been partitioned are shaded; so two-thirds of the region is shaded.

Throughout the work with fractional numbers, both the number line and regions have wide applications, including their application in developing the concept of equivalent fractions.

In the study of measurement, arithmetic aspects have traditionally received the major emphasis. However, a careful development of the con-

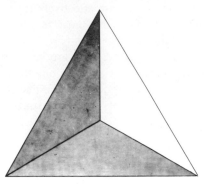

Fig. 2.16

cepts of linear, area, and volume measure has its basis in geometric concepts.

The width of a room, as in figure 2.17, is considered to be the length of a line segment perpendicular to the two longer walls. To find the measure of line segment XY, its length is compared with that of a unit segment. In the actual measuring process, segments congruent to the unit segment can be laid off along $\overline{XY}$; more frequently, however, the measurement is done by means of a ruler or a yardstick.

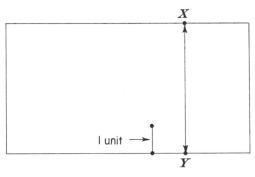

Fig. 2.17

Similarly, finding the area of the table top in figure 2.18 involves comparing the rectangular region with that of a given unit region. Many informal activities can be used prior to a formal study of area. One early experience deals with which kinds of regions have the property of covering a surface, leaving no gaps. Pupils are given a variety of regions, such as those shown in figure 2.19, and are asked to determine which of them will cover a given surface. As further preparation for the study of area,

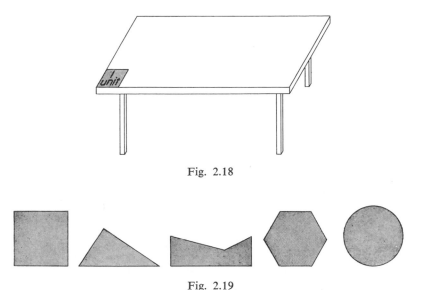

Fig. 2.18

Fig. 2.19

pupils can be asked to determine the number of regions enclosed by figures drawn on grids, like those shown in figures 2.20 and 2.21.

A highly significant example of the connection between number and geometry occurs in the development of the coordinate plane. In this work, pairs of numbers are associated with points of the plane. Graphing work is enjoyed by elementary school pupils and has many applications at this level. Again, through this work with the coordinate plane, a strong foundation in geometry can be provided by using the powerful tool of graphing.

Geometric and number ideas are also related in the use of bar, line, and circle graphs. These types of graphs use a geometric approach for organizing and presenting data.

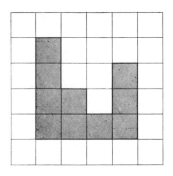

Fig. 2.20

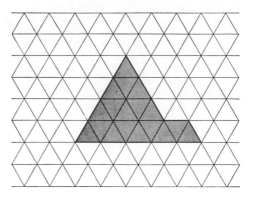

Fig. 2.21

In this section, geometry has been justified as an area of study appropriate for children on the basis of psychological, pedagogical, and mathematical considerations alone. Other cases, as evidenced in early writings, can be built for the study of geometry. Among these are the cultural, historical, and aesthetic values derived from such work.

Perspectives on the Classroom Use of Geometry

Geometry is a new area in the elementary school mathematics curriculum. As such, many curricular considerations are raised regarding its proper role and the experiences that are the most productive for pupils to study throughout their educational program. The attempt to develop direction, perspective, and balance is of appropriate concern for mathematics educators and curriculum developers.

Geometry also poses problems for the classroom teacher. Many teachers may feel insecure about their competency with geometric ideas. Some recall unsuccessful experiences associated with their own study of high school geometry, or they may equate the study of geometry in the elementary school with approaches used in the secondary school. Teachers are also faced with a curriculum that they may perceive as being already too extensive for the amount of time available. Nonetheless, all teachers share the common desire to have enjoyable, successful learning experiences with children. They are well aware of the powerful effect of such experiences on motivation. On this point the message is clear: Geometry is a most effective source of ideas that children will enjoy and with which they will find success. This source of ideas is very extensive. In basal and enrichment textbook materials, professional journals, state mathematics guides, and activity programs, a wide assortment of topics is presented. Many geometry experiences do not require an extensive knowledge on the part of the teacher. Rather, a teacher needs only the willingness to explore new ideas with children and the initiative to seek appropriate experiences for them.

Geometry has other characteristics that recommend it for pupil and teacher use. Although some aspects of number and operation work require a careful sequencing of learning with specific prior knowledge, such is not true of many geometric ideas. Geometry experiences that can be explored independently of each other exist, requiring little previous knowledge on the part of the learner. Such experiences can be brief in duration or can be extended over a period of several days.

Today, many teachers desire to engage their learners in direct learning experiences involving physical activities. It is again being emphasized that the effective learning of mathematics often involves exploration and discovery through the use of materials. Geometry fits well into this renewed emphasis on learning through active, direct experience. This aspect of geometry may have been one factor in its rapid implementation in the curriculum. Many geometry experiences that involve cutting, folding, drawing, constructing, and other uses of materials can be found. Some of these experiences were suggested in an earlier section of this chapter. Since geometry, with its close connection with the physical world, lends itself to activity approaches, such techniques should be maximized in the study of geometry and take precedence over solely visual or paper-and-pencil approaches. For example, the topic of the intersections of planes with closed surfaces is interesting to pupils. Although a visual experience, such as that in figure 2.22, showing possible figures formed by the intersection of a plane and a cone, is helpful, clay models of cones also can be constructed, with a moving wire, representing a plane, being passed through the model. Many kinds of direct, physical experiences can be provided both in the group development of central ideas and in the individual exploration of interesting concepts and relationships.

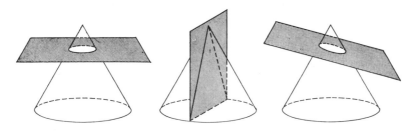

Fig. 2.22

Today, many teachers have found geometry to be a most profitable addition to their mathematics program. Initially, they undoubtedly felt hesitant about the advisability of including geometry. Yet the response of their pupils encouraged them to expand the use of geometry and to begin to develop a planned series of geometry experiences. In an increasing number of classrooms, geometry, then, having been found enjoyable to learn and to teach, has gained acceptance. As educators seek to maximize the value of a program of geometric ideas, it is important to develop a sense of perspective while enjoying the fun of using new ideas.

Some further considerations regarding the use of geometry are now presented:

1. It was previously suggested that one advantage of geometry is that many activities can be developed without an extensive background on the part of the learner. Yet it is important that geometry eventually be presented as more than a disjoint collection of facts about figures or a series of interesting discoveries. Productive learning ultimately involves the fitting of individual pieces of information into a larger context through the careful consideration of concepts, their relationships to each other, and the combining of concepts to generate new ideas. To emphasize only the bits of geometry in which little organization is evident or to use it primarily as a recreation is to reduce the opportunity for a more productive comprehension of geometry throughout a child's schooling. However, presenting the need for a systematic development of important ideas is not to deny the motivational value that lies in many of the independent experiences that can be presented in geometry—rather it argues that a series of such experiences alone does not constitute a sound geometry program.

2. Many ideas of geometry seem obvious, at least from an adult perspective. Thus there may be a temptation to try to "nail down" learning too soon with children, that is, to teach for closure. It has been recognized that mathematical understandings develop over a period of time as a result of many experiences and are not a matter of instantaneous insight. This awareness has received renewed emphasis in recent years through the work of those investigating children's thinking and learning. Geometric ideas, too, should be developed over a period of time in different contexts and at different levels of abstraction before a full understanding can be assumed. First-grade pupils understand the idea of a straight path. Yet the understanding of a line segment as an infinite set of points (with all that statement implies) or the distinction between the drawing of a line segment and the idea of a segment comes much later in the child's schooling.

There is also a tendency to teach for closure by emphasizing the ability to match names with figures and relationships rather than focusing on the ideas themselves. Although the naming of figures and relationships is a valid objective of teaching geometry, an early emphasis on this may not contribute to productive learning.

3. The study of geometry ultimately involves many new terms and symbols. As previously noted, the tendency to focus on precise terminology at early levels should be avoided—the early use of the accepted symbols for such ideas as line, line segment, ray, parallelism, and perpendicularity incurs the risk that the pupils will sense that the use of symbols is more important than the ideas themselves. Of graver consequence is the fact that a pupil's ability to recognize and use the symbols correctly may be considered evidence that he understands the ideas and relationships.

There is a close connection between geometry and elementary set ideas. The study of geometry involves the study of space—the set of all points—and the subsets of space. However, this relationship should not be abused by making geometry an exercise in set symbolism gymnastics. At some point it is helpful to indicate that $\triangle ABC$ can be considered as the union of three line segments with common endpoints, but whether it is helpful to symbolize this at an early stage as $\overline{AB} \cup \overline{BC} \cup \overline{CA}$ is doubtful. Similarly, the fact that two lines may intersect in a single point need not be symbolized as $\overline{AB} \cap \overline{CD} = \{x\}$ for the concept to be understood by pupils. It may even be that symbolizing the relationship in this fashion will obscure the ideas one wishes to communicate.

There is a place in the learning of mathematics for learning how to communicate mathematical ideas clearly and concisely through appropriate symbolism. However, perceptive teaching recognizes that ideas are primary, and the comprehension of ideas precedes and extends beyond mere adroit performance with symbols.

The Search for Direction in Future Programs

The description of elementary school geometry programs in an earlier section of the chapter raised the significant question, "What should the program in elementary school geometry be?" This question is the focus of current discussion among mathematicians and educators concerned with elementary school mathematics curriculum development. The discussion is the result of concern that some content included for study may not be appropriate for particular age levels. The concern is further articulated in the form of questions about the whole thrust of the geometry

curriculum, or at least certain emphases found at the elementary school
level. At present, there is consensus neither on what constitutes valid
content for study nor on how it is to be taught. Thus it is not surprising
that mathematics educators are in agreement on the need for continued
study in the search for an appropriate mathematical and pedagogical
direction for the geometry program at this level.

In order to bring the issue into sharper focus, some key elements in the
development that led to the geometry program of the early 1970s will now
be examined. Throughout the decade of the 1960s to the present time,
the geometry program could be broadly characterized as an intuitive development of standard Euclidean geometry. Axioms were not stated as
axioms, theorems were not stated as theorems, and no formal proof was
presented. Nonetheless, the programs frequently first introduced the notions of point, line, and plane, next examined relationships between these
elements (that is, the incidence relationships), and then continued to build
logically and systematically up to the notion of polygon before investigating properties of particular classes of polygons. Even though the program in geometry has continued to evolve from its inception, the above
description still characterized the major direction after ten years of development.

Speculating on factors that caused this emphasis to occur can provide
insight into why this direction was taken. Extended work in informal
geometry was first introduced at the junior high school level and from there
implemented in intermediate grade levels, where much of the elementary
school emphasis in geometry was concentrated. In addition, it had long
been advocated that pupils needed an extended informal exposure to geometric ideas prior to a formal deductive treatment in high school. Furthermore, curriculum developers, many drawn from secondary and collegiate
levels, felt that not only would an early, informal, yet systematic, development have a definite preparatory value but also that pupils could do such
work and would likely find it enjoyable. Finally, it was recognized that
elementary teachers had some familiarity with standard Euclidean geometry and the task of training teachers would be lessened if such an approach
were used. Because of a combination of factors, it appeared natural that
the direction of geometry content take the form of an elementary version
of what would later be treated in a more formal fashion.

As the curriculum revision movement gained in momentum, the creation of a geometry curriculum for the primary grades was an inevitable
step. The willingness to try new ideas and the acceptance found for
geometry in the upper grades were among the factors that created support
for attempts to develop curricular experiences for the early grades also.
However, at this level the direction to be taken was far less clear. At one

extreme was the attempt to move the systematic point-line-plane develop-ment into the primary grades, while at the other extreme the work was limited to the recognition of common polygons and physical models of them, augmented with a few additional random experiences. Questions began to be raised regarding the appropriateness of seven- and eight-year-old pupils trying to determine the number of lines that may pass through one point.

The need to find a framework for primary-grade geometry was a key factor that led educators to question the purpose and values of studying geometry. The questions were stimulated by the desire to find a pedagogi-cal and conceptual framework appropriate to the thinking of young chil-dren. It was only natural that educators eventually asked, "Why is the program so closely aligned with a traditional Euclidean development?" and "Why should a standard Euclidean approach be the basis on which to construct programs?"

Thus, toward the end of the 1960s a variety of new approaches were suggested, many of which differed markedly from the earlier work. One approach that has interested many is based on ideas of transformational geometry. Lessons were developed centering on topics of congruence, symmetry, and similarity. The lessons were described as ones in "motion geometry" rather than as lessons in transformational geometry. Motion geometry was found to be an appropriate term for young children, since movements of figures and shapes in the plane and in space were studied. Because the discussion of transformational approaches is of current inter-est, the following brief description will help orient the reader to some basic elements of such work.

The notion of congruence is central to the study of geometry. Under a transformational approach, congruence receives a strong emphasis and is treated in a different context. Congruence can be presented to young learners using the intuitive ideas of "same size and shape" and "exactly match." The active involvement of children with materials provides a means by which the congruence of two regions can be verified. For ex-ample, pupils might be involved in sorting triangular disks into equivalence classes. The activity is planned so that a pupil is challenged to flip some of the disks to determine whether they are congruent to others. (See fig. 2.23.)

The sliding, turning, and flipping of these triangular disks on a desk top provides an important intuitive experience for the more formally described motions of translation, rotation, and reflection. The classification of other polygonal forms provides similar readiness experiences. The later work using motion geometry builds on the intuitive work done in the early years. The learner is led to consider whether two polygons are congruent by

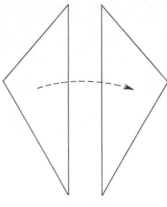

Fig. 2.23

rotating (fig. 2.24), translating (fig. 2.25), or reflecting (fig. 2.26) one to fit "exactly" on the other.

In the case of the motion of reflection, the early concepts developed in symmetry are especially appropriate. In particular, the line of symmetry takes on a deeper meaning. The child recognizes that the line of symmetry used in the motion of reflection is the mirror he has used in earlier experiences. The study of the distance relationship between the point and its image and the line of symmetry formalizes the learner's intuitive feeling about the object, the mirror, and the image.

In determining whether two polygons are congruent by using motions on

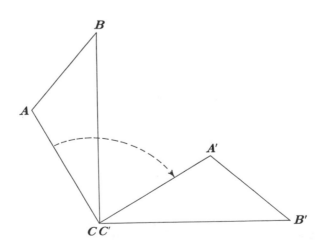

Fig. 2.24

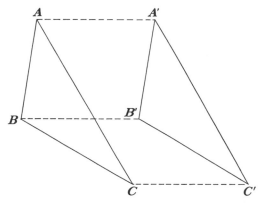

Fig. 2.25

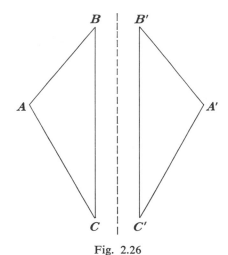

Fig. 2.26

the plane, the learner may be challenged to use a composition of two or more motions, as shown in figure 2.27.

The motions described are rigid motions—that is, the shape and size of the polygon remain the same after the transformation. Approaching congruence from the motions in the plane seems to be natural and lends itself to activities appealing to both teachers and children.

The development of programs using some experiences in rigid motion transformation has led to a discussion of the appropriateness of an introduction to topics through other transformations. A transformation that preserves shape but not size can be used to generate triangles that are

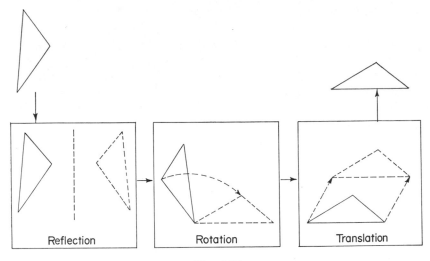

Fig. 2.27

similar. Again, a certain transformation might explain what happens to an image on a piece of sheet rubber as it is stretched.

The fact that some elements of transformational geometry are being used as an approach to the study of such familiar topics as congruence, symmetry, and similarity is interesting. Yet it remains a question whether this is the appropriate direction for future elementary school geometry programs. It again appears evident that some of the interest is the result of work in this direction at the secondary school level, as well as less formal approaches for junior high school pupils. There again remains the temptation to justify elementary school curriculum work primarily on its preparatory value for later work.

From the preceding discussion, it is evident that the search for direction in elementary school geometry is still a central focus in mathematics education in the 1970s. Even the proponents of innovative and promising approaches (such as the motion geometry approach) recognize that their efforts have not yet resulted in total elementary school programs.

This continuing lack of a clear direction for the geometry program is due to several reasons or obstacles. One of the obstacles in establishing a geometry program that is truly a program for children and not just a reflection of the secondary program is the lack of agreement on the goals for elementary school geometry. Geometry is unlike arithmetic in that there is no consensus of social necessity for geometric skills as there has been for arithmetic skills. In a culture whose emphasis has been on utilitarian skills, the purpose of a geometry program in the elementary school has not been uniformly recognized. Hence, without consensus on goals for geometry, the prospect of agreement in the direction for content and pedagogy of the elementary program remains remote. However, some progress in fashioning a geometry program has evolved from recent efforts: curriculum developers have produced some activities in elementary school geometry that pupils seem to enjoy. Clearly, these efforts in developing activities and lessons in geometry that are appropriate for young children should continue even while discussion of the goals for geometry continues.

Another obstacle to the possibility of agreement in the programs is the lack of evidence about what young children can and cannot do in geometry. Although articles on experiences in teaching geometry have frequently appeared in professional journals, the messages gained are diverse and often add little insight into how children think or what they can indeed understand. There is only a meager history of research in mathematics education related to geometry and the elementary school child. Fortunately, a consensus among mathematics educators on the need for basic research on how children learn mathematics is growing. This emphasis on basic research, along with a balancing effort of classroom applications of the research, is a vital and missing link in the search for direction in elementary school geometry.

A third obstacle in this search for direction is the diffuse nature of geometry as compared to arithmetic. The pedagogical development of the arithmetic program is structured to a great extent by reasons of the internal organization and structure of the content itself. For example, the study of rational numbers would usually not precede the study of whole numbers; likewise, multiplication of whole numbers generally follows addition. In geometry, however, the study of polygons does not necessarily follow the study of incidence relations; a child needs neither an understanding of, nor a set of skills in, the point-line-plane development to understand an intuitive development of coordinate geometry. Whatever the organization of topics in the programs in the 1970s, the basis for this organization was not apparent to the teachers and children using it. The overall impression was that the geometry program consisted of a diffuse set of topical ideas. Whether this apparent diffuseness must be inherent in the elementary geometry program or whether a truly organized and structured program can be developed remains for the future to determine.

A geometry program conceived and developed for the elementary school child has not yet been fashioned. It is clear that although it would be desirable to outline the elementary geometry program of the future, such a task is not possible now, given the state of present knowledge; however, some aspects of an appropriate program for elementary school geometry can be described here. The following factors should be present in any program developed for the elementary school:

1. *The program should be mathematically cohesive.* In the early days of the reform period, the geometry program often appeared to consist of a disconnected set of bits and pieces of concepts. This approach has not been entirely without merit in the sense that several sets of good lessons about a particular topic have been developed. However, these sets do not fit together, and thus the geometry curriculum has not been presented as a cohesive entity either to children or to teachers. Teaching demonstrations related to geometry often emphasized the aspects that would be *enjoyed* by children—clearly, children should find the lessons and experiences in geometry interesting, but the program should not be relegated to a role of fun and games. Although geometry does not have to be as tightly structured as the arithmetic program (in fact, it probably should not be), the child should be able to relate new learnings to previous learnings, and—what is perhaps equally important—the teacher should understand why he is teaching a particular idea in geometry and how this idea fits into a geometry sequence.

2. *The program should be informal rather than formal.* The view that the elementary school program serves only as a preparation for high school and college is one that promoted a fair amount of formality in some of the programs in the sixties. Consequently, some programs merely included secondary school topics that were modified for the elementary school pupil only by the augmentation of explanations, illustrations, and examples. In some cases, the attention devoted to definitions and symbolism seemed to be greater than the emphasis on the geometric concepts themselves. Certainly a geometry program in the elementary school should not exist in ignorance of more advanced concepts, but it is important that the initial presentation of concepts be informal, using the child's language to assure communication. Thus the distinctions between a rectangle and a rectangular region or between a line and a line segment may be inappropriate for the very young child. As valid reasons for such distinctions arise, they can be made in a natural setting so that refinements of concepts develop within the young learner. An appropriate geometry program will consist of informally presented concepts that, in turn, are used as building blocks for more formally developed ideas. Initially, then, the child should handle objects, discuss their properties in his own language, and discover some of the spatial relations himself or through interaction with other children. These informal experiences can then be gradually translated into more formal organizing aspects embodied in definitions and symbols.

3. *The geometry program should relate geometric ideas to the real world.* One of the reasons continually given by educators for the inclusion of geometry in the elementary school is that it provides numerous opportunities to explain spatial relations to the ever-curious minds of children. One focus of direction in the elementary school program, then, should be to encourage children to investigate the spatial relations around them. The child should be encouraged to observe the shapes of objects, to compare and contrast these shapes, to articulate and discuss the properties he has observed, and to begin to organize these observations. Conversely, the child should be encouraged to seek out in the physical world examples of ideas and concepts presented in a classroom lesson in geometry. Laboratory activities involving experiences with geoboards and other manipulative devices should be a part of every elementary program in geometry. Relating geometry to the physical world (and vice versa) provides an opportunity for the child to see mathematics as something really useful in interpreting his surroundings.

4. *The geometry program should fit the thinking and learning patterns of the elementary child.* The need for basic research in the learning of mathematics, and of geometry in particular, has been mentioned as a widespread belief among mathematics educators. This research is necessary to help determine both the appropriate content and the appropriate pedagogical development for elementary school geometry. In addition to the basic research aimed at how and what children can learn, classroom studies must continue to examine the effectiveness of varied sequences of topics. This intermeshing of basic research and classroom applications is a prerequisite for any definitive direction in geometry. Clearly, a determination of a logical order of development is not enough, for the logical order is not always the appropriate pedagogical order. Thus the development of geometry programs must rest equally as well on considerations of child development as on consideration of content. In past years, it appears that the latter consideration often received undue emphasis in the development of elementary school geometry. Aspects of child development—his interest in objects, his natural attraction to activity-oriented tasks, his natural curiosity, his early limitations in vocabulary—must be taken into account in the construction of a geometry program.

The four factors described in the preceding paragraphs are essential for any geometry program developed for the elementary child. Although such a program does not yet exist, it is encouraging to note the many serious efforts that have been made in the decade of the sixties and early seventies. Research and curriculum development in elementary school geometry is now legitimate scholarly activity for mathematics educators. With the continued cooperative efforts of scholars, experimental projects, curriculum writers, and classroom teachers, a program truly developed for the elementary child will evolve.

In summary, as considerations are raised about the appropriate direction for future programs in elementary school geometry, they should not diminish the recognition of the substantial gains of the past decade. Geometry has been established as an integral and vital component in the elementary school curriculum. It has already been recognized as an effective means of enhancing and broadening the learning of mathematics for children. The recommendations and hopes of mathematicians and educators of the past 100 years were put into effect in the 1960s. In spite of shortcomings in the program, geometry has enjoyed widespread success among teachers and children. The promise of a more appropriate program is encouraged by the continuing efforts of educators through the sixties up to the present. While the search for direction in the elementary school geometry program continues, one fact remains: Geometry is and will remain a part of the elementary school mathematics program.

REFERENCES

1. Betz, William. "The Teaching of Intuitive Geometry." In *The Teaching of Mathematics in the Secondary School*. Eighth Yearbook of the National Council of Teachers of Mathematics. New York: Bureau of Publications, Teachers College, Columbia University, 1933.

2. Cambridge Conference on School Mathematics. *Goals for School Mathematics*. Boston: Houghton Mifflin Co., 1963.

3. Cambridge Conference on Teacher Training. *Goals for Mathematical Education of Elementary School Teachers*. Boston: Houghton Mifflin Co., 1967.

4. Coleman, Robert, Jr. *The Development of Informal Geometry*. Contributions to Education, no. 865. New York: Teachers College, Columbia University, 1942.

5. "Final Report of the National Committee of Fifteen on Geometry Syllabus." *Mathematics Teacher* 5 (December 1912): 46–131.

6. Moore, Eliakim H. "On the Foundations of Mathematics." In *A General Survey of Progress in the Last Twenty-five Years*. First Yearbook of the National Council of Teachers of Mathematics. New York: Bureau of Publications, Teachers College, Columbia University, 1926.

7. National Education Association. "Report of Committee on Elementary Education." In *Journal of Proceedings and Addresses, Session of the Year 1890*. Topeka, Kans.: Kansas Publishing House: Clifford C. Baker, 1890.

8. ———. "Report of the Committee on Secondary School Studies." Washington, D.C.: Government Printing Office, 1893.

9. Reeve, William D., ed. *Significant Changes and Trends in the Teaching of Mathematics throughout the World since 1910*. Fourth Yearbook of the National Council of Teachers of Mathematics. New York: Bureau of Publications, Teachers College, Columbia University, 1929.

10. Reeve, William D. "The Teaching of Geometry." In *The Teaching of Geometry*. Fifth Yearbook of the National Council of Teachers of Mathematics. New York: Bureau of Publications, Teachers College, Columbia University, 1930.

11. School Mathematics Study Group. *Geometry Units for Elementary School*. Edited by J. Fred Weaver. Stanford, Calif.: SMSG, 1971.

12. Shibli, Jabir. *Recent Developments in the Teaching of Geometry*. State College, Pa.: The author, 1932.

13. Smith, David Eugene. *The Teaching of Elementary Mathematics*. New York: Macmillan Co., 1900.

3

Informal Geometry
in Grades 7–14

JOHN C. PETERSON

FOR MANY years informal geometry was the exclusive domain of the junior high school curriculum. Then, in the 1950s, a revolution in school mathematics began sweeping through the schools. These revolutionary ideas included an emphasis on the structure of mathematics and an emphasis on "discovery teaching." The emphasis on discovery teaching should have been a boon to the teaching of informal geometry in secondary schools. However, many teachers were reluctant to use discovery methods in their teaching, and informal geometry relies quite heavily on a discovery philosophy toward teaching. As a result, informal geometry was de-emphasized.

Fortunately, some of the decrease in informal geometry at the secondary school level can be compensated for by the increase in informal geometry in the elementary school curriculum. Yet, there is a place for informal geometry in the secondary school curriculum. As educators come to realize how people learn mathematics, many are deciding that the study of geometry as a deductive system can be enhanced by the use of inductive reasoning. Informal geometry provides an excellent vehicle with which to use inductive reasoning.

This chapter will include (1) a historical development of informal geometry in grades 7–14, (2) a discussion of the possible role of informal geometry in grades 7–9, and (3) the possible continuing role of informal geometry in grades 10–14. Appropriate methods to use in teaching informal geometry and how informal geometry is related to other mathematics will be discussed.

Historical Background

Courses in geometry were not a part of the secondary school curriculum until after geometry was made a college entrance requirement. In 1865 Yale began to require geometry for entrance; Princeton, Michigan, and Cornell included geometry as an entrance requirement in 1868; and Harvard required geometry and logarithms for entrance in 1870 (20, p. 27). Thus when geometry entered the secondary school curriculum, it entered as essentially a college subject—a formal, demonstrative subject.

At the same time, attempts were made to introduce informal geometry into the elementary schools. (Since there were no junior high schools before 1910, the elementary schools included grades 7 and 8.) In 1854 Thomas Hill wrote his *First Lessons in Geometry* (23, p. 10). He intended this book for children of from six to twelve years of age. Hill's philosophy, aptly stated in the preface of his book, can also be seen in the motto on the title page: "Facts Before Reasoning." In 1863 Hill published his *Second Book in Geometry*. The motto on the title page of his second book was "Reasoning Upon Facts." He intended the second book for children of from thirteen to eighteen years of age.

Another effort to institute geometry in the elementary school was made by Bernhard Marks, principal of the Lincoln School in San Francisco. His text, *First Lessons in Geometry,* was published in 1871 (5, pp. 63–64).

Throughout New England more and more experiments were conducted in intuitive geometry. Cambridge, Massachusetts, began to include the elements of intuitive geometry in its grammar schools by 1893. "In that year, Professor Paul H. Hanus of Harvard University gave a course of lectures 'to the teachers of the seventh, eighth, and ninth grades' " (5, pp. 64–65).

More and more people were advocating reforms in mathematics teaching and were in favor of the introduction of geometry into the elementary schools. At this time the mathematics curriculum of the eight years of elementary school consisted of arithmetic, arithmetic, and still more arithmetic. Most students completed their formal education with nothing more than an elementary school education, and so the majority of people had no knowledge of geometry of any kind.

The 1890s saw many additional suggestions for reforms in the secondary school curricula. Among these was the *Report of the Committee of Ten on Secondary School Studies,* published in 1894. The recommendations of the Committee of Ten included one favoring a reduction of the elementary school from eight years to six years. This would allow many secondary school subjects to begin two years earlier. The subcommittee on mathematics suggested that demonstrative geometry should begin at the end of the first year's study of algebra. However, the mathematics subcommittee also recommended that

> a course of instruction in concrete geometry, with numerous exercises, be introduced into the grammar school. The object of this course would be to familiarize the pupil with the facts of plane and solid geometry, and with those geometrical conceptions to be subsequently employed in abstract reasoning. During the early years the instruction might be given informally, in connection with drawing, and without a separate appointment in the school calendar; after the age of ten years, one hour per week should be devoted to it. [21, p. 106]

The recommendation specifically called for a reduction in the teaching of arithmetic along with the introduction of concrete geometry. Concrete geometry was to occupy one forty-to-forty-five–minute recitation period a week during the fifth, sixth, seventh, and eighth grades.

The following year, 1895, the Committee of Fifteen presented their report to the Department of Superintendence of the National Education Association. Here, also, was the call for a reduction in arithmetic teaching:

> With the right methods, and a wise use of time in preparing the arithmetic lesson in and out of school, five years are sufficient for the study of mere arithmetic—the five years beginning with the second school year and ending with the close of the sixth year. [11, p. 24]

But in conjunction with the reduction of "mere arithmetic," the Committee of Fifteen promoted algebra instead of geometry. In fact, they were quite vehement that informal geometry should not be included in the curriculum (11, p. 44).

So we see that there was disagreement between the Committee of Ten and the Committee of Fifteen. Yet there also was disagreement among the members of the Committee of Fifteen: Charles B. Gilbert, in his dissent from the report, commented, "I desire to suggest that geometry may be so taught as to be a better mathematical study than algebra to succeed or accompany arithmetic in the seventh and eighth grades" (11, p. 126).

Four years later, in 1899, the Committee on College Entrance Requirements attempted to bring about better articulation between the secondary schools and the colleges. In their recommendations they advised that, assuming the length of the recitation period to be at least forty-five minutes, the seventh grade include four periods a week of concrete geometry and introductory algebra and that the eighth-grade program include four periods a week of introductory demonstrative geometry and algebra (22).

For the next dozen years no new recommendations were forthcoming. Then in 1911, just as the junior high school was becoming established, a new committee, the National Committee of Fifteen on Geometry Syllabus, submitted their preliminary report (12). The final report was published in 1912 (13). Commenting on the poor results that previous recommendations had brought, the committee made the following statement:

> In spite of all the discussion about constructive geometry (intuitive, metrical, etc.) in the first eight grades, carried on in the past half century, no generally accepted plan has been developed to replace the old custom of teaching the most necessary facts of mensuration in connection with arithmetic. We have, therefore, at this time, algebra in the ninth school year, plane geometry in the tenth, and algebra and geometry in the eleventh and sometimes in the twelfth. [13, pp. 83–84]

Hoping that their suggestion would replace the "old custom," they made the following proposals:

> A special course in geometry in the graded school is desirable, if at all, only in the last grade or the last two grades [seventh and eighth]. In such a course no work of demonstrative character should be undertaken, though work may be done to *convince* the student of the truth of certain facts; for example, by paper folding or cutting a variety of propositions may be made evident, such as the sum of the angles of a triangle is 180°, etc. . . .
> Emphasis should be laid upon the facts with which the student is already familiar through the work described above. . . .
> Emphasis should also be laid on other work of a concrete nature which involves direct use of geometric facts. Thus the propositions concerning the similarity of triangles should be introduced by means of the drawing of figures to scale. . . .
> Contrary to the traditional procedure, the forms of solid geometry should be emphasized even more than those of plane geometry, for they are more real and more capable of concrete illustration. [13, pp. 90–92]

Junior high schools were becoming more and more prevalent, and the curriculum of the junior high schools became one of the major issues. At their spring meeting in 1917, the president of the Association of Teachers of Mathematics in New England appointed the seven-member Committee to Recommend a Suitable Program in Mathematics for the Junior High School. The committee included the following geometry recommendations among the topics that should appear in the junior high school course in mathematics. For the seventh grade, mensuration was the only geometric topic. Included in the realm of mensuration were "(a) formulas for the areas of rectilinear figures and circles; (b) rectangular and circular graphs" (1, pp. 136–37). Eighth grade would include "plane geometric figures—construction and classification of lines, angles, triangles, and quadrilaterals; graphical representation of statistics using squared paper" (p. 137). Ninth graders would study the "properties of plane geometric figures informally developed (lines, angles, triangles, and quadrilaterals)" (p. 138).

In 1916 the National Committee on Mathematical Requirements was established by the Mathematical Association of America. The committee's final report, *The Reorganization of Mathematics in Secondary Education*, was published in 1923 and is often referred to as *The 1923 Report* (19).

The 1923 Report suggested that junior high school mathematics include the topics of arithmetic, intuitive geometry, algebra, numerical trigonometry, and demonstrative geometry. A detailed list of recommended intuitive geometry topics was included in *The 1923 Report* along with a general philosophy toward intuitive geometry:

> The work in intuitive geometry should make the pupil familiar with the elementary ideas concerning geometric forms in the plane and in space with respect to shape, size, and position. Much opportunity should be provided for exercising space perception and imagination. The simpler geometric ideas and relations in the plane may properly be extended to three dimensions. The work should, moreover, be carefully planned so as to bring out geometric relations and logical connections. Before the end of this intuitive work the pupil should have definitely begun to make inferences and to draw valid conclusions from the relations discovered. In other words, this informal work in geometry should be so organized as to make it a gradual approach to, and provide a foundation for, the subsequent work in demonstrative geometry. [19, pp. 22–23]

The committee then offered five plans suggestive of the ways in which the work might be completed. None of the five was recommended as superior to the others. Two of the plans completely omitted demonstrative geometry, and four of the five plans recommended the inclusion of intuitive geometry.

A preliminary report of the National Committee on Mathematical Requirements entitled *Junior High School Mathematics* was published in 1920 (18). The recommendations for informal geometry in this preliminary report were essentially the same as the recommendations in *The 1923 Report*.

The January 1921 issue of the *Mathematics Teacher* (the first issue published under the auspices of the newly formed National Council of Teachers of Mathematics) included "A Symposium of Discussion on the

National Committee Report on Junior High School Mathematics." The symposium consisted of several mathematics educators giving their reactions to the report: Marie Gugle, assistant superintendent of the Columbus, Ohio, schools and fourth president of the NCTM, stated that "there is no other phase of elementary mathematics that makes such an appeal to all types of pupils [as does intuitive geometry]" (6, p. 27), and William Betz, mathematics specialist for the Rochester, New York, schools and seventh president of the NCTM, cautioned that "it might be well to inform inexperienced teachers that intuitive geometry is not primarily a textbook subject, that it requires constant contact with concrete material, and that the method of procedure is all-important" (p. 33).

The 1923 Report contained no actual recommendations for including informal geometry in the high school curriculum. It did, however, contain some hints that could be interpreted as advocating the inclusion of informal geometry. (See 19, pp. 36–37.)

What effect were the recommendations of these various committees having on the curriculum? One of the several studies conducted in the 1920s gives some indication. William Betz reported on a 1924 study by James M. Glass on "Curriculum Practices in the Junior High School and Grades 5 and 6," in which Glass made a detailed study of the junior high school curriculum in fourteen representative centers. His results indicated that intuitive geometry was being taught an average of 90, 75, and 85 minutes each week in grades 7, 8, and 9, respectively (4, pp. 154–55). Yet Betz, the most outspoken advocate of intuitive geometry, was quick to point out that "intuitive geometry has not yet received the attention to which it is entitled" (p. 162).

In 1932 the NCTM appointed the Committee on Geometry. This committee originally consisted of six members, but on seeing the immensity of their task, they quickly expanded to twenty-six members. The final report of the Committee on Geometry was published in 1935. (See 2.) The committee surveyed 101 teachers and found that informal geometry was almost entirely thought of as a junior high school subject. A few teachers indicated that they preferred to use the result of any inductive proofs. J. Shibli disagreed with his fellow committee members on the placement of informal geometry. He felt that the material then being taught in grades 7 and 8 should be transferred to grades 5 and 6. "Most of this material is too easy to be interesting and challenging to children of grades 7 and 8; and it is easy enough for grades 5 and 6." He then proposed an inductive-deductive, intuitive-demonstrative geometry course for junior high school (2, p. 342).

In 1940 the reports of two more committees were published. One committee, the Joint Commission of the Mathematical Association of America and the National Council of Teachers of Mathematics, divided their specific recommendations for informal geometry into four sections: Basic Concepts, Basic Skills or Techniques, Important Geometric Facts and Relations, and Discovering and Testing Geometric Relationships (17, pp. 83–85). The only grades indicated for teaching informal geometry were the seventh and eighth. The ninth grade was reserved for the review and extension of concepts, skills, facts, and relations and applications, preferably in relation to algebra.

The next year World War II enlarged, and United States troops were directly involved. The inadequacies of the mathematical programs of our schools became very conspicuous in the light of the mathematics deficiencies discovered among the inductees. Several committees were formed by the U.S. Office of Education in cooperation with the NCTM. Many recommendations were made, but they were primarily concerned with refresher courses for high school students not studying mathematics.

At its annual meeting in 1944, the NCTM created the Commission on Post-War Plans. This commission was to plan for effective programs in secondary mathematics in the postwar period. A checklist of twenty-eight items that the commission considered essential for functional competence in mathematics was given. Among these were only four that would seem to apply to informal geometry in the junior high school (10, pp. 197–98).

The next ten years were void of any new recommendations, but in 1955 the College Entrance Examination Board (CEEB) appointed the Commission on Mathematics "to study the present secondary school mathematics curriculum with a view toward the recommendation of a realignment of the entire secondary mathematics program to provide adequate preparation for the study of present-day college and advanced mathematics" (7, p. 19). In 1956 the Board of Directors of the NCTM appointed the Secondary School Curriculum Committee. The committee was "to make a comprehensive and critical study of the curriculum and instruction in mathematics in secondary schools with relation to the needs of contemporary society" (7, p. 35). Both committees published their reports in 1959.

The Secondary School Curriculum Committee gave a detailed list of the elements of informal geometry it felt were appropriate for grades 7 and 8. This material might well constitute from 20 to 30 percent of that presented in grades 7 and 8. However, the committee's only recommendation for informal geometry in the ninth grade was that a liberal use should be made of geometric drawing in illustrating problems wherever possible (25, p. 406).

The CEEB Commission on Mathematics did not feel that its task included a careful study of the mathematics program of grades 7 and 8.

However, the commission did give some comments on junior high school mathematics and the essential subject matter that should be included. The geometry that the commission felt was essential for junior high school students was divided into two sections: (1) measurement and (2) relationships among geometric elements. All this could be classified as informal geometry.

In the section entitled "Specific proposals for geometry" are these comments concerning the role of informal geometry in the secondary school curriculum:

> As has already been set forth, it is felt that a substantial introduction to geometry on an intuitive and informal basis should be accomplished in the seventh and eighth grades. A sense of geometric form, a knoweldge of simple geometric facts, and skill in simple geometric constructions are all appropriate achievements for junior high school or upper grade-school pupils. . . . In a sense, this intuitive geometry may be thought of as the physical geometry of the space in which we live, rather than as an abstract mathematical system. [9, pp. 24–25]

Several organizations were being formed at this time to improve the teaching of mathematics. Among the most notable of these were the School Mathematics Study Group (SMSG) and the University of Illinois Committee on School Mathematics (UICSM). Most of the curriculum organizations were interested in improving the mathematics curriculum for the college-bound student. As a result, informal geometry did not gain in importance, since the emphasis was placed on the axiomatic nature of mathematics. Thus, although SMSG believed that the curriculum for grades 7 and 8 "should include a sound intuitive basis for algebra and geometry courses to follow" (27, p. 455), they did not follow the recommendation of the CEEB commission to teach geometry "as a series of deductive sequences linked by more informal and intuitive sections to lay a stress on postulational systems" (20, p. 278). However, the emphasis that was being placed on discovery teaching, if used to its best advantage, would have compensated for this lack of informal geometry.

In 1963, a group of twenty-five professional mathematicians reviewed school mathematics and established goals for mathematics education. Their goals, known as the Cambridge report, included some recommendations for the elementary school, which were given in the previous chapter of this book, but presented two somewhat different approaches to a curriculum for the secondary schools. The first approach consisted of algebra and probability in grades 7 and 8 and included in grade 9 a section on "intuitive and synthetic geometry to the Pythagorean theorem" (8, p. 44). The second approach included the axiomatic development of Euclidean geometry of two and three dimensions in grades 7 and 8. In the ninth grade the only geometry that might be considered informal was that of the volumes of figures. Informal geometry, according to the Cambridge report, was to become almost the exclusive property of the elementary school curriculum.

In the last eighty years, many committees have been formed and have made many recommendations. All the recommendations for including informal geometry in grades 7–14 limited it to grades 7 and 8. Thus historically there is little justification for including informal geometry topics later than grade 8. But informal geometry is more than just a list of specific topics—it is a method of teaching geometry. Informal geometry at its best makes use of discovery methods of teaching, inductive reasoning, and the student's inquisitiveness. Thus, whereas the topics normally included in informal geometry have not been advocated for high school students, the philosophy of teaching that pervades informal geometry has been in the mainstream of mathematics education the last fifteen years.

Informal Geometry in Grades 7–9

Informal geometry is an established part of the secondary school curriculum. However, most of the informal geometry in the curriculum has been traditionally confined to grades 7 and 8. Until recently there was relatively little informal geometry in the elementary school curriculum. The fact that informal geometry is becoming more and more a part of the elementary school curriculum was pointed out in the previous chapter. Even with the increase of informal geometry in the elementary school, there is still a great deal of informal geometry that can be, and should be, included in the secondary school curriculum.

Some of this informal geometry is a *continuation and extension* of the informal geometry currently being taught in elementary school. Other aspects of this informal geometry involve completely *different approaches* to some concepts. One example of this is the University of Illinois Committee on School Mathematics (UICSM) textbook series *Motion Geometry* (26), which provides slow learners at about the eighth-grade level with a look at informal geometry through geometric transformations. Another way in which informal geometry can be used in secondary schools is to provide *inductive lessons* that will enable students to form conjectures. These conjectures can then be either proved or disproved using deductive reasoning.

This use of informal geometry emphasizes the philosophical idea that informal geometry is a method of teaching as well as a list of specific topics. In many instances this use of informal geometry is more important than the topical issue. The remainder of this chapter will include examples of each of these three ways in which informal geometry is, or should be, included in the secondary school curriculum.

Continuation and extension

Many geometric ideas are begun in elementary school, but most of these ideas need to be reinforced and then extended. Reinforcement usually includes a reintroduction and a review of the concepts. These activities are necessary, since many students will have forgotten much of what they had previously learned. Extension, or continuation, of the concepts involves a more sophisticated look at the concepts and uses the previously learned ideas to investigate some problems that, in the student's experiences, are unsolved. A more sophisticated look at a concept can be accomplished in several different ways: the concept could be embedded in a structure, more information could be provided, applications could be presented, more precision could be used, or an in-depth study could be undertaken.

For example, suppose the students had previously learned how to determine the area of a rectangle. An extension of this concept would be to use it to help determine the areas of parallelograms and triangles. Similarly, the student's knowledge of how to determine the volume of a right rectangular prism is used to help determine first the volumes of other specific types of prisms (right triangular prisms, rectangular prisms, and so on) and then the volume of any prism. Related to these concepts are the concepts of conservation of area and volume, which are begun in the elementary school but which need to be reinforced in the junior high school.

Much of the work in the elementary schools includes the definition of many geometric ideas. Many, if not all, of these definitions are reviewed in junior high school. An investigation of some junior high school textbooks being used in 1970 indicated that the following concepts were included: point, line, plane, space, segment, ray (half-line), half-plane, half-space, angles (acute, right, obtuse, straight, vertical, adjacent, complementary, supplementary, corresponding, alternate interior, alternate exterior), perpendicularity, parallelism, skewness, collinearity, coplanarity, coincidence, concurrence, betweenness, separateness, perimeter and circumference, area, volume, polygons, circles, prisms, pyramids, cones, cylinders, spheres, congruence, symmetry, and similarity. The following constructions were included: copying an angle, bisecting an angle, copying a segment, copying a triangle, drawing a perpendicular to a line through a point either on or not on the line, bisecting a segment, drawing polygons (such as rectangles, squares, parallelograms, and regular hexagons), and drawing incircles and circumcircles.

The introduction of informal geometry in the elementary school has provided mathematics educators with an excellent opportunity to make better use of the spiral approach when teaching geometric concepts. A geometric concept can now be taught during three or more school years instead of the previously allotted two years. This extra time allows for additional reinforcement and extension of the concepts. Students should thus be able to understand a concept better and retain it longer. Examples of how students can improve their understanding of a concept can be seen in some approaches toward the measurement concepts of length, perimeter, area, and volume. Approaches vary, but some of the best approaches begin in the elementary school by focusing attention on the basic measurement procedure of selecting a unit, dividing the object to be measured into the chosen units, and then counting the number of units. This approach enables the student to gain an understanding of the measurement process before he begins to generalize formulas. Figures 3.1 through 3.4 illustrate how different units can be used to measure the same object. For example, in figure 3.1 the student is asked to determine the distance around the object shown in the figure by using each of the indicated units; in the next three figures (3.2–3.4) he must determine the measure of the interior of the objects shown using the units indicated for each object. This process is extended in the junior high school. Using the understanding of the measurement concept that was established in elementary school, students are now able to generalize their ideas and develop formulas.

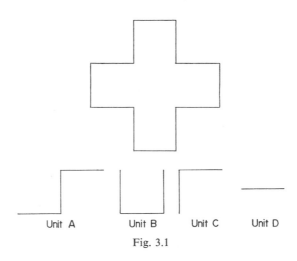

Fig. 3.1

Exercises such as those provided in figures 3.1–3.4 should enable the student to realize that the unit used to find the length, area, or volume is primarily a unit of convenience. He should also realize that the measure

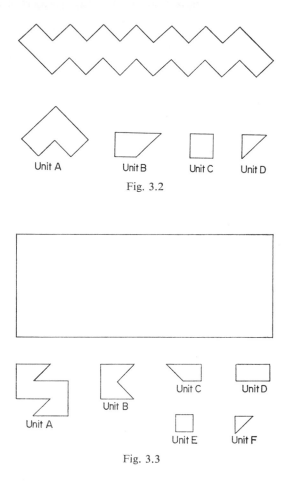

Unit A   Unit B   Unit C   Unit D

Fig. 3.2

Unit A

Unit B

Unit C   Unit D

Unit E   Unit F

Fig. 3.3

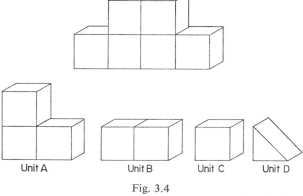

Unit A   Unit B   Unit C   Unit D

Fig. 3.4

of any object depends on the unit that is used to measure the object. Finally, the student should realize that although the unit of measure is arbitrary, men have selected a few units that are usually more convenient to use than other units.

Once the student has accepted the idea that convenient units for measuring the areas of plane figures are square units, he is ready to use this unit to develop formulas for the areas of these plane figures. If the teacher is alert in preparing problems, the counting procedure often yields an exact answer and is available and appropriate for many simple problems. As the problems become more difficult, the student gets the feeling that there ought to be a better way to compute the area than counting, even though counting may be an appropriate method. Figure 3.5 shows how to create in the student the need for a better method of computing the area of a rectangle. The area of rectangles A and B in figure 3.5 can easily be found by counting, but finding the area of rectangle C by counting is a very tedious task and should instill in the student a need for an easier method to determine the area of a rectangle.

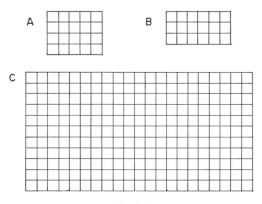

Fig. 3.5

Counting is not always an exact procedure for finding the areas of figures. To determine the area of each of the figures in figure 3.6, it will be necessary for students to estimate. Some procedure must be agreed on for counting the squares that are not entirely contained within the geometric figure. The presence of incomplete squares means that students will experience difficulty in determining the area of these figures. This difficulty, plus the tedium of counting squares in such a figure as E in figure 3.6, should again arouse in students the need for an easier method for determining the area of plane geometric figures. Yet it should also create an understanding of what is meant by area.

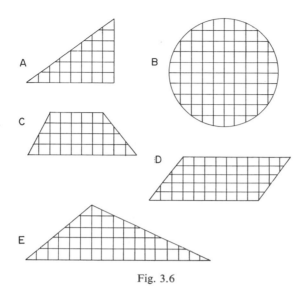

Fig. 3.6

There is some ambiguity about what is meant by length, area, and volume. Some mathematicians assert that these concepts are numbers. Hence it is odd to speak of measuring length, area, or volume. Other mathematicians are willing to equivocate in the use of these terms and find it reasonable to speak about measuring length, area, and volume. Thus *area* refers to a measure of the region. This understanding of area is as important as, if not more important than, getting the student to use a formula to determine the area.

One other important concept that is necessary for developing many of the area formulas is the concept of conservation of area. Even though conservation of area is taught in the elementary school, it needs to be reintroduced and reinforced in the junior high school. Thus the student must realize that the area of shape A in figure 3.7 is not altered if it is cut and reassembled into a different shape such as shape B. Similarly, the areas of shapes C and D in figure 3.7 are not altered if they are cut and reassembled into the different shapes indicated by the dotted lines. Unless the student understands conservation of area, he will not be able to understand some of the demonstrations for determining the area of geometric figures.

Returning once again to the concept of the area of a rectangle, the student should realize by now that there certainly must be an easier method of finding the area of a rectangle than by counting squares. One approach would be to have the student work on exercise sets such as the one in figure 3.8. Again the student is asked to count, but this time he is asked

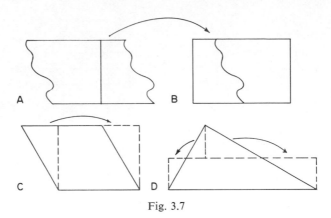

Fig. 3.7

to find not only the area but also the length and width of each rectangle in figure 3.8 and to enter his findings in the following table:

Rectangle	Length	Width	Area
A			
B			
C			
D			
E			

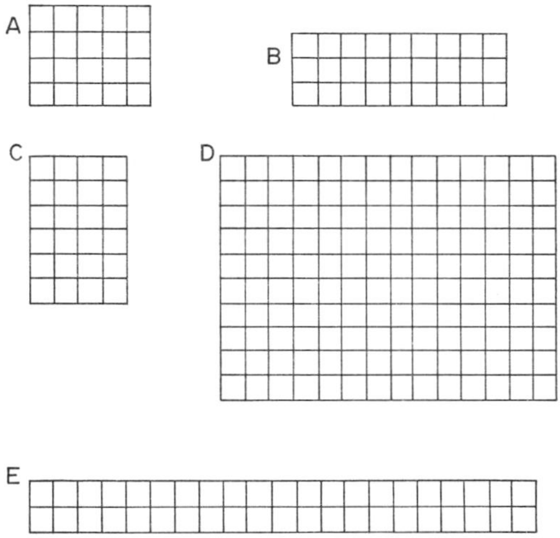

Fig. 3.8

The geoboard and its substitute, dot paper, have recently become very popular manipulative devices for teaching the area of polygons. For example, if the shaded region (A) in figure 3.9 is used to represent one unit of area, then the area of each polygon (B through H) in figure 3.9 can be determined. One very interesting relation, known as Pick's theorem, can be investigated by first having the student complete the following table and then studying its data. The student is then asked if he can see a pattern in the table.

	Number of pegs on the perimeter	Number of pegs in the interior	Area
A			
B			
C			
D			
E			
F			
G			
H			

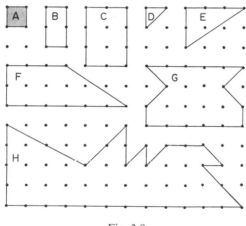

Fig. 3.9

The unit selected to measure the area of a polygon can be altered when the geoboard is used in a manner similar to the one used earlier. (See, for example, figs. 3.1–3.4.) Thus the children can determine the area of shapes A–E in figure 3.10 in terms of each of the shaded units of area in the figure, recording their data in a table such as the one that follows.

Figure	Area using	
	unit 1	unit 2
A		
B		
C		
D		
E		

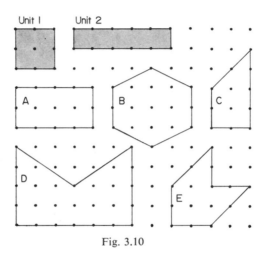

Fig. 3.10

Is Pick's theorem still true when these units of area are used?

Similar approaches could be taken for developing the other formulas for the area of various polygons. Yet for many of those formulas it is important that the student understand conservation of area. This enables the student to understand that the area of a parallelogram is the same as the area of a rectangle with the same length and height.

Another concept that is begun in the elementary school and that should be extended into the junior high school is similarity. Similar polygons have two important properties that students should realize: (1) corresponding angles are congruent, and (2) corresponding sides are proportional. It is very easy for either the teacher or the textbook to tell the students that if two polygons are similar then they possess both these properties. Figures 3.11 and 3.12 give examples of inductive lessons that could be used as an alternative approach to teaching these properties. In figure 3.11, students are told that the three pairs of polygons are similar. They are asked to measure each of the angles of the six polygons with their protractors and to record these measurements in the following table. Then they are asked if they see any pattern.

Polygon	$m \angle A$	$m \angle B$	$m \angle C$	$m \angle D$	$m \angle E$
I					
II					
III					
IV					
V					
VI					

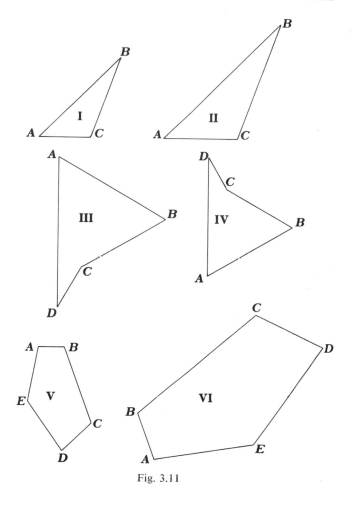

Fig. 3.11

In figure 3.12 the students are told that polygon I is similar to polygon II and that polygon III is similar to polygon IV. Then they measure with their rulers the lengths of corresponding sides of the similar polygons and complete the ratios of these lengths in tables such as the following. Finally, they are asked if they see a pattern.

AB	BC	AC	DE	EF	DF	$\dfrac{AB}{DE}$	$\dfrac{BC}{EF}$	$\dfrac{AC}{DF}$

RS	ST	TU	UR	WX	XY	YZ	ZW	$\dfrac{RS}{WX}$	$\dfrac{ST}{XY}$	$\dfrac{TU}{YZ}$	$\dfrac{UR}{ZW}$

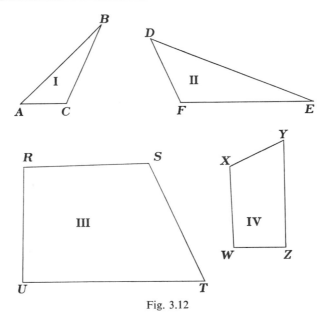

Fig. 3.12

Different approaches

Recently there have been some innovations in the teaching of informal geometry at the junior high school level. To a large extent, this can probably be attributed to some of the many innovations that have been suggested for the traditional formal geometry course.

The largest change in the informal geometry described in the previous section has been a result of a program developed by the UICSM for slow learners enrolled in the eighth grade. Entitled "Motion Geometry," the program essentially is a course in geometry through isometric mappings. The topics and concepts in "Motion Geometry" are no different from the topics listed in the previous section. What makes "Motion Geometry" so new and different is the approach used to teach the topics.

At the present stage of development of the mathematics curriculum, it would be safe to say that few students have been introduced to any transformational geometry. For this reason, a transformational approach provides a fresh approach to teaching many geometric concepts. Since few secondary school students have studied transformational geometry, the slow learners using this approach are given the opportunity to learn something that the non–slow learners do not know either. For once, the slow learners are taught something new instead of getting just a rehash of mathematical concepts that they have seen—and failed to learn—many times before.

However, perhaps what is more important than the fresh approach to geometry provided through geometric transformations is the fact that it is a very manipulative approach toward teaching geometry. This provides the teacher more opportunity to get the students physically involved in the course material. Physical involvement by itself, however, is not significant. Unless students find this physical involvement meaningful or unless the physical involvement helps to make the mathematics meaningful, it will not benefit a mathematics program. In order for physical involvement to be significant, it must aid students in concept attainment. This physical involvement makes transformational geometry well suited for the laboratory approach to teaching mathematics, which is gaining in popularity with educators.

One of the most useful tools for a student taking UICSM "Motion Geometry" is a plentiful supply of tracing paper. (In 1970, the late Max Beberman stated [3, p. 2], "Teachers now using the program have found that they use approximately $40 worth of expendable supplies such as ditto masters, ditto paper, transparencies, tracing paper, index cards, etc. for each class that they teach.") Since intuitively two geometric figures are congruent if they are the same size and shape, students are encouraged to see if two geometric figures are congruent by tracing one of the figures and seeing if the tracing exactly matches the second figure. In this way students are able to verify physically if two figures are congruent or not.

"Motion Geometry" uses three basic motions: (1) *slide*—a motion along a straight line for a specified distance and direction without any accompanying twisting or turning; (2) *turn*—a motion along a circle with fixed center for a specified amount and and direction; and (3) *flip*—a "turnover" motion about a line that serves somewhat the same purpose as a mirror (26, p. T4).

An excellent introduction to "Motion Geometry" at the junior high school level can be found in Sanders and Dennis (24).

Another recent innovation in the teaching of informal geometry at the junior high school level approaches the key concepts of nonmetric geometry by using a finite geometry. For example, unit 4 of Immerzeel, Wiederanders, and Scott's *Ginn Mathematics, Level 7* (15) develops the basic concepts of "parallelism, perpendicularity, congruence, similarity, and the classification of angles, triangles, and quadrilaterals" within a set of thirty-six points (16, p. 40).

The thirty-six points are arranged in a 6×6 array, and each point is given a "letter" as its coordinate (see fig. 3.13). A line consists of 2, 3, 4, 5, or 6 points. For example, in figure 3.14, $\overleftrightarrow{GB}$ contains exactly 2 points, whereas $\overleftrightarrow{TD}$, $\overleftrightarrow{\text{Ⴉ}F}$, $\overleftrightarrow{\text{Ɔ}L}$, and $\overleftrightarrow{\text{Ⴉ}Y}$ contain 4, 6, 5, and 3 points, respectively. Students are reintroduced to the concepts of parallel, intersecting, and skew lines in this novel manner, which enables old ideas to be studied from a fresh point of view. All the lines in figure 3.14 are parallel. In figure 3.15, $\overleftrightarrow{AN}$ and $\overleftrightarrow{EQ}$ intersect at Ⴉ, and thus are intersecting lines. However, $\overleftrightarrow{AN}$ and $\overleftrightarrow{B\text{Ⴈ}}$ are examples of skew lines: they are not parallel and they do not intersect.

Fig. 3.13

The unit continues with the investigation of parallelism and perpendicularity through the idea of slope, two parallel lines and a transversal, congruent and similar triangles, and the classifications previously men-

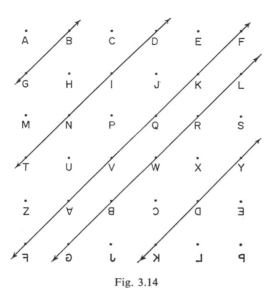

Fig. 3.14

tioned. With the exception of skew lines, everything is analogous to Euclidean geometry. Skew lines are deceptive. They are not parallel, and they do not intersect at a point. However, they can "cross," as in the example of $\overleftrightarrow{AN}$ and $\overleftrightarrow{B\daleth}$ in figure 3.15.

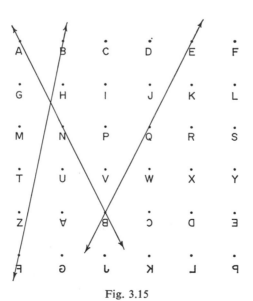

Fig. 3.15

Inductive lessons

A third aspect in which informal geometry can be used in the junior high school is in inductive lessons, which assist students in forming conjectures. Some of these examples were given in the discussion of continuation and extension. These examples were the ones dealing with the development of the area of some geometric figures.

Many of the inductive lessons in the junior high school will lead to conjectures. Some of these conjectures the students will be able to prove; some they will not be able to prove either because they are false or because the students do not, as yet, have the necessary mathematical tools. Others they will disprove.

It is only fair to point out that a proof to a junior high school student does not usually mean a deductive proof. A proof can consist of testing more examples that seem to verify the conjecture being tested until the students are satisfied that the conjecture is proved. This is not a mathematical proof, but it is certainly much more reasonable, and often more logical, to the majority of children of junior high school age. (This method of proof is sometimes called a "proof by exhaustion" because examples are tested until everyone is exhausted.) The teacher and the class must always remember that whenever they employ a "proof by exhaustion," the conjecture may be incorrect, and they may later find a counterexample.

Six examples of geometry problems that most junior high school students can solve are given here:

1. Determine the area and perimeter of each geometric figure shown in figure 3.16 and enter your answers in the table.

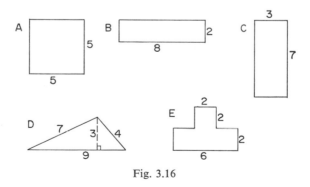

Fig. 3.16

What do you notice about the perimeters? What do you notice about the areas? Do you see any relationship between the areas and the perimeters?

Figure	Area	Perimeter
A		
B		
C		
D		
E		

2. Determine the number of diagonals in each of the polygons in figure 3.17 and enter your answers in the following table:

Polygon	No. of diagonals	No. of vertices	No. of diagonals from each vertex
Triangle			
Quadrilateral			
Pentagon			
Hexagon			
Septagon			
Octagon			

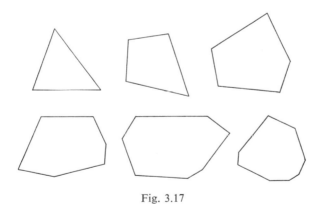

Fig. 3.17

Do you see any pattern? How many diagonals do you think a nonagon (9 sides) would have? How many diagonals do you think a dodecagon (12 sides) would have? How many diagonals do you think an n-gon (n sides) would have?

3. In figure 3.18, $\overleftrightarrow{AB} \parallel \overleftrightarrow{CD}$ and $\overleftrightarrow{QR}$ is the transversal of $\overleftrightarrow{AB}$ and $\overleftrightarrow{CD}$. Complete the table.

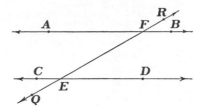

Fig. 3.18

Angle	Measure	Angle	Measure	Angle	Measure	Angle	Measure
∠QED		∠AFR		∠RFB		∠AFE	
∠EFB		∠CEF		∠FED		∠CEQ	

Here is another drawing (fig. 3.19) of two parallel lines intersected by a transversal. Complete the table.

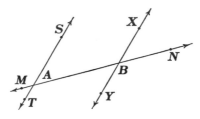

Fig. 3.19

Angle	Measure	Angle	Measure	Angle	Measure	Angle	Measure
∠YBN		∠MAT		∠MAS		∠BAS	
∠TAB		∠ABY		∠ABX		∠NBX	

Do you see any pattern? If so, what is it?

4. Fill in the corresponding row in the table for each of the polyhedrons shown in figure 3.20. Let V equal the number of vertices, E equal the number of edges, and F equal the number of faces.

Polyhedron	V	E	F
1	4	6	4
2			
3			
4			
5			
6			

Polyhedron	V	E	F
7			
8			
9			
10			
11			
12			

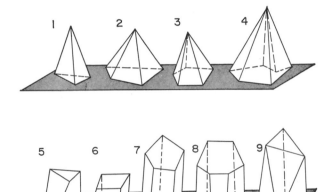

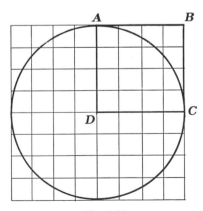

Fig. 3.20

Draw three more polyhedrons, different from the previous nine, and fill in the corresponding entries on the table. Do you see any patterns in the table?

5. Find the area of the circle in figure 3.21 and then complete the table.

Fig. 3.21

Area of circle	Area of $ABCD$	$\dfrac{\text{Area of circle}}{\text{Area of } ABCD}$

Repeat the procedure with the drawing shown in figure 3.22.

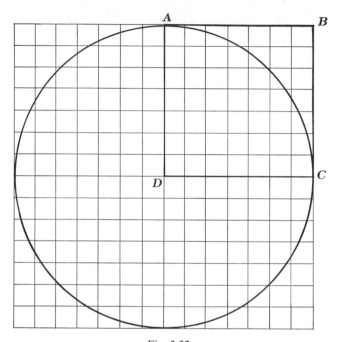

Fig. 3.22

Area of circle	Area of $ABCD$	$\dfrac{\text{Area of circle}}{\text{Area of } ABCD}$

6. Four triangles are shown in figure 3.23. Using your protractor, measure the angles indicated and list your findings in the table.

$m\angle A =$	$m\angle C =$	$m\angle A + m\angle C =$	$m\angle x =$
$m\angle D =$	$m\angle F =$	$m\angle D + m\angle F =$	$m\angle y =$
$m\angle H =$	$m\angle G =$	$m\angle H + m\angle G =$	$m\angle z =$
$m\angle K =$	$m\angle J =$	$m\angle K + m\angle J =$	$m\angle w =$

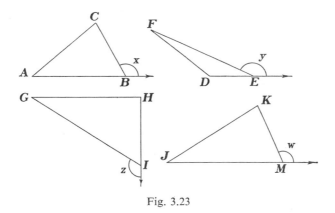

Fig. 3.23

The Continuing Role in Grades 10–14

The use of informal geometry in grades 10–14 has been greatly over-looked. Some informal geometry has been included in general mathematics and technical mathematics and has been of the same caliber as the informal geometry in grades 7–9. However, the use of informal geometry as an integral part of formal geometry has been ignored by all but a few.

Geometry courses have been courses in deductive reasoning. Students were given theorems and asked to prove them. Axioms, definitions, and theorems were arranged in a more or less logical order. Students were seldom given a situation, asked to use inductive reasoning to form some conjectures, and then asked to attempt either to prove or to disprove these conjectures using deductive reasoning. One of the main purposes of this section will be to give examples of how informal geometry can be used to provide opportunities for students to form conjectures.

Several reasons can be cited for advocating that informal geometry should become an integral part of the formal geometry course. One reason is that some students who have difficulty with deductive proofs will excel at reaching conjectures. However, probably the most important reason is that it allows students some of the excitement of generating their own con-jectures and of being a mathematician. In this section examples will be given of how informal geometry can be used to help stimulate the study of formal geometry in some of the approaches discussed in the following chapters.

Transformations

The transformational approach to geometry is one area in which efforts have been made to encourage informal investigations of geometric ideas.

For example, the first five chapters of Coxford and Usiskin's *Geometry: A Transformation Approach* (14) is devoted to an informal investigation of geometry and geometric transformations. Much of this, however, consists of a review of algebra and of giving definitions. No formal proof is begun until chapter 6.

The first geometric transformation introduced is a reflection in a line. A reflection in a line can be introduced by considering the techniques used in UICSM's "Motion Geometry" program. Consider some geometric figure and a line of reflection such as in figure 3.24. Mark a reference dot on the line, place a sheet of tracing paper over the geometric figure and the line, and trace the geometric figure, the line, and the reference dot, as in figure 3.25. Flip the tracing face down so that the tracing of the line and the reference dot coincides with the original line and reference dot, as in figure 3.26. The tracing of the geometric figure is the reflection of the original figure in the line (26, pp. T4–T5).

Fig. 3.24

Fig. 3.25

Fig. 3.26

The UICSM "Motion Geometry" technique might be considered too unsophisticated by many tenth graders. However, it can be used to investigate what happens to points that are either on or not on the line of reflection. Results of this investigation can be used to motivate students to produce the formal definition of the reflection image of a point in a line. Once the definition is given, then two alternative techniques for finding the reflection image of a point are explained. One technique uses geometric construction with a compass; the other uses a protractor. Although the protractor technique is not as mathematically "pure" as the compass-only technique, it is quick, easy, and probably highly preferred by students (14, pp. 70–71).

The following example illustrates how to use the "protractor technique" to find the image A', of a point A over a reflecting line l, given A and l. Place the protractor so that its straight edge is perpendicular to l. See figure 3.27. Move the protractor until this edge is touching point A. Measure the distance from l to A and locate A' this same distance from l but on the opposite side of l and touching the edge of the protractor. A protractor that has a ruler of some kind along the bottom edge simplifies measuring from l to A and from l to A'.

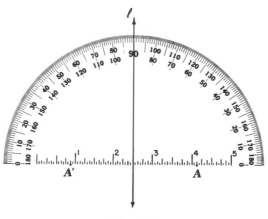

Fig. 3.27

Now that the student knows what a reflection is and has some techniques enabling him to determine reflection images, he is able to begin some informal investigations of reflections. For example, given $\triangle ABC$ and lines l and m, what is the result of reflecting $\triangle ABC$ in l and then reflecting this image in m?

The awkwardness of the last question should help create an urge for some notation. Let r_l denote a reflection in line l. Then $r_l\,(\overline{PQ})$ denotes the reflection of segment PQ in line l. Also, composition of transformations will be denoted by a "∘." Then $r_y \circ r_x\,(\overline{PQ})$ means reflect $\overline{PQ}$ in line

x and then reflect this result in line y. Now the last question can be re-phrased as follows:

Given $\triangle ABC$ and lines l and m, what is $r_m \circ r_l (\triangle ABC)$?

Investigations into this question should lead students to two more trans-formations: (1) The slide when $l \parallel m$ or $l = m$ and (2) the rotation when $l \cap m$ is exactly one point.

Other questions that can be investigated informally, if l, m, n denote lines, include the following twelve problems:

1. Is it true that $r_l \circ r_m = r_m \circ r_l$?

2. What is $r_l \circ r_m \circ r_n (\triangle ABC)$?

3. Is it true that $r_l \circ (r_m \circ r_n) = (r_l \circ r_m) \circ r_n$?

4. If $A' = r_l \circ r_m (A)$, $l \parallel m$, and $d(l, m)$ is the distance between l and m, is there any relationship between AA' and $d(l, m)$?

5. If $A' = r_l \circ r_m (A)$ and $l \cap m = \{P\}$, is there any relationship between $m \angle APA'$ and the angle between l and m?

6. If $R_{A,\theta}$ is a rotation around point A through an angle of θ, then what is $R_{P,\theta} \circ R_{Q,\phi}$? (Does it make any difference if $P = Q$? If $\theta = \phi$? If $\phi = -\theta$?)

7. Let T_{AB} be a slide in the direction of $\overrightarrow{AB}$ and for a distance equal to AB. Is it true that $T_{AB} \circ r_l = r_l \circ T_{AB}$?

8. Is it true that $T_{AB} \circ R_{P,\theta} = R_{P,\theta} \circ T_{AB}$?

9. Is it true that $l_m \circ R_{P,\theta} = R_{P,\theta} \circ l_m$?

10. If $l \cap m \cap n = \{P\}$ and A is any point in the plane, what is $r_n \circ r_m \circ r_l \circ r_n \circ r_m \circ r_l (A)$?

11. A billiard ball bounces off a side of a table in such a manner that the two lines along which it moves before and after hitting the sides are equally inclined to the sides. If a rectangular billiard table is bordered by lines l, m, n, p, will a ball placed at point A and hit so that it bounces consecutively off l, m, n, and p return to point A? What is the sum of the lengths of the diagonals of the billiard table? What is the total length the ball traveled before it returned to point A?

12. Given the billiard table in the previous exercise, a ball at point A, and a different ball at point B, in what direction should the ball at point A be hit so that it will hit the ball at point B after striking (1) exactly one side of the table, (2) exactly two sides of the table, (3) exactly three different sides of the table, or (4) all four sides of the table exactly once?

The twelve problems above are just a few of the questions in transforma-tional geometry that can be investigated inductively. Many students will be more interested in proving a theorem such as

If $l = m$, $r_m \circ r_l (A) = A'$, then $AA' = 2d(l, m)$

after investigating problem 4 above and conjecturing that $AA' = 2d(l, m)$.
Students find that it is more enjoyable to investigate their own hypotheses
instead of proving something that they are told is true.

Other problems

There are many interesting problems that students can intuitively investigate, and some examples of these problems follow. In most cases it does
not matter whether students are taking a coordinate geometry course, a
transformational geometry course, a vector geometry course, an affine
geometry course, or whatever. The problem will still be the same. Only
the method of proof will differ.

13. Shown in figure 3.28 is a triangle, $\triangle ABC$. On each side of the triangle construct an equilateral triangle external to $\triangle ABC$, as in figure 3.29. What happens

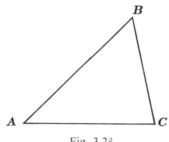

Fig. 3.28

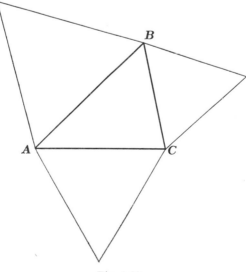

Fig. 3.29

when you connect the centroids of the three equilateral triangles? Label this new triangle $\triangle DEF$.

Repeat the exercise, only this time construct the equilateral triangles so they intersect the interior of $\triangle ABC$, as in figure 3.30. Let the centroids of these equilateral triangles be G, H, I. What can you say about $\triangle GHI$?

Compute the area of $\triangle ABC$, $\triangle DEF$, and $\triangle GHI$. Do you see any relationship?

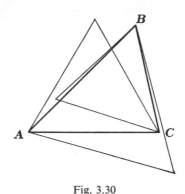

Fig. 3.30

14. Triangle ABC in figure 3.31 is an equilateral triangle. The three medians $\overline{AD}$, $\overline{BE}$, and $\overline{CF}$, are concurrent at P. Are the six triangles APE, APF, BPF, BPD, CPD, and CPE congruent? Do the six triangles have equal areas?

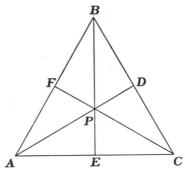

Fig. 3.31

Let $\triangle ABC$ be an isosceles triangle with $\overline{AB} \cong \overline{AC}$. Draw the three medians to form six smaller triangles the sum of whose areas is the area of $\triangle ABC$. Are there any pairs of congruent triangles? Are there any pairs of triangles which are not congruent but which have equal areas?

Now draw a triangle, $\triangle ABC$, with no two of its sides congruent. Again, draw its medians to intersect at P and form six triangles the sum of whose areas is the area of $\triangle ABC$. Are there any pairs of congruent triangles? Are there any pairs of triangles which are not congruent but which have equal areas?

15. Let C be any circle and P a point in the interior of C. Draw a chord containing point P. Label one endpoint of the chord A_1 and the other endpoint B_1. (See fig. 3.32.) Measure $\overline{A_1P}$ and $\overline{B_1P}$ and let these distances be a_1 and b_1, respectively. (Note: Students will usually find it more convenient to use the metric system than the English system.) Draw several other chords, $\overline{A_2B_2}$, $\overline{A_3B_3}$, and so on, containing point P. Let $a_i = A_iP$ and $b_i = B_iP$. Complete the following table. Do you see any relationship?

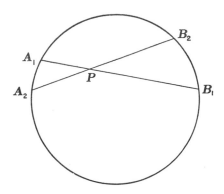

Fig. 3.32

i	a_i	b_i	$(a_i)(b_i)$
1			
2			
3			
4			

16. Let $ABCD$ be any quadrilateral. Let E, F, G, H be the midpoints of $\overline{AB}$, $\overline{BC}$, $\overline{CD}$, $\overline{DA}$, respectively. What can you say about quadrilateral $EFGH$? How does the area of $ABCD$ compare to the area of $EFGH$?

17. Given a circle with center O. Let $\angle ABC$ be an inscribed angle. Is there a relationship between $m\angle ABC$ and $m\angle AOC$?

18. Given a circle with center O. Let P be a point in the exterior of the circle. Let $\overrightarrow{PA}$ intersect circle O in points A and B and let $\overrightarrow{PC}$ intersect circle O in points C and D. (Note: If $A = B$, then $\overrightarrow{PA}$ is tangent to circle O. Also, if $C = D$, then $\overrightarrow{PC}$ is tangent to circle O.) Is there any relationship between $m\angle APC$, $m\,\widehat{AC}$, and $m\,\widehat{BD}$?

19. Let O and P be noncongruent circles, line l tangent to O at point A and to P at point B, and line m tangent to O at point C and to P at point D. If $l \cap m = \{E,\}$ is there any relationship between $\overline{AE}$, $\overline{BE}$, $\overline{CE}$, and $\overline{DE}$?

20. Let $ABCD$ be any convex quadrilateral. On each side of the quadrilateral construct a square external to $ABCD$, as in figure 3.33. Let M_1, M_2, M_3, M_4, respec-

tively, be the centers of these squares. What <u>kind</u> of <u>figure</u> is quadrilateral $M_1M_2M_3M_4$? Is there any relationship between $\overline{M_1M_3}$ and $\overline{M_2M_4}$? Let $ABCD$ be a parallelogram and repeat the exercise.

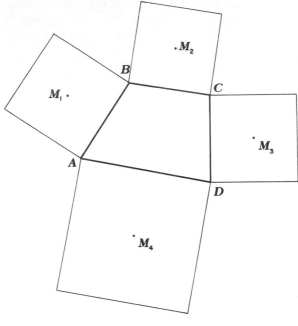

Fig. 3.33

Repeat the exercise, only this time construct the squares so that they intersect the interior of $ABCD$ as in figure 3.34. Let N_1, N_2, N_3, N_4, respectively, be the centers of these squares. What kind of figure is quadrilateral $N_1N_2N_3N_4$? Is there any relation-

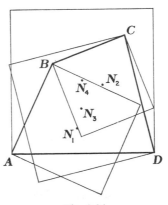

Fig. 3.34

ship between $\overline{N_1N_3}$ and $\overline{N_2N_4}$? Let $ABCD$ be a parallelogram and repeat the exercise.

21. Let $\triangle ABC$ be any triangle. Draw the medians of $\triangle ABC$. Does anything seem to happen?

22. Let $\triangle ABC$ be any triangle. Construct the angle bisectors of $\triangle ABC$. Does anything seem to happen?

23. Let $\triangle ABC$ be any triangle. Let D, E, and F be points on $\overline{BC}$, $\overline{CA}$, and $\overline{AB}$, respectively, such that D is halfway around the triangle from A, E is halfway around from B, and F is halfway around from C. Draw $\overline{AD}$, $\overline{BE}$, and $\overline{CF}$. Does anything appear to happen?

24. Let $\triangle ABC$ be any triangle. Let l_1, l_2, and l_3 be the perpendicular bisectors of $\overline{AB}$, $\overline{BC}$, and $\overline{CA}$, respectively. What seems to be true about l_1, l_2, and l_3?

25. Let $\triangle ABC$ be any triangle. Let D be a point on $\overline{BC}$ such that $\overrightarrow{AD}$ is the bisector of $\angle BAC$. Is there any relationship between $\overline{AB}$, $\overline{BD}$, $\overline{DC}$, and $\overline{CA}$?

26. Let $\triangle ABC$ be any triangle. Let D and E be the midpoints of $\overline{AB}$ and $\overline{BC}$, respectively. What seems to be true about $\overline{DE}$? Measure $\overline{DE}$, $\overline{AB}$, $\overline{BC}$, and $\overline{AC}$. Do you see any relationship?

27. Let $ABCD$ be any square. Join the midpoints of the sides of $ABCD$ to form a square, $EFGH$. Determine the areas of $ABCD$ and $EFGH$. Do you see any relationship?

28. Let $\triangle ABC$ and $\triangle DEF$ be any two noncongruent triangles such that $\overline{AB} \parallel \overline{DE}$, $\overline{BC} \parallel \overline{EF}$, and $\overline{AC} \parallel \overline{DF}$. Draw $\overleftrightarrow{AD}$, $\overleftrightarrow{BE}$, and $\overleftrightarrow{CF}$. What can you say about $\overleftrightarrow{AD}$, $\overleftrightarrow{BE}$, and $\overleftrightarrow{CF}$?

29. Let $\triangle ABC$ be any triangle and let D, E, and F be the midpoints, respectively, of $\overline{AB}$, $\overline{BC}$, and $\overline{AC}$. How is $\triangle ABC$ related to $\triangle DEF$? How does the area of $\triangle ABC$ compare with the area of $\triangle DEF$?

30. Given in figure 3.35 $\triangle ABC$ and point P not on $\overleftrightarrow{AB}$, $\overleftrightarrow{AC}$, or $\overleftrightarrow{BC}$. Let $\overleftrightarrow{PA} \cap \overleftrightarrow{BC} = \{X\}$, $\overleftrightarrow{PB} \cap \overleftrightarrow{AC} = \{Y\}$, $\overleftrightarrow{PC} \cap \overleftrightarrow{AB} = \{Z\}$. Measure each of the indicated

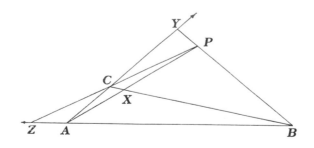

Fig. 3.35

segments and place your results in the following table. Draw two more triangles and repeat the experiment. Do you see any pattern?

Triangle	AZ	ZB	BX	XC	CY	YA
I						
II						
III						

31. Draw three concurrent lines. (See figure 3.36.) On one of these lines, choose points A and A'. On one of the other lines, choose B and B', and on the third line choose C and C'. Draw $\triangle ABC$ and $\triangle A'B'C'$. Draw $\overleftrightarrow{AB}$ and $\overleftrightarrow{A'B'}$. If they intersect, call this point of intersection P. Draw $\overleftrightarrow{AC}$ and $\overleftrightarrow{A'C'}$. If they intersect, call this point of intersection Q. Draw $\overleftrightarrow{BC}$ and $\overleftrightarrow{B'C'}$. If they intersect, call this point of intersection R. What seems to be true about P, Q, and R?

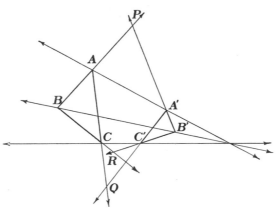

Fig. 3.36

32. Let C be any simple closed curve. If the area of C is 25 square units, what is the perimeter?

33. Let S be any simple closed curve. If the perimeter of S is 100 units, what is the area?

34. Given $\triangle ABC$, let m be a line that intersects $\overleftrightarrow{BC}$, $\overleftrightarrow{CA}$, $\overleftrightarrow{AB}$ in points L, M, N, respectively, as in figure 3.37. Measure each of the segments indicated in the

Triangle	AN	NB	BL	LC	CM	MA
I						
II						
III						

table and write in the measurements. Draw two more triangles and repeat the procedure. Do you see any pattern?

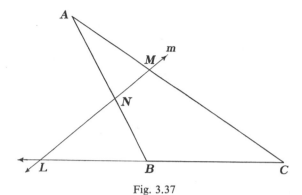

Fig. 3.37

35. Let $\triangle ABC$ be any triangle. Draw a line m so that $m \parallel \overline{BC}$ and m intersects $\overline{AB}$ and $\overline{AC}$ in points D and E, respectively, as in figure 3.38. Measure each of the segments indicated in the following table and write the measurements in the table. Complete the table. Do you see any pattern? Repeat the procedure with another triangle. Do you get the same results?

Triangle	AB	AC	BC	AD	AE	DE	$\dfrac{AD}{AB}$	$\dfrac{AE}{AC}$	$\dfrac{DE}{BC}$
I									
II									
III									

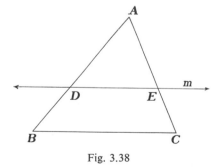

Fig. 3.38

Most of the problems above were taken from theorems that are stated and proved in many current high school geometry texts. Some of the problems are relatively easy, and others are quite difficult. The easy problems can be given to students one or two days before the teacher plans to prove the theorem. The more difficult problems can be given a week or two in advance. This allows ample time for the teacher occasionally to give a hint that may help the students if they do not see any relationships.

Assigning interesting problems that are not in the text prevents students from looking ahead in the text until they find the theorem (or answer) that states the relationship. For a few of the problems no relationship exists—but students often think that a relationship does exist. Once the students believe that they have found a relationship, then they should attempt to prove their conjecture. Not until the relationship is proved can the conjecture be considered a theorem. The use of informal geometry in what is usually considered a formal geometry course should make the study of geometry more interesting.

REFERENCES

1. Association of Teachers of Mathematics in New England. "Report of the Committee to Recommend a Suitable Program in Mathematics for the Junior High School." *Mathematics Teacher* 11 (March 1919): 133–40.

2. Beatley, Ralph. "Third Report of the Committee on Geometry." *Mathematics Teacher* 28 (October and November 1935): 329–79, 401–50.

3. Beberman, Max. "Memo to Potential UICSM Resource Team Participants." Mimeographed. Urbana, Ill.: UICSM, 1970.

4. Betz, William. "The Development of Mathematics in the Junior High School." In *A General Survey of Progress in the Last Twenty-five Years.* First Yearbook of the National Council of Teachers of Mathematics. New York: Bureau of Publications, Teachers College, Columbia University, 1926.

5. ———. "The Teaching of Intuitive Geometry." In *The Teaching of Mathematics in the Secondary School.* Eighth Yearbook of the National Council of Teachers of Mathematics. New York: Bureau of Publications, Teachers College, Columbia University, 1933.

6. Breslich, E. R., Marie Gugle, William Betz, C. B. Walsh, and Raleigh Schorling. "A Symposium of Discussion on the National Committee Report on Junior High School Mathematics." *Mathematics Teacher* 14 (January 1921): 16–41.

7. Butler, Charles H., and F. Lynwood Wren. *The Teaching of Secondary Mathematics.* 4th ed. New York: McGraw-Hill Book Co., 1965.

8. Cambridge Conference on School Mathematics. *Goals for School Mathematics.* Boston: Houghton-Mifflin Co., 1963.

9. College Entrance Examination Board, Commission on Mathematics. *Program for College Preparatory Mathematics.* New York: The Board, 1959.

10. Commission on Post-War Plans. "The Second Report of the Commission on Post-War Plans." *Mathematics Teacher* 38 (May 1945): 195–221.

11. Committee of Fifteen. *Report of the Committee of Fifteen.* Boston: New England Publishing Co., 1895.

12. Committee of Fifteen on Geometry Syllabus. "Provisional Report on Geometry Syllabus." *School Science and Mathematics* 11 (1911): 330, 434–60, 509–31.

13. ———. "Final Report of the National Committee of Fifteen on Geometry Syllabus." *Mathematics Teacher* 5 (December 1912): 46–131.

14. Coxford, Arthur F., and Zalman P. Usiskin. *Geometry: A Transformation Approach.* River Forest, Ill.: Laidlaw Bros., 1971.

15. Immerzeel, George, Don Wiederanders, and Lloyd F. Scott. *Ginn Mathematics, Level 7.* Lexington, Mass.: Ginn & Co., 1972.

16. ———. *Teacher's Manual, Ginn Mathematics, Level 7.* Lexington, Mass.: Ginn & Co., 1972.

17. Joint Commission of the Mathematical Association of America and the National Council of Teachers of Mathematics. *The Place of Mathematics in Secondary Education.* Fifteenth Yearbook of the National Council of Teachers of Mathematics. New York: Bureau of Publications, Teachers College, Columbia University, 1940.

18. National Committee on Mathematical Requirements. *Junior High School Mathematics.* A preliminary report by the NCMR. Department of the Interior, Bureau of Education, Secondary School Circular no. 6. Washington, D.C.: Government Printing Office, 1920.

19. National Committee on Mathematical Requirements of the Mathematical Association of America. *The Reorganization of Mathematics in Secondary Education.* Buffalo, N.Y.: The Association, 1923.

20. National Council of Teachers of Mathematics. *A History of Mathematics Education in the United States and Canada.* Thirty-second Yearbook. Washington, D.C.: The Council, 1970.

21. National Education Association. *Report of the Committee of Ten on Secondary School Studies.* New York: American Book Co., 1894.

22. Nightingale, A. F. "Report of the Committee on College Entrance Requirements." In *Journal of Proceedings and Addresses of the Thirty-eighth Annual Meeting.* Chicago: National Education Association, 1899.

23. Reeve, William David. "The Teaching of Geometry." In *The Teaching of Geometry.* Fifth Yearbook of the National Council of Teachers of Mathematics. New York: Bureau of Publications, Teachers College, Columbia University, 1930.

24. Sanders, Walter J., and J. Richard Dennis. "Congruence Geometry for Junior High School." *Mathematics Teacher* 61 (April 1968): 354–69.

25. Secondary School Curriculum Committee of the National Council of Teachers of Mathematics. "The Secondary Mathematics Curriculum." *Mathematics Teacher* 52 (May 1959): 389–417.

26. University of Illinois Committee on School Mathematics. *Motion Geometry, Book 1: Slides, Flips, and Turns.* Teacher's ed. New York: Harper & Row, 1969.

27. Wagner, John. "The Objectives and Activities of the School Mathematics Study Group." *Mathematics Teacher* 53 (October 1960): 454–59.

Formal Geometry in the Senior High School

4

Conventional Approaches Using Synthetic Euclidean Geometry

CHARLES BRUMFIEL

IN THIS chapter we shall consider options available to the teacher who prefers to teach a somewhat conventional geometry course, one whose roots extend back more than two thousand years to the work of the Greek geometers. Many things have changed since students studied Euclid in Plato's Academy, but many things have remained much the same. It is difficult to improve on Socrates' techniques for involving students in the educational process. Today we continue to present many of Euclid's proofs in geometry classrooms for the simple reason that no one has found better proofs. For example, Euclid's proof of the triangle inequality shown in figure 4.1 is beautiful in its simplicity and elegance:

$\overline{BA}, \overline{AD}$ are together greater than $\overline{BC}$ because in $\triangle DBC$, $\angle C$ is greater than $\angle D$, and it has been established earlier that in any triangle the greater of two angles lies opposite the greater side.

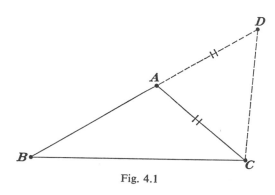

Fig. 4.1

In organizing a conventional geometry course, we have essentially three sources for a teaching philosophy—Euclid, Descartes, and Hilbert. Euclid's treatment of geometry is wholly a development of geometric intuition. There is no measurement in Euclid, that is, no assignment of measures (numbers) to segments, angles, or regions. Euclid develops the pre-measurement concepts of *greater than, less than,* and *same size*—the triangle inequality is concerned, not with *how much* greater are the two sides of a particular triangle than the third, but with the fact that they *are* greater. Euclid's treatment of geometry has an inherent appeal to the aesthetic and philosophic sides of our natures. If it were possible to sort children into classes, separating out those who are not going to make a serious study of science, then for these future housewives, clerks, musicians, lawyers, writers, politicians, social workers, and so on, a Euclidean approach to geometry, with its emphasis on the beauty of form, clear verbal arguments, and purity of logical reasoning, would undoubtedly be far more valuable than a computational approach.

During the sixteenth century, ideas of analytic geometry were created and very quickly had a profound influence on mathematical development. This marriage of algebra and geometry led quickly to the invention of the calculus, the tool of the practical man and the scientist. We begin to teach these ideas to children when we introduce the number line. It is essential that in our teaching of future scientists and technicians, we bring geometric, arithmetic, and algebraic modes of reasoning together. The arithmetical formulation of limit concepts in calculus is really too difficult to teach to college freshmen, but if these students have well-developed geometric intuitions and substantial skill in the mechanics of algebra, then most of them can master enough of the calculus to serve their professional needs.

Authors of conventional geometry texts today are attempting to present a mixture of Euclid and Descartes. We can disagree with particular decisions made by an author but not with his intent, which is to combine the purity of Euclid's geometric approach with the practicality we associate with numerical methods.

The work of Hilbert can be viewed as a reconciliation of Euclid with Descartes. Hilbert did two things to clarify our understanding of geometry, thereby showing how to "remove every blemish" from Euclid: first, he corrected Euclid's postulates in a wholly geometric fashion, with no reliance on numerical concepts, and second, he showed that the geometric structure defined by the corrected postulates is a mirror image of the analytic geometry of Descartes. We can either begin with Hilbert's geometric postulates and construct analytic concepts, or we can begin with postulates characterizing the real numbers and construct geometry.

Threads of geometry should run through most of the mathematics we teach. In order to make wise decisions in our selection of geometric material and in our choice of approaches to geometric concepts, we should understand clearly the specific contributions that each of these three philosophic approaches—that of Euclid, Descartes, and Hilbert—can make to the teaching of geometry: Euclid for the dreamer or the artist, who wants to perceive beauty without dissecting it; Descartes for the practical man, who must feed, clothe, and shelter the rest of us; Hilbert for the mathematician or the scientist, who has a fierce thirst for knowledge and must understand everything.

Combining Analytic and Synthetic Concepts

Analytic and synthetic geometry sit at two extremes: a geometry is analytic if a coordinate system (employing real or complex numbers) is imposed on it, thus enabling the geometer to use the machinery of real or complex analysis; a geometry is synthetic if it foregoes the use of a coordinate system. If the analytic choice is made, no geometric postulates are needed; rather, definitions are used. For example, in plane analytic geometry, the plane can be defined as $R \times R$, the set of all ordered pairs of real numbers. Similarly, a distance function can be defined; for example, the Euclidean distance from $P(x_1, y_1)$ to $Q(x_2, y_2)$ is given by

$$d(P,Q) = \sqrt{(x_1 - x_2)^2 + (y_1 - y_2)^2}.$$

Lines in Euclidean geometry are defined as sets of points satisfying linear equations of the form $Ax + By + C = 0$. With the analytic approach to Euclidean geometry, all geometric concepts—betweenness, sides of a line, congruence, and so on—are precisely defined, and all theorems follow from calculations. The tools are the properties of real numbers. Euclid's *Elements* is a presentation of synthetic geometry. Descartes is generally credited with the conception of analytic geometry. Since the work of Hilbert, we have known that the two approaches are equivalent to each other in the sense that one can either begin with axioms characterizing the real number system and construct all geometric concepts or begin with the Hilbert axioms for geometry and construct the real number system. Actually, more than two thousand years before the work of Hilbert, the Greek mathematician Eudoxus, in developing his theory of ratio and proportion, showed how to construct the real number system. Implicit in his work is the concept of a Dedekind cut. As Eudoxus defines equality of ratios, each equivalence class of ratios establishes a Dedekind cut in the set of positive rational numbers.

Conventional approaches to school geometry are neither purely synthetic

nor purely analytic. The line is coordinatized, lengths are assigned to segments, and measures are assigned to angles; but the plane is not coordinatized for the purpose of proving theorems, a practice that seems wise.
The successful study of analytic geometry requires a high degree of algebraic sophistication. Analytic geometry leads directly into the calculus
and is well reserved for those students whose professional interests require
the serious study of university mathematics. As an example of the difficulty of an analytic development, a class of university juniors and seniors
(all mathematics majors or minors who plan to teach school mathematics)
was given the following problem:

> If lines are defined as sets of points satisfying linear equations, $Ax +$
> $By + C = 0$, prove that for each pair of distinct points (x_1,y_1), (x_2,y_2)
> there is one and only one line that contains them.

Not one student could handle this problem, although all twenty-five had
had three semesters of calculus and analytic geometry, some had taken
advanced calculus, and many had studied modern abstract algebra.

   In teaching geometry today it is sensible to use any analytic techniques
that facilitate the development of the ideas. We have an algebraic symbolism that was not available to Euclid, and it would be ridiculous not to
use this symbolism. But it is just as ridiculous to force the use of algebraic
notation and numerical computation in situations where intuitive geometric reasoning is more easily applicable. A good geometry course should
develop both geometric and algebraic modes of reasoning. In many situations the geometric point of view seems superior to the algebraic one,
and in other situations the opposite is the case. Often the two viewpoints
supplement each other nicely. For example, Euclid shows how to construct

Geometric solution   Algebraic solution

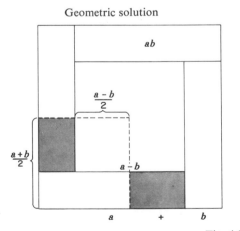

$$(a + b)^2 - (a - b)^2 = 4ab$$
so
$$\left(\frac{a + b}{2}\right)^2 - \left(\frac{a - b}{2}\right)^2 = ab$$

Fig. 4.2

a square "equal" to a given rectangle without employing the theory of similarity. He does this by first constructing two squares whose difference is equal to the rectangle and then using the Pythagorean theorem to construct a square equal to the difference of the squares. The problem of expressing the rectangle as a difference of squares is solved both geometrically and algebraically in figure 4.2. Note that one can look at the geometric diagram and "see" the algebra.

Student Understanding of Geometric Structure

During the elementary and junior high school years, students accumulate a great many geometric facts. Most of the better students learn during these years to distinguish between the geometry of the real world and mathematical geometry. They know that mathematical geometry is an abstraction; they have made a few informal proofs. Traditionally, the tenth-grade mathematics program has been dedicated to a systematization and extension of this geometric knowledge with primary emphasis on the concept of formal proof based on an axiomatic structure. No one objects to organizing and extending the knowledge of geometry during the high school years; however, the effectiveness of current approaches, which use rather complicated axiom systems, is being more and more sharply questioned. Critics of present practice argue that a concern with axiomatics, logic, and formal proof slows the development of geometry and actually has only superficial impact on students. It would be illuminating to run a series of studies and find out what understanding of the structure of geometry students retain at three stages in their education: (a) at the end of their geometry course, (b) at high school graduation, and (c) some years after their departure from high school.

In 1954 I started gathering bits of information on the understanding of axiomatic structure that students carry away from their high school geometry courses. Forty freshmen in a college-algebra class were asked to name some *theorems* they had proved in school geometry. Here are the only responses I received:

1. We proved that if equals are added to equals the sums are equal.

2. We proved that an isosceles triangle has two equal sides.

I asked for a theorem and what did I get? First an axiom, then a definition! These students in 1954 had no conception of the structure of geometry.

Recently, similar questions were posed to a group of twenty-five juniors and seniors at the University of Michigan. These were mathematics majors and minors who will themselves soon be teaching high school geometry.

Nearly all had studied an SMSG-influenced geometry based on the postulate of a distance function, the ruler postulate, the protractor postulate, and so on. The only axiom they could recall was that two points determine a line. Almost none of the twenty-five could classify correctly these two statements:

1. An isosceles triangle has two congruent sides (a definition, of course).
2. An isosceles triangle has two congruent angles (a theorem they all had seen proved).

Twelve of the twenty-five called both of these statements theorems. Not one of the twenty-five could say which of the following two statements is a postulate of special importance in Euclidean geometry:

1. Through a point P not on a line l, there is a line parallel to l.
2. Through a point P not on a line l, there is at most one line parallel to l.

Most of these students had used a text that states the following two postulates:

P_1. Each line contains at least two points.

P_2. There is exactly one line through any two different points.

When asked if either of these statements could be used to prove the other, every one insisted that P_2 implies P_1, and several tried to write out proofs. Of course, some of the students also claimed that P_1 implies P_2.

Naturally, my examination of these university students left me a bit dismayed. However, I related their confusion to the fact that they were at least five years removed from their study of geometry; so quite recently I submitted questionnaires to fifty-two high school students early in the fall of their junior year. These were students headed for advanced-placement calculus in their senior year. They had taken an accelerated geometry course during the preceding year and had had only a summer to forget their geometry.

The text these fifty-two students had used is a conventional one based on the currently popular ruler-protractor postulates. In the back of this text are listed 39 postulates and 215 theorems. The students are excellent students. Their school is one of the finest high schools in the United States. I daresay that in their college work most of them will not be handicapped by a lack of knowledge of the facts of geometry.

The most important items on the questionnaire were requests that students—

1. list as many postulates as they could remember;

2. list as many theorems as they could remember;
3. choose some one interesting theorem and write out a proof in any style.

A summary of responses follows. (The students had approximately fifty minutes to complete the questionnaire.)

Postulates

Twenty-six students left this page blank. They listed no postulates at all.

Sixteen students listed between them a total of thirty-four statements that they thought were postulates, but none of them were. Of these thirty-four statements, eighteen were theorems, five were definitions, one was false, three were nonsense, and seven defied classification.

Ten students listed one or more postulates each. Between them they named a total of nine of the thirty-nine postulates their text lists. Of course, along with these postulates they also listed theorems, definitions, and false statements.

Among the ten students who listed some postulates, not one student mentioned the ruler or protractor postulate. Only one space postulate was mentioned—the postulate that *space exists*. The student who listed this postulate also listed three other existence postulates: points exist, lines exist, and planes exist.

Theorems

The group performed better in the listing of theorems. Still, twenty-one of the fifty-two students could not think of a single theorem. Of course, mixed in with the theorems were axioms, definitions, and false statements. Theorems about transversals of parallel lines tended to dominate. Interestingly, only a single student mentioned a space theorem: A line perpendicular to a plane is perpendicular to every line of the plane through its foot. This fact suggests that space geometry plays an insignificant role in the school geometry program.

Proofs

I was most of all interested in seeing what theorems these students would single out for proof. What had they considered beautiful and impressive?

Forty-two students attempted no proof.

Ten students tried to make proofs, but only one presented a satisfying theorem whose proof was conceptually correct. His proof as he wrote it follows. (Note the abuse of the symbols $\Rightarrow$ and $\wedge$.)

THEOREM. *In a circle a chord "a" that is closer to the center than chord "b"* $\Rightarrow$ *ma* > *mb*. (See fig. 4.3.)

1. $r^2 = x^2 + a^2$
2. $r^2 = y^2 + b^2$
3. $x^2 < y^2$
4. $a^2 > b^2$
5. $a \wedge b > 0$
6. $a > b$

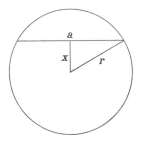 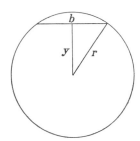

Fig. 4.3

Another student "proved" that the measure of two right angles is 180° by writing $90 + 90 = 180$. Another "proved" that triangle congruence implies similarity by appealing to the AA condition for similarity. Yet another "proved," If $ab = bc$, then $a = c$, by writing the following:

$$a\cancel{b} = \cancel{b}c$$
$$a = c$$

What can I conclude? Students of 1954 who studied an old-fashioned hodgepodge geometry had no conception of geometric structure. Students of today who have studied a tight axiomatic treatment also have no conception of geometric structure.

It would seem that I am arguing that any emphasis on axioms in school geometry is a waste of time. For the great majority of students, this is undoubtedly the case, as geometry is now taught. Certainly the responses described above suggest that we should look carefully and honestly at our geometry students and describe accurately their grasp of geometric concepts. We need to listen to students and learn what they really think. If we do so listen, what we hear will provide useful guidance as we experiment in the years ahead with various approaches to the teaching of geometry.

Let me make it clear that I am not arguing for an abandonment of the teaching of axiomatic structure in geometry. Certainly we are not teaching it effectively today, but this is no reason to quit trying. If we are committed to the goal of teaching students something about logical reasoning and proofs, we have no choice. At some stage, we must agree on the bases of our arguments. We have no alternative but to list some assumptions. The moment we start questioning statements of "fact" in the classroom, we begin to press students back on their fundamental beliefs. It is good for each generation of young people to learn that undefined terms are a necessity in every logical development of ideas. It is good for them to learn that the validity of an argument can be determined from the *logical structure* of the argument. How can we teach this better than in geometry? (A few will answer, "By teaching formal logic," but I choose to ignore this reply at this time.)

The single most important long-range goal in the teaching of mathematics is not to teach students to perform algorithmic calculations (we have computers for these tasks) but to teach students how to carry along chains of deductive reasoning—how to think in the "if . . . then . . ." style. Of course, we must have widely different objectives for different students. We do need a variety of choices in geometry, a range of textbooks, but above all, we wish students to learn to make concise, meaningful arguments.

We shall undoubtedly continue to teach synthetic geometry in many high school classrooms, and we shall continue to call attention to the fact that it is possible to describe an axiomatic structure for this geometry. I urge that there be no unseemly haste to do this. For the average class, it seems reasonable to focus the sharpest attention on the axiomatic structure near the end of the course—this could be the capstone of the year's work. Naturally, all through the course proofs will have been clearly based on assumptions, but there is no great hurry to examine the relationships among these assumptions. For a while, we may wish to assume all four of the triangle congruence conditions, SAS, SSS, ASA, and AAS; but if we do not at some stage in the course pause and make it clear that we can establish the other three conditions by assuming SAS alone, *and without dependence on the parallel postulate,* then we shall have taught a miserable course indeed.

Possible Choices for Axioms

It is easy to argue that it makes no difference what axioms are chosen for the development of a geometry text, since students will not remember any of them, but the selection of basic axioms does affect the way a text unfolds. The axioms influence authors and teachers and thus indirectly affect students.

Most texts today use axioms patterned after those introduced by Moise and used in the geometry of the School Mathematics Study Group. These axioms show the influence of Euclid, Hilbert, and George Birkhoff (a mathematician at Harvard during the first half of this century who suggested using a small set of axioms, two of which are essentially the familiar ruler and protractor postulates). In this section we shall look briefly at the axiomatic approaches of Euclid, Hilbert, and Moise.

Every geometry teacher should realize how faithfully the ten axioms of Euclid describe our intuitive feelings for geometric relationships. Euclid postulated that—

1. a line (segment) can be drawn from any point to any other point;
2. a line (segment) can be extended indefinitely;
3. given any first point as center and any second point, a circle can be drawn containing the second point.

These first three postulates are the ruler-compass construction postulates of Greek geometry. They reflect the Greek geometers' desire to understand exactly what configurations arise in the plane if one begins with only two points, draws the *one* line and *two* circles that these two points determine, then uses the *four new points* that are obtained as intersections to draw the *nine* new lines and *sixteen* new circles (count them!) that are determined, and so on ad infinitum. (See fig. 4.4.)

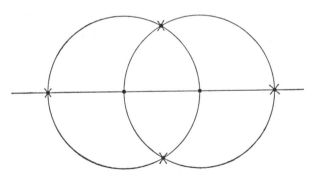

Fig. 4.4

Euclid knew that every angle and every line segment that appear as this network of lines and circles spreads over the plane will eventually be bisected. Moreover, every segment will be trisected, quadrisected, and so on. Triangles, parallelograms, rectangles, rhombi, squares, and regular pentagons appear in this expanding configuration. Many geometric figures, however, are not generated by this infinite sequence of construction—for example, no 20° angle is constructed, although angles of 3° are; no 7-sided regular polygon occurs, but 17-sided ones do; and no square and circle with equal areas are produced. Some of the above facts Euclid recognized; others are of more recent discovery.

Euclid's remaining seven axioms enabled him to prove that figures brought into existence by these first three axioms have certain properties. Continuing, Euclid postulated that—

4. all right angles are equal;

5. if a transversal of two lines forms interior angles on one side of the transversal and the sum of these angles is less than two right angles, then the lines meet on that side.

This last axiom is the famous fifth postulate, which characterizes Euclidean geometry. Note that a statement logically equivalent to postulate 5 above is the following:

If two lines cut by a transversal are parallel, then on each side of the transversal the sum of the interior angles is not less than two right angles.

It follows at once from this reformulation of Euclid's fifth postulate that for parallel lines cut by a transversal, the sum of interior angles on one side is two right angles and that alternate interior angles are equal. These consequences of the fifth postulate are probably more familiar to the reader than is the fifth postulate itself.

Continuing, Euclid further postulated that—

6. things equal to the same are equal;

7. equals added to equals yield equals;

8. equals subtracted from equals yield equals;

9. things that coincide are equal;

10. the whole is greater than each of its parts.

The unfolding of geometry from the Euclidean postulates is beautiful. Euclid proves a sequence of almost thirty theorems in Book 1 of his *Elements* before he finally uses the fifth postulate. Hence, these theorems are valid in the hyperbolic non-Euclidean geometry of Gauss, Bolyai, and Lobachevski. These theorems deal with the following:

1. The SAS, SSS, ASA, and AAS triangle congruence theorems

2. Congruence of base angles of an isosceles triangle (and the converse)

3. Bisection of segments and angles

4. Construction of perpendiculars

5. Congruence of vertically opposite angles

6. The exterior angle theorem and its important consequences, namely, uniqueness of the perpendicular from a point to a line, sum of any two angles of a triangle less than two right angles, greater side opposite the greater angle, the triangle inequality, and existence of parallel lines.

Without appealing to the parallel postulate, one cannot even prove the existence of a rectangle. Use of the fifth postulate leads quickly to the familiar theorems concerning parallelograms. The Pythagorean theorem becomes a key tool in the development of some important construction theorems.

School texts today tend to fall into two categories: those that respect Euclid's judgment and withhold the parallel postulate for some time, thus roughly following the Euclidean order of theorems (e.g., 6); and those that introduce a parallel postulate early in the development, thus blurring the distinction between Euclidean and non-Euclidean geometries (e.g., 5).

The spirit of Euclid's treatment is lost in the usual high school course. Anyone who reads carefully Book 1 of *The Elements* cannot help but be impressed by the way the ideas hang together. Students would form a much clearer picture of the structure of geometry than they presently obtain if at an appropriate time they could see a few of Euclid's sequences of theorems. The following diagram presents one such sequence:

SAS → Base angles of an isosceles triangle are congruent (and the converse) → SSS → Angles can be bisected → Segments can be bisected and perpendiculars can be drawn

(For details, see any translation of *The Elements*.)

Another remarkable sequence of theorems in Euclid follows the exterior angle theorem for triangles. Figures 4.5 through 4.12 suggest some of these theorems and their proofs.

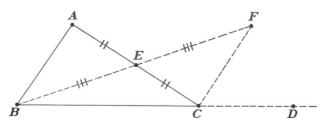

Fig. 4.5.   The exterior angle theorem

Proof. $\angle A \cong \angle ACF < \angle ACD$

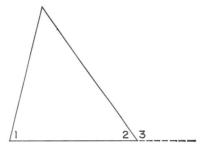

Fig. 4.6.   Any two angles of any triangle are together less than two right angles.

Proof. ∠ 1 + ∠ 2 < ∠ 2 + ∠ 3

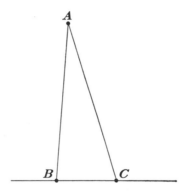

Fig. 4.7.   Uniqueness of perpendicular from a point to a line

Proof. If ∠ *B* and ∠ *C* were both right, this would contradict the preceding theorem.

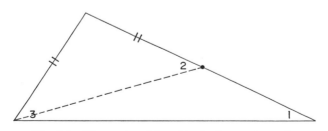

Fig. 4.8.   The greater side subtends the greater angle.

Proof. ∠ 1 < ∠ 2 < ∠ 3

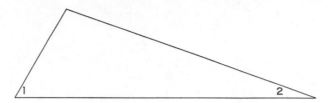

Fig. 4.9. The greater angle lies opposite the greater side; ∠1 > ∠2.

Proof. No geometrical argument is needed. The proof is a logical consequence of the preceding theorem.

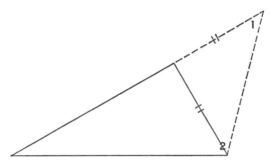

Fig. 4.10. The triangle inequality

Proof. ∠ 1 < ∠ 2, . . .

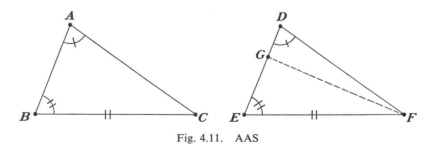

Fig. 4.11. AAS

Proof. If $\overline{DE} > \overline{AB}$, then on cutting off segment $\overline{EG}$ congruent to $\overline{BA}$ we should have ∠ EGF ≅ ∠ A ≅ ∠ D. But ∠ EGF > ∠ D.

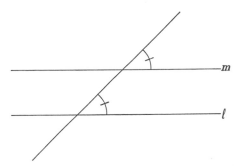

Fig. 4.12.   Existence of parallel lines

Proof. If lines *l* and *m* intersected at a point *P,* a triangle having an exterior angle congruent to a remote interior angle would be formed.

Much has been written about the errors that come to light when one scrutinizes the Euclidean development from the standpoint of present-day mathematical knowledge, but it seems a bit presumptuous to criticize a body of mathematics that stood without major improvements for over two thousand years. It was not until David Hilbert published in 1899 his Göttingen lectures on *The Foundations of Geometry* that it became clear exactly what errors Euclid had made and just how those errors could be corrected. A brief description of Hilbert's revision of Euclid's axioms follows.

Euclid's postulates 1 and 2 speak of drawing lines. Hilbert replaces these by *incidence* postulates, which include basic existence postulates. Modern texts that state such existence postulates as "In each plane are three noncollinear points" and "In each line are two points" are copying Hilbert.

Of course, Hilbert drops Euclid's third postulate. One needs only a definition of circle. Hilbert proves Euclid's fourth postulate, which states that all right angles are congruent. His parallel postulate is essentially Euclid's.

Euclid's postulate 6, "Things equal to the same are equal," comes in for more careful treatment by Hilbert. He postulates enough about segment congruence and angle congruence that he is able to prove each to be an equivalence relation.

Euclid's postulates 7 through 10 reflect his greatest oversights. Addition and subtraction of segments, angles, and regions cannot be treated correctly without introducing the concept of *betweenness.* Therefore, Hilbert replaces these postulates by a group that characterizes betweenness for points on a line. By postulating "addition of segments" and SAS, Hilbert is able to prove theorems concerning subtraction of segments (hence

Euclid's postulate 8 is not needed) and addition and subtraction of angles. The betweenness axioms make it possible to prove that each point on a line separates the line into two half-lines and that each line separates the plane. Now the interiors of angles and triangles can be defined and their properties proved.

Hilbert's postulate set for plane geometry is completed by listing two more, a *completeness postulate* and the *axiom of Archimedes*. This latter is the postulate that for any two segments there is a multiple of the first that exceeds the second. This postulate is really due to Eudoxus, who introduced it in order to develop a theory of ratio and proportion. Euclid employs the axiom but does not state it formally. The axiom of Archimedes, together with the congruence and betweenness postulates, enables one to define a measurement process that (given any unit segment) associates each segment to a real number. The completeness postulate operates in reverse, enabling one to prove that given a unit segment, for each real number r there exists a segment with length r.

Following is a version of the Hilbert postulates listed side by side with axioms presented in the Moise-Downs text mentioned earlier (6):

Hilbert   Moise-Downs text (SMSG postulates)

Incidence postulates

1. There are three noncollinear points in each plane.

2. There are two points on each line.

3. There is a unique line on each two points.

1. There are three noncollinear points in each plane.

2. There are as many points on each line as there are real numbers.

3. There is a unique line on each two points.

Betweenness postulates

4. If a point is between two points, the three points are collinear.

5. For any two points A, B, there is a point C such that B is between A and C.

6. For any three points on a line, at most one is between the other two.

7. If a line cuts one side of a triangle, it cuts a second side.

4. Each line separates the plane.

Congruence and measurement postulates

8. For each segment and each ray AX, there is a point B on the ray such that $\overline{AB}$ is congruent to the given segment.

5. To every pair of distinct points, there corresponds a positive number, the distance between the points.

Hilbert   Moise-Downs

Congruence and measurement postulates—continued

9. If $\overline{AB} \cong \overline{CD}$ and $\overline{A'B'} \cong \overline{CD}$, then $\overline{AB} \cong \overline{A'B'}$.

6. The points of a line can be placed in one-to-one correspondence with the real numbers so that the absolute value of the difference between any two numbers is the distance between their points.

10. If $\overline{AB} \cong \overline{A'B'}$, $\overline{BC} \cong \overline{B'C'}$, B is between A and C, and B' is between A' and C', then $\overline{AC} \cong \overline{A'C'}$.

7. Given two points P, Q of a line, the coordinate system can be chosen so that the coordinate of P is zero and the coordinate of Q is positive.

11. If ray AB is on the edge of a half-plane H and A' is any angle, then there is exactly one ray AP, with P in H, such that $\angle PAB \cong A'$.

8. To each angle there corresponds a real number between 0 and 180.

12. Every angle is congruent to itself.

13. SAS

14. Axiom of Archimedes

15. A completeness postulate

9. Let ray AB be on the edge of the half-plane H. For every number r between 0 and 180, there is exactly one ray AP, with P in H, such that $m \angle PAB = r$.

10. If D is in the interior of BAC, then $m \angle BAC = m \angle BAD + m \angle DAC$.

11. If two angles form a linear pair, then they are supplementary.

12. SAS

Parallel postulate

16. Through a point not on a given line, there is no more than one line parallel to the given line.

13. Through a point not on a given line, there is no more than one line parallel to the given line.

(Note that this is a *uniqueness,* not an *existence,* postulate.)

Using the Moise axioms, one defines betweenness and congruence. Hence, many of Hilbert's betweenness and congruence statements become theorems. Of course, with the Hilbert axioms one can prove that segments and angles have measures satisfying the statements included in the Moise axioms. It is possible to interest fine students in the difficult proofs that are involved in developing geometry meticulously from the Hilbert axioms. For example, figure 4.13 indicates how Hilbert proves that supplements of congruent angles are congruent.

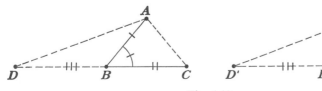

Fig. 4.13

Given $\angle B \cong \angle B'$, one uses SAS to prove $\triangle ABC \cong \triangle A'B'C'$, and continues to prove, using SAS, that both $\triangle ACD \cong \triangle A'C'D'$ and $\triangle ADB \cong \triangle A'D'B'$.

Figure 4.14 suggests the Hilbert proof of Moise's axiom 10, which states that sums of congruent angles are congruent.

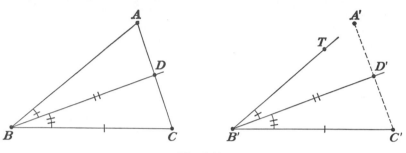

Fig. 4.14

By SAS, $\triangle DBC \cong \triangle D'B'C'$. Choosing A' as indicated, so that $\overline{D'A'} \cong \overline{DA}$, it follows by SAS that $\triangle A'D'B' \cong \triangle ADB$, hence A' falls on ray $B'T$. Now applying SAS to triangles $A'C'B'$ and ACB, it follows that $\triangle ABC \cong \triangle A'B'C'$.

Comparing the two sets of axioms above, it is clear that the Hilbert axioms have more geometric content. The SMSG axioms are deceptive. Students probably think they understand them, when pictures of number lines are drawn, but they really do not. There is a psychological circularity in the axioms that deceives nearly all students and teachers. The axioms pretend to lead to definitions of betweenness and congruence in terms of a distance function, but it is a certainty that students think of the distance between two points as something obtained by choosing a unit segment and then laying it off on the segment the points determine. That is, in their intuitive understandings of SMSG geometry, students rely on the concepts of betweenness and congruence, axiomatized by Hilbert, to justify the SMSG postulates 5, 6, and 7. Later in the course, they impassively watch "proofs" of properties of betweenness and congruence. Yet these are proofs of theorems without which the SMSG postulates would have made no sense to them in the first place!

Of course, the SMSG postulates are mathematically sound. The trouble is that almost no one understands their implications. An article by Clemens (2) nicely pinpoints some of the misconceptions related to the SMSG postulates. As Clemens points out clearly, a key postulate that forces the SMSG distance function to be the familiar Euclidean distance is the SAS postulate.

One way to gain insight into the significance of the SMSG postulates is to discard all geometric intuition and to subject postulates 5, 6, 7, 8, and 11 to numerical scrutiny alone. Postulate 5 would be satisfied if to every pair of points our distance function assigned the number 1. But, of course, this distance function will not satisfy postulate 6; so, we begin with postulate 6. Simply think of each line as a set with as many elements as has the set of real numbers, and for each line imagine an arbitrary one-to-one correspondence from the reals to its points. Now go back and define a distance function. Each pair of points P,Q lies on one and only one line. On that line these points have unique coordinates, p,q. We assign the number $|p - q|$ to this point pair as the "distance" from P to Q. Now postulates 5 and 6 are in perfect harmony, and we can draw pictures that violate our geometric intuition (see fig. 4.15). Note the conclusions: $d(T,Q) = 1$, $d(Q,R) = 10^{10} - 1$, Q is *between* T and R, $d(T,R) = 10^{10}$; $d(S,T) = 1/3$, $d(T,Q) = 1$, T is *between* Q and S, $d(S,Q) = 4/3$.

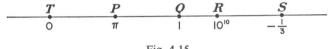

Fig. 4.15

We have defined betweenness and congruence numerically, and as far as postulates 5 and 6 are concerned, there is nothing wrong with the above picture.

Postulate 7 is easily satisfied. To get a coordinate system for the line above that has coordinate 0 for R and a positive coordinate for S and still gives the same distances between points on the line, simply multiply each coordinate by -1 and then add 10^{10} to each coordinate.

It should be obvious that in much the same way we have infinitely many choices for coordinate systems on pencils of rays, choices which when pictured would violate our intuitive feeling of what it means for one ray to be between two other rays. There are innumerably many distance functions and angle measure functions that satisfy these five postulates, 5, 6, 7, 8, and 11. It follows that other of Moise's postulates put restrictions on our choices for these functions. These are, of course, postulates 4, 9, 10, and 12. In studying figure 4.16, we see that postulate 4 (the plane separation postulate) takes away much of our freedom to slap arbitrary coordinate systems on lines.

Note that if P and Q are on opposite sides of l, and P and R are on the same side, then since postulate 4 requires that the point S in which lines l and m intersect be *between* Q and R, we do not have complete freedom in the assigning of coordinates to points on line m.

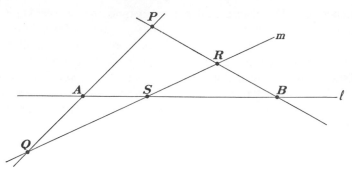

Fig. 4.16

In his article referred to earlier, Clemens shows how to construct a distance function and an angle measure function that together satisfy all of Moise's postulates except SAS. This is reminiscent of Hilbert's achievement in his *Foundations of Geometry*. Hilbert established the independence of several of his postulates from the others by exhibiting for each one a geometry in which that particular axiom is false but all other axioms are true. Clemens has shown the independence of SAS in the SMSG collection of axioms.

I am sure all the above, and much more, was known to Moise before he made the selection of axioms that appeared in the SMSG geometry and that since then have been copied imperfectly by many authors of school texts. One cannot fault the Moise axioms from a mathematical standpoint, but imitations of these axioms often have objectionable characteristics. One text may first postulate coordinate systems on lines and then postulate a distance function that agrees with them. This is putting the cart before the horse with a vengeance. Once coordinate systems are on lines, as we have seen, a distance function is defined.

Of course, criticisms of axiom choices are purely academic. Whatever axioms are printed in the text, students will not be disturbed, for they will cheerfully ignore them all. The goal should be to keep the axioms from getting in the way too much and to move students as efficiently as possible into the important work of geometry—the accumulation of useful facts and the development of skills in making proofs.

Summary

We are unrealistic today in our view of geometry. Concern with the particular axioms that shall be used in developing a geometry course seems hardly justified when we look candidly at our students and recognize exactly what impact the axioms have on them. Any approach to geometry that hopes to achieve spectacular success by some reorganization of the axiomatic structure is doomed to founder on the rocks of student indifference.

With all our present-day understanding of geometry, we have little more facility in imparting understanding and appreciation of geometric concepts than we had twenty years ago. The currently popular ruler-protractor postulates, although mathematically sound, have a built-in phoniness. Students misinterpret the mathematical significance of these two axioms, which are of little help in proving the respectable theorems that need to be covered in a serious geometry course. Employment of these axioms does little for students' algebraic skills and contributes even less to their ability to reason geometrically.

However, teaching has never been easy. We can look forward pleasurably to interesting experimentation as we try to do better. Good students accumulate quite a bit of geometrical knowledge during their junior high school years, and it should be possible to build a substantial geometry course on this foundation. It will be interesting to watch our struggles to do so.

REFERENCES

1. Birkhoff, George David, and Ralph Beatley. *Basic Geometry*. 2d ed. Chicago: Scott, Foresman & Co., 1941.
2. Clemens, Stanley R. "A Non-Euclidean Distance." *Mathematics Teacher* 64 (November 1971): 595–600.
3. Euclid. *The Thirteen Books of Euclid's "Elements."* Vol. 1. Translated from the text of Heiberg by Thomas L. Heath. Cambridge: Cambridge University Press, 1908.
4. Hilbert, David. *The Foundations of Geometry*. Translated by Edgar Jerome Townsend. Chicago: Open Court Publishing Co., 1902.
5. Jurgensen, Ray C., Alfred J. Donnelly, and Mary P. Dolciani. *Geometry*. Modern School Mathematics series. Boston: Houghton Mifflin Co., 1969.
6. Moise, Edwin E., and Floyd L. Downs. *Geometry*. Reading, Mass.: Addison-Wesley Publishing Co., 1967.

5

Approaches Using Coordinates

LAWRENCE A. RINGENBERG

THIS CHAPTER is concerned with high school geometry courses in which coordinates are introduced early and used extensively. These courses develop elementary geometry synthetically and analytically as a deductive system. One textbook designed for such a course is *Geometry with Coordinates* (4), a School Mathematics Study Group (SMSG) text.

The 1959 report of the Commission on Mathematics of the College Entrance Examination Board included a recommendation that coordinate geometry be a part of the high school geometry course. The commission believed that an early introduction of coordinates would add flexibility to the student's attack on geometrical problems, simplify the proofs of many theorems, and strengthen the student's preparation in algebra.

The first SMSG geometry textbook for high school students (3) includes an introduction to coordinates late in the book. The commission's report played an important part in the decision of the SMSG Panel on Textbooks to promote another experimental textbook on formal geometry. The result was *Geometry with Coordinates,* a book in which coordinates play a key role. The 1962 revised edition includes numerous improvements suggested by the teachers and consultants who used the 1961 preliminary edition in seven centers scattered throughout the United States. Since first published, *Geometry with Coordinates* has sold about 71,000 copies. In what follows we are primarily concerned with courses using an approach similar to that in *Geometry with Coordinates.* We shall refer to such courses as GWC courses and to *Geometry with Coordinates* specifically as SMSG-GW.

Some two thousand years ago Euclid first developed geometry as a deductive system. His development has some logical shortcomings, primarily due to his failure to recognize the need for undefined terms and his failure to formulate an adequate set of postulates. GWC courses develop geometry as a deductive system. Although complete proofs for all the theorems may not be included in the course, the postulational basis is complete. Teachers of such courses should be able (sometimes with the help of a hint from a teacher's edition) to supply missing proofs or to help interested students write them.

Some of the gaps in Euclid's logical development are plugged rather easily by postulates that capitalize on properties of real numbers. These postulates may be thought of as formal statements about abstract rulers and protractors. By means of these postulates, measurement and betweenness properties suggested by physical rulers and protractors are built into formal geometry developed as a deductive system. The ruler and protractor postulates play a key role in GWC courses.

In the first SMSG geometry there is a distance postulate that, roughly speaking, fixes the unit of distance in any given consideration and requires different coordinate systems on a line to have the same scale. In the SMSG-GW text, the distance and ruler postulates permit any two distinct points on a line to be the points with coordinates 0 and 1 in some coordinate system on that line. The SMSG-GW postulational basis is a bit more elaborate than the one for the SMSG geometry, but it permits coordinates to be used more effectively in the deductive development.

Characteristics of GWC Courses

GWC courses are college-preparatory courses suitable for classes of college-capable students and have two main objectives. One objective is to help students learn a body of important facts about geometrical figures. The facts of Euclidean geometry, interpreted physically, are facts about the space in which we live. These facts are important for intelligent citizenship and for success in careers. The other main objective is to help students attain a degree of mathematical maturity. In the elementary and junior high schools, students are encouraged to learn by observing and manipulating physical objects. There is considerable emphasis on intuitive and inductive reasoning. The power and the beauty of mathematics, however, are due primarily to its abstractness. The generality of its theorems makes possible a variety of applications. Understanding mathematics in the abstract is tantamount to understanding the deductive method in mathematics. Students will attain a degree of mathematical maturity when they experience the development of elementary geometry as a deductive system.

GWC courses are adaptable to students with varying backgrounds. It is expected that most students entering the tenth grade will have a background of experience in informal geometry and algebra that includes some exposure to the properties of real numbers and the basic concepts and terminology of sets. Sections on sets and equations may be included. The preparation of the students entering GWC courses should be considered in determining the amount of time to be devoted to algebraic topics.

In GWC courses geometric figures are considered as sets of points. *Point* is the basic undefined term. *Line* and *plane* are also undefined terms. Every line is a set of points, and every plane is a set of points. Points, lines, and planes have no properties except those stated in the postulates and those that can be proved using the postulates.

The postulates and some of the key definitions of the SMSG-GW text follow:

DEFINITION. *Space* is the set of all points.

POSTULATE 1. Space contains at least two distinct points.

POSTULATE 2. Every line is a set of points and contains at least two distinct points.

POSTULATE 3. If P and Q are two distinct points, there is one and only one line that contains them.

POSTULATE 4. No line contains all points of space.

POSTULATE 5. Every plane is a set of points and contains at least three noncollinear points.

POSTULATE 6. If P, Q, R are three distinct noncollinear points, then there is one and only one plane that contains them.

POSTULATE 7. No plane contains all points of space.

POSTULATE 8. If two distinct points of a line belong to a plane, then every point of the line belongs to that plane.

POSTULATE 9. If two planes intersect, then their intersection is a line.

POSTULATE 10. If A and A' are distinct points, there exists a correspondence that associates with each pair of distinct points in space a unique positive number such that the number assigned to the given pair $\{A, A'\}$ is one.

DEFINITION. The set consisting of the two points A, A' mentioned in postulate 10 is called the *unit-pair*.

DEFINITION. The number that corresponds, by postulate 10, to a pair of distinct points is called the *measure of the distance between the points,* relative to the unit-pair $\{A, A'\}$.

DEFINITION. The *measure of the distance between any point and itself,* relative to $\{A, A'\}$, is the number zero.

POSTULATE 11. If $\{A, A'\}$ is any unit-pair and if B and B' are two points such that BB' (relative to $\{A, A'\}$) $= 1$, then for every pair of points, the distance between them relative to $\{B, B'\}$ is the same as the distance between them relative to $\{A, A'\}$.

DEFINITION. Let $\{A, A'\}$ be any unit-pair and let l be any line. A *coordinate system on* l, relative to the unit-pair $\{A, A'\}$, is a one-to-one correspondence between the set of all points on l and the set of all real numbers with the following property: If numbers r and s correspond to points R and S on l and if $r > s$, then $r - s$ is the same as the number RS (relative to $\{A, A'\}$).

DEFINITIONS. The *origin of a coordinate system on a line* is the point that corresponds to the number 0. The *unit-point of a coordinate system on a line* is the point that corresponds to the number 1. The number that a given coordinate system assigns to a point is the *coordinate* of that point in that coordinate system.

POSTULATE 12 (the ruler postulate). If $\{A, A'\}$ is any unit-pair, if l is any line, and if P and Q are any two distinct points on l, then there is a unique coordinate system on l relative to $\{A, A'\}$ such that the origin of the coordinate system is P and the coordinate of Q is positive.

DEFINITION. If A and B are distinct points, then in the coordinate system on $\overleftrightarrow{AB}$ with origin A and unit-point B, the subset of $\overleftrightarrow{AB}$ consisting of all points whose coordinates x satisfy $0 \leq x \leq 1$ is called the *segment* joining the points A and B.

DEFINITION. The distance between two distinct points is called the *length* of the segment joining the two points.

DEFINITION. Two segments (whether distinct or not) that have the same length are *congruent* segments, and each is said to be *congruent* to the other.

POSTULATE 13. Let A and A' be any two distinct points and let B and B' be any two distinct points. Then for every pair of distinct points P and Q,

$$\frac{PQ \text{ (in } \overline{AA'} \text{ units)}}{PQ \text{ (in } \overline{BB'} \text{ units)}} \text{ is a constant.}$$

DEFINITION. A set containing more than one point is said to be a *convex set* if and only if for every two points of the set the segment joining the points is in the set. Every set of points that contains no more than one point is also said to be a convex set.

POSTULATE 14 (the plane separation postulate). For any plane and any line contained in the plane, the points of the plane that do not lie on the line form two nonempty sets such that (1) each of the two sets is convex and (2) every segment that joins a point of one of the sets to a point of the other intersects the given line.

POSTULATE 15. For any plane, the points of space that do not lie on the plane form two sets such that (1) each of the two sets is convex and (2) every segment that joins a point of one of the sets and a point of the other intersects the given plane.

DEFINITIONS. An *angle* is the union of two rays that have a common endpoint but do not lie on the same line. Each of the two rays is called a *side* of the angle, and their common endpoint is called the *vertex* of the angle.

POSTULATE 16. There is a correspondence that associates with each angle a unique number between 0 and 180.

DEFINITION. The number that corresponds to an angle, by postulate 16, is called the *measure of the angle*.

DEFINITION. Two angles, whether distinct or not, that have the same measure are called *congruent* angles, and each is said to be *congruent to* the other.

DEFINITION. Let V be any point in a plane P. A *ray-coordinate system* in P relative to V is a one-to-one correspondence between the set of all rays in P with endpoint V and the set of all real numbers x such that $0 \leq x < 360$ with the the following property: If numbers r and s correspond to rays $\overrightarrow{VR}$ and $\overrightarrow{VS}$ in P and if $r > s$, then

$m\angle RVS = r - s$ if $r - s < 180$;

$m\angle RVS = 360 - (r - s)$ if $r - s > 180$;

$\overrightarrow{VR}$ and $\overrightarrow{VS}$ are opposite rays   if $r - s = 180$.

DEFINITIONS. The number assigned to a ray in a ray-coordinate system is called its *ray-coordinate*. The ray whose ray-coordinate is zero is called the *zero-ray* of the coordinate system.

POSTULATE 17 (the protractor postulate). If P is a plane and if $\overrightarrow{VA}$ and $\overrightarrow{VB}$ are noncollinear rays in P, then there is a unique ray-coordinate system in P relative to V such that $\overrightarrow{VA}$ corresponds to 0 and such that every ray $\overrightarrow{VX}$ with X and B on the same side of $\overrightarrow{VA}$ corresponds to a number less than 180.

POSTULATE 18 (the interior of an angle postulate). If $\angle AVB$ is any angle, and (1) R is the set of all interior points of rays between $\overrightarrow{VA}$ and $\overrightarrow{VB}$, (2) I is the set of all points that belong both to the half-plane with edge $\overleftrightarrow{VA}$ and containing B and to the half-plane with edge $\overleftrightarrow{VB}$ and containing A, (3) S is the set of all interior points of segments joining an interior point of $\overrightarrow{VA}$ to an interior point of $\overrightarrow{VB}$, then R and I are the same set, and this set contains S.

DEFINITION. A one-to-one correspondence between the vertices of one triangle and the vertices of another triangle in which corresponding parts (sides and angles) are congruent is called a *congruence* between the two triangles.

DEFINITION. Two triangles are *congruent* if and only if there exists a one-to-one correspondence between their vertices that is a congruence.

POSTULATE 19 (the SAS postulate). Given a one-to-one correspondence between the vertices of two triangles (not necessarily distinct triangles), if two sides and the included angle of the first triangle are congruent to the corresponding parts of the second triangle, then the correspondence is a congruence.

POSTULATE 20 (the ASA postulate). Given a one-to-one correspondence between the vertices of two triangles, if two angles and the included side of one triangle are congruent to the corresponding parts of the other, then the correspondence is a congruence.

POSTULATE 21 (the SSS postulate). Given a one-to-one correspondence between the vertices of two triangles, if the three sides of one triangle are congruent to the corresponding sides of the other, then the correspondence is a congruence.

DEFINITION. Two coplanar lines (whether distinct or not) whose intersection is not a set consisting of a single point are called *parallel lines,* and each is said to be *parallel to* the other.

POSTULATE 22 (the parallel postulate). There is at most one line parallel to a given line and containing a given point not on the given line.

POSTULATE 23 (the proportional segments postulate). If a line is parallel to one side of a triangle and intersects the other two sides in interior points, then the lengths of one of those sides and the two segments into which it is cut are proportional to the lengths of the three corresponding segments in the other side.

DEFINITION. A line and a plane are *perpendicular* to each other if and only if they intersect and every line lying in the plane and passing through the point of intersection is perpendicular to the given line.

POSTULATE 24. There is a unique plane that contains a given point and is perpendicular to a given line.

DEFINITION. A line and a plane whose intersection does not consist of exactly one point are *parallel to* each other.

DEFINITION. Two planes (whether distinct or not) whose intersection is not a line are *parallel planes,* and each is *parallel to* the other.

POSTULATE 25. Two lines that are perpendicular to the same plane are parallel.

DEFINITION. A *dihedral angle* is the union of a line and two half-planes having this line as edge and not lying in the same plane.

POSTULATE 26. If R is any given polygonal region, there is a correspondence that associates to each polygonal region a unique positive number, such that the number assigned to the given polygonal region is one.

POSTULATE 27. Suppose that the polygonal region R is the union of two polygonal regions R_1 and R_2 such that the intersection of R_1 and R_2 is contained in the union of a finite number of segments. Then, relative to a given unit-area, the area of R is the sum of the areas of R_1 and R_2.

POSTULATE 28. If two triangles are congruent, then the respective triangular regions consisting of the triangles and their interiors have the same area relative to any given unit-area.

DEFINITION. Given a unit-pair for distance, a square region is called a *unit-*

square if and only if the square region consists of a square and its interior and the length of each side of the square is one.

POSTULATE 29. Given a unit-pair for measuring distance, the area of a rectangle relative to a unit-square is the product of the lengths (relative to the given unit-pair) of any two consecutive sides of the rectangle.

DEFINITION. The *degree measure, $m\overarc{AXB}$,* of an arc of a circle is given by the following:

1. If $\overarc{AXB}$ is a minor arc, then $m\overarc{AXB}$ is the measure of the associated central angle.

2. If $\overarc{AXB}$ is a semicircle, then $m\overarc{AXB} = 180$.

3. If $\overarc{AXB}$ is a major arc and if $\overarc{AYB}$ is the corresponding minor arc, then

$$m\overarc{AXB} = 360 - m\overarc{AYB}.$$

POSTULATE 30. If $\overarc{AB}$ and $\overarc{BC}$ are arcs of the same circle having only the point B in common and if their union is an arc $\overarc{AC}$, then $m\overarc{AB} + m\overarc{BC} = m\overarc{AC}$.

POSTULATE 31. The lengths of arcs in congruent circles are proportional to their degree measures.

The scope of the SMSG-GW text is suggested by its postulates and the definitions listed above as well as by its chapter titles, which are as follows:

1. Introduction to Formal Geometry
2. Sets; Points, Lines, and Planes
3. Distance and Coordinate Systems
4. Angles
5. Congruence
6. Parallelism
7. Similarity
8. Coordinates in a Plane
9. Perpendicularity, Parallelism, and Coordinates in Space
10. Directed Segments and Vectors
11. Polygons and Polyhedrons
12. Circles and Spheres

The formal geometry developments in most high school textbooks are overpostulated in the sense that one or more of the postulates follow logically from the others. Thus statements that could be proved and included as theorems are unproved and included as postulates in the logical structure. The ultimate in mathematical efficiency is achieved if the structure is not overpostulated, that is, if every postulate is independent of the other postulates in the system. High school mathematics texts are overpostulated to achieve pedagogical efficiency. Some statements with involved proofs that would appear in a more advanced textbook are accepted as postulates in a high school textbook. In some instances statements are accepted as postulates to expedite the development and hence to provide time for building a more extensive geometrical structure. In most instances statements accepted as postulates and that would be proved in a more sophisticated treatment are plausible statements that students are willing to accept without fuss or bother. Some examples of overpostulation in the SMSG-GW text will now be noted.

Postulate 15 can be proved. See, for example, theorem 9 in *Foundations of Geometry* (5, pp. 65–66). It is appropriate, however, in a high school textbook to accept a space-separation postulate along with a plane-separation postulate.

Postulate 18 can be proved. Given $\angle AVB$, then three sets, R, I, and S, associated with this angle are defined in the postulate. See theorems 5.12–5.15 in *College Geometry* (1, pp. 90–91) for proofs that R and I are the same set and that I contains S.

Postulates 19, 20, and 21 are the triangle-congruence postulates. Assuming one of these statements as a postulate, the other two can be proved as theorems. See postulate 17, theorem 6.14, and theorem 6.17 in *College Geometry* (1, pp. 105–6).

Postulates 30 and 31 can be proved. See theorem 13.15 (p. 580) and theorem 14.15 (p. 648) in *Geometry* (2). The textbook by Ringenberg and Presser (2) is a GWC textbook for the high school level. On the subject of properties of circular arc measure, it is less overpostulated than the SMSG-GW text; in other respects it is more overpostulated. The extent to which a particular textbook is overpostulated is determined by the author. The teacher of a particular textbook may increase the overpostulation by treating some of the theorems as postulates.

Examples of GWC Content

Example 1

An exercise following postulates 1–4 in the SMSG-GW text asks, Which postulates imply that given a point, there is a line that contains it? The students would not be required to write a proof to support the answer. A student might defend his answer orally as follows: A point is given. Call it P. Since space contains at least two distinct points, there must be (by postulate 1) some point Q besides P. Then by postulate 3 there is a line l containing P and Q. Therefore, there is a line containing the given point.

Example 2

A theorem following the incidence, distance, and ruler postulates.

THEOREM (the two coordinate systems theorem). *Let a line l and two coordinate systems, C and C', on l be given. There exist two numbers a, b, with a ≠ 0, such that for any point on l, its coordinate x in C is related to its coordinate x' in C' by the equation x' = ax + b.*

Proof. The postulates regarding distances and coordinate systems imply that a ratio of distances, such as $\dfrac{PX}{PQ}$, is independent of the coordinate system and that betweenness for points agrees with betweenness for coordinates (see fig. 5.1). It follows that if P and Q are distinct points on a line with coordinates p and q in a coordinate system C and with coordinates p' and q' in a coordinate system C', then for every point X on $\overset{\leftrightarrow}{PQ}$, its coordinate x in C is related to its coordinate x' in C' by the equation

$$\frac{x - p}{q - p} = \frac{x' - p'}{q' - p'},$$

or

$$x' = \frac{q' - p'}{q - p}\, x + \frac{p'q - pq'}{q - p},$$

which is the form of $x' = ax + b$.

Fig. 5.1

A useful special case of the theorem is obtained if p, q, x, p', q', x' are replaced by 0, 1, k, x_1, x_2, x, respectively (see fig. 5.2). In this case the equation relating x and k is $x = x_1 + k(x_2 - x_1)$. Also, X is a point of the ray $\overset{\rightarrow}{PQ}$ if $k \geq 0$, and X is a point of the ray opposite to $\overset{\rightarrow}{PQ}$ if $k \leq 0$.

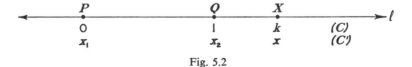

Fig. 5.2

Several exercises that illustrate the use of this theorem follow.

Exercise. Find the x-coordinate of X on $\overleftrightarrow{PQ}$ if X lies on the ray $\overrightarrow{PQ}$, $PX = 5 \cdot PQ$, the x-coordinate of P is 2, and the x-coordinate of Q is -4. (See fig. 5.3).

Fig. 5.3

Solution. Let k be the coordinate of X in the system with P as origin and Q as unit-point (the coordinate of P is 0 and the coordinate of Q is 1). Then $x = 2 + k(-4 - 2)$, $x = 2 - 6k$. Since $PX = 5 \cdot PQ$, $k = 5$ or -5. Since $X \in \overrightarrow{PQ}$, $k \geq 0$. Therefore $k = 5$, and it follows that $x = -28$.

Exercise. Find the x-coordinate of X on $\overleftrightarrow{PQ}$ if X lies on the ray opposite to $\overrightarrow{PQ}$, $PX = 5 \cdot PQ$, the x-coordinate of P is 2, and the x-coordinate of Q is -4. (See fig. 5.4.)

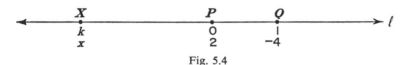

Fig. 5.4

Solution. Let k be the coordinate of X in the coordinate system on $\overleftrightarrow{PQ}$ with P as origin and Q as unit-point. Then $x = 2 + k(-4 - 2)$, $x = 2 - 6k$. $PX = 5 \cdot PQ$ implies that $k = 5$ or $k = -5$, and $X \in$ opp $\overrightarrow{PQ}$ implies that $k \leq 0$. Therefore $k = -5$, and it follows that $x = 32$.

Alternate Solution. Let k be the coordinate of X in the coordinate system on $\overleftrightarrow{PQ}$ with Q as origin and P as unit-point. Then $x = -4 + k(2 + 4)$, $x = -4 + 6k$. Set $k = 6$. Then $x = 32$. (See fig. 5.5.)

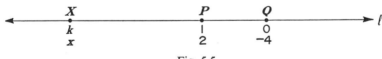

Fig. 5.5

Example 3

A theorem regarding additivity of angle measure.

THEOREM (the betweenness angles theorem). *Let* $\overrightarrow{VE}$, $\overrightarrow{VF}$, $\overrightarrow{VG}$ *be rays such that* $\overrightarrow{VF}$ *is between* $\overrightarrow{VE}$ *and* $\overrightarrow{VG}$. *Then* $m\angle EVF + m\angle FVG = m\angle EVG$.

Proof. The definition of betweenness for rays and the protractor and angle measure postulates imply that there is a ray-coordinate system with center V in which the coordinates of $\overrightarrow{VE}$, $\overrightarrow{VF}$, $\overrightarrow{VG}$ are 0, f, g, respectively, with $0 < f < g < 180$. Then $m\angle EVF + m\angle FVG = (f - 0) + (g - f) = g = m\angle EVG$. (See fig. 5.6.)

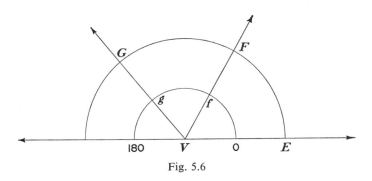

Fig. 5.6

Example 4

An exercise following the introduction of coordinates in a plane asks the students to plot the graphs of the sets

$$A = \{(x, 7): x = 7\}, B = \{(x, y): y > 5\},$$
$$C = \{(x, y): x = 7 \text{ and } y > 5\}.$$

The graph of A is the line parallel to the y-axis and 7 units to the right of the y-axis (fig. 5.7). The graph of B is a half-plane (fig. 5.8). The graph of C is a half-line; it is the intersection of the line that is the graph of A and the half-plane that is the graph of B (fig. 5.9). See figures 5.7–5.9 for the solution.

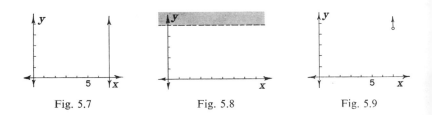

Fig. 5.7   Fig. 5.8   Fig. 5.9

Example 5

A theorem regarding parametric linear equations.

THEOREM. *If $A(x_1, y_1)$ and $B(x_2, y_2)$ are any two distinct points then*

$$\overleftrightarrow{AB} = \{(x, y): x = x_1 + k(x_2 - x_1), y = y_1 + k(y_2 - y_1), k \text{ is real}\}.$$

If k is a real number and P is the point (x, y) where $x = x_1 + k(x_2 - x_1)$ and $y = y_1 + k(y_2 - y_1)$, then

$$k = \frac{AP}{AB} \text{ and } P \in \overrightarrow{AB} \text{ if } k \geq 0,$$

$$-k = \frac{AP}{AB} \text{ and } P \in opp \ \overrightarrow{AB} \text{ if } k \leq 0.$$

Proof (plan). Let $\overleftrightarrow{AB}$ be a line in an xy-plane not parallel to either axis (see fig. 5.10). It follows from the definition of an xy-coordinate system that the one-to-one correspondence that matches each point on the x-axis (y-axis) with its x-coordinate (y-coordinate) is a coordinate system on the x-axis (y-axis). The x-coordinate (y-coordinate) of a point on $\overleftrightarrow{AB}$ is the x-coordinate (y-coordinate) of its projection on the x-axis (y-axis). It may be proved that the one-to-one correspondence that matches each point on $\overleftrightarrow{AB}$ with its x-coordinate (y-coordinate) is a coordinate system on $\overleftrightarrow{AB}$. The assertions of the theorem for the case of a line parallel to either axis as well as for the oblique case follow from preceding results regarding coordinate systems on a line.

There is a similar theorem regarding parametric linear equations and the representation of a line in an xyz-space.

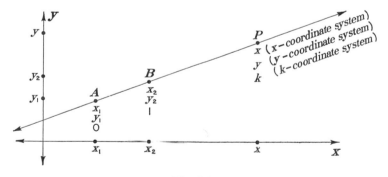

Fig. 5.10

Example 6

An exercise illustrating the use of parametric equations states, Given $A = (3, -7)$ and $B = (-5, -12)$, find the points of trisection of $\overline{AB}$.

Solution. $\overline{AB} = \{(x, y): x = 3 - 8k, y = -7 - 5k, 0 \leq k \leq 1\}$. Set $k = 1/3$. Then $x = 1/3$ and $y = -26/3$. Set $k = 2/3$. Then $x = -7/3$ and $y = -31/3$. The points of trisection are $(1/3, -26/3)$ and $(-7/3, -31/3)$.

Example 7

A classical theorem relating slopes and perpendicularity.

THEOREM. *Two nonvertical lines are perpendicular if and only if the product of their slopes is* -1.

Proof. Yet p_1 and p_2 be nonvertical lines in an xy-plane with slopes m_1 and m_2, respectively. Let q_1 and q_2 be lines through $O = (0, 0)$ parallel to p_1 and p_2, respectively. Let $R(1, m_1)$ and $S(1, m_2)$ be points on q_1 and q_2, respectively. Then p_1 is perpendicular to p_2 if and only if $\overline{OR}$ is perpendicular to $\overline{OS}$. (See fig. 5.11.)

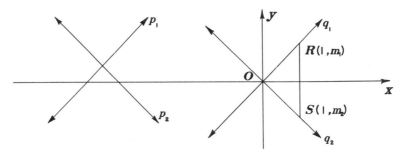

Fig. 5.11

Using the Pythagorean theorem and its converse, it follows that p_1 is perpendicular to p_2—

1. if and only if $(OR)^2 + (OS)^2 = (RS)^2$;
2. if and only 1 $+ m_1{}^2 + 1 + m_2{}^2 = (m_1 - m_2)^2$;
3. if and only if $m_1 m_2 = -1$.

Example 8

A classical theorem regarding parallelograms.

THEOREM. *A quadrilateral is a parallelogram if and only if its diagonals bisect each other.*

Proof. Given quadrilateral $ABCD$, set up an xy-coordinate system so that $A = (0, 0)$, $B = (a, 0)$, $C = (x, y)$, $D = (b, c)$. Then (by a preceding theorem) $ABCD$ is a parallelogram if and only if $x = a + b$ and $y = c$. But $x = a + b$ and $y = c$ if and only if the midpoint $\left(\dfrac{x}{2}, \dfrac{y}{2}\right)$ of $\overline{AC}$ is the same as the midpoint $\left(\dfrac{a + b}{2}, \dfrac{c}{2}\right)$ of $\overline{BD}$. Therefore $ABCD$ is a parallelogram if and only if $\overline{AC}$ and $\overline{BD}$ bisect each other. (See fig. 5.12).

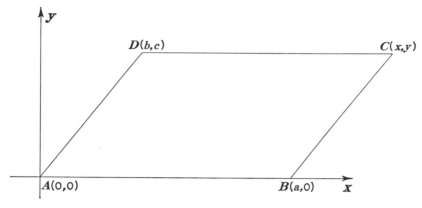

Fig. 5.12

Example 9

A theorem whose proof illustrates the power of the coordinate method.

THEOREM. *If a line in the plane of a circle intersects the interior of the circle, then it intersects the circle in two distinct points.*

Proof. Suppose point P lies on line l and in the interior of circle C. Set up an xy-coordinate system so that C, l, and P are given by $C = \{(x, y): x^2 + y^2 = r^2\}$, $l = \{(x, y): y = b\}$, $P = (a, b)$. (See fig. 5.13.) Then $a^2 + b^2 < r^2$, and the intersection of l and C is the set $\{(x, y): x^2 + b^2 = r^2$ and $y = b\}$. Since $a^2 \geq 0$, it follows that $r^2 - b^2 > 0$, and the equation

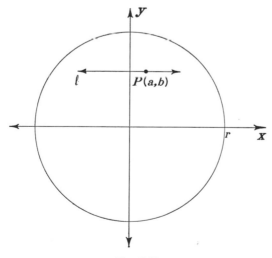

Fig. 5.13

$x^2 + b^2 = r^2$ has two real roots. Therefore l intersects C; the two distinct points of intersection are $(\sqrt{r^2 - b^2}, b)$ and $(-\sqrt{r^2 - b^2}, b)$.

Example 10

Another theorem with a proof illustrating the power of the coordinate method.

THEOREM. *The medians of a triangle are concurrent in a point that is two-thirds of the distance from each vertex to the midpoint of the opposite side.*

Proof. Let $\triangle ABC$ be given. Select an xy-coordinate system in the plane of this triangle with the origin at A, with $\overleftrightarrow{AB}$ the x-axis, with the abscissa of B positive, and with the ordinate of C positive. Let a, b, c be numbers such that $B = (6a, 0)$ and $C = (6b, 6c)$. Let D, E, F be the midpoints of $\overline{BC}, \overline{CA}, \overline{AB}$, respectively. Then $\overline{AD}, \overline{DE}, \overline{EF}$ are the medians of $\triangle ABC$. Then $F = (3a, 0)$ and $\overline{CF} = \{(x, y): x = 6b + (3a - 6b)k, y = 6c + (0 - 6c)k, 0 \le k \le 1\}$. (See fig. 5.14.) The point P on $\overline{CF}$ such that $CP = \frac{2}{3} \cdot CF$ is obtained by setting $k = \frac{2}{3}$. Then $P = (x, y) = \left(6b + [3a - 6b]\frac{2}{3}, 6c + [0 - 6c]\frac{2}{3}\right) = (2a + 2b, 2c)$.

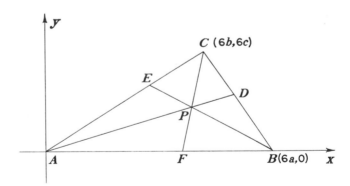

Fig. 5.14

Similarly it may be shown that the point P' that is two-thirds of the distance from B to E is $P' = (2a + 2b, 2c)$ and that the point P'' that is two-thirds of the distance from A to D is $P'' = (2a + 2b, 2c)$. Therefore $P = P' = P''$, and the proof is complete.

Example 11

An exercise regarding internal and external points of division of a segment states, Given two points A and B on a line l with x-coordinates 3 and -21, respectively, find the x-coordinates of the points P and Q on $\overleftrightarrow{AB}$ that divide $\overrightarrow{AB}$ (the directed segment from A to B) internally and externally, respectively, in the ratio 7/8.

Solution. Let S and T be two coordinate systems on l with coordinates of several points as marked in figure 5.15. Then

$$\frac{x-3}{-21-3} = \frac{k-0}{1-0}$$

and $x = -24k + 3$ for every point X on l. Since $\dfrac{AQ}{QB} = \dfrac{7}{8}$, AQ is less than QB, and Q lies on the ray opposite to $\overrightarrow{AB}$. The coordinates x and x' of P and Q are computed as follows:

$$AP = k - 0 \qquad\qquad AQ = 0 - k'$$
$$PB = 1 - k \qquad\qquad QB = 1 - k'$$
$$\frac{k}{1-k} = \frac{7}{8} \qquad\qquad \frac{-k'}{1-k'} = \frac{7}{8}$$
$$k = \frac{7}{15} \qquad\qquad\quad k' = -7$$
$$x = -8.2 \qquad\qquad\quad x' = 171$$

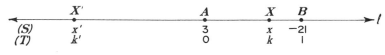

Fig. 5.15

Example 12

A "locus" theorem that illustrates the use of set language.

THEOREM. *The set of all points equidistant from the endpoints of a segment is the perpendicular bisecting plane of the segment.*

Proof. Let α be the perpendicular bisecting plane of $\overline{AB}$ and let C be the midpoint of $\overline{AB}$. Let P be a point. (See fig. 5.16.) There are two statements to prove:

1. If $P \in \alpha$, then $AP = PB$.
2. If $AP = PB$, then $P \in \alpha$.

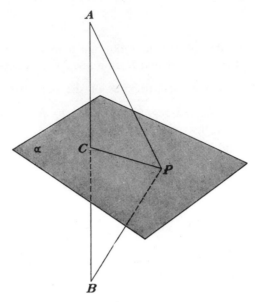

Fig. 5.16

To prove statement 1, use the SAS postulate to prove that $\triangle ACP \cong$ $\triangle BCP$; then $AP = PB$. To prove statement 2, use SSS postulate to prove that $\triangle ACP \cong \triangle BCP$; then $\angle ACP \cong \angle BCP$. It follows easily from preceding theorems that $\overleftrightarrow{CP} \perp \overleftrightarrow{AB}$ and that P lies in α.

The Role of Coordinates
in Selected High School Geometry Textbooks

Comments on the role coordinates play, along with pertinent bibliographic information, are given for the following textbooks:

Geometry, A. Wilson Goodwin, Glen D. Vannatta, and F. Joe Crosswhite. Columbus, Ohio: Charles E. Merrill Publishing Co., 1970.

Linear coordinate systems are introduced in chapter 2, and rectangular coordinate systems for a plane are introduced in chapter 5. Coordinate geometry is included as an enrichment topic near the ends of most chapters following chapter 5. Coordinates appear also in chapter 20 (Introduction to Trigonometry).

Geometry, Edwin E. Moise and Floyd L. Downs, Jr. Reading, Mass.: Addison-Wesley Publishing Co., 1964, 1971.

On the role of coordinates, this book is similar to the 1961 SMSG *Geometry.*

Geometry, Lawrence A. Ringenberg and Richard S. Presser. Benziger-Wiley Mathematics Program series. Beverly Hills, Calif.: Benziger, 1971.

This book also features an approach using coordinates similar to that of the SMSG-GW text.

Geometry, Myron F. Rosskopf, Harry Sitomer, and George Lenchner. Morristown, N.J.: Silver Burdett Co., 1971.

This book features an approach using coordinates similar to that of the SMSG-GW text.

Geometry, School Mathematics Study Group. New Haven: Yale University Press, 1960, 1961.

This is the first SMSG high school geometry textbook. The postulates provide for a unique distance function. Postulate 2 (the distance postulate) asserts that "to every pair of different points there corresponds a unique positive number." Coordinates on a line are introduced in chapter 3. Postulate 3 (the ruler postulate) asserts that "the points of a line can be placed in correspondence with the real numbers in such a way that (1) to every point of the line there corresponds exactly one real number, (2) to every real number there corresponds exactly one point of the line, and (3) the distance between two points is the absolute value of the difference of the corresponding numbers." Betweenness for points is defined in terms of distance, that is, B is between A and C if $AB + BC = AC$. The relationship of betweenness for coordinates to betweenness for points is established as a theorem. Set-builder symbols involving coordinates are not introduced to represent subsets of lines, planes, and space. Coordinates are not used in developing angle measure. Coordinates do not appear in chapters 3 through 16. The final chapter of the book (chapter 17) is devoted to plane coordinate geometry.

Geometry, James F. Ulrich and Joseph N. Payne. New York: Harcourt, Brace & World, 1969.

A coordinate system on a line and a coordinate system for rays are two of the topics in chapter 2. Parallels and coordinates are the subject of chapter 6. Coordinates are used to a considerable extent in chapter 9 (Introduction to Locus) and chapter 13 (Introduction to Trigonometry) and to a lesser extent in chapter 11 (Pythagorean Theorem) and chapter 12 (Polygons and Area).

Lines, Planes, Space, H. Vernon Price, Philip Peak, and Phillip S. Jones. Book 2 of Mathematics: An Integrated Series. New York: Harcourt, Brace & World, 1965.

Coordinates on a line and in a plane are introduced in chapter 2. Coordinate geometry is the subject of chapter 8. Coordinates are used to varying extents in chapters 9 (Functions), 10 (Exponents and Logarithms), 11 (Quadratic Functions), and 12 (Circles and Spheres).

Modern Geometry, Eugene D. Nichols, William F. Palmer, and John F. Schacht. New York: Holt, Rinehart & Winston, 1968.

Coordinates appear in three of the sixteen chapters of this book. Coordinates on a line are mentioned in chapter 1. Chapter 10 is a formal introduction to coordinate geometry—line, plane, and space. Equations and their graphs are the subject of chapter 15.

Modern Geometry: Structure and Function, 2d ed., Kenneth B. Henderson, Robert E. Pingry, and George A. Robinson. New York: McGraw-Hill Book Co., 1968.

A ruler postulate and a protractor postulate are introduced in chapter 4. The first postulate in chapter 7 coordinatizes a plane. Coordinates are used in some proofs.

School Mathematics Geometry, Richard D. Anderson, Jack W. Garon, and Joseph G. Gremillion. Boston: Houghton Mifflin Co., 1966.

Again, on the role of coordinates, this book is similar to the 1961 SMSG *Geometry.*

The Role of Coordinates
in Selected College Geometry Textbooks

Teachers whose college background in formal geometry is limited to a course in analytic geometry (preceding, or integrated with, the calculus) and a course in advanced Euclidean geometry using a synthetic approach should have some lead time to prepare to teach a GWC course. Many teachers with such a background who have taken courses in the foundations of geometry as a part of a summer institute program or an in-service institute for teachers are enthusiastic about GWC courses and prefer them to courses using the traditional synthetic approach. Many younger mathematics teachers have had a college course that includes coordinates in developing the foundations of geometry and are prepared for an assignment to teach a GWC course. A teacher who feels his background is inadequate to teach a GWC course should be encouraged to study a textbook for such a course. With the aid of a teacher's edition he should be able to complete it as an independent study assignment. Having completed it, he is likely to be enthusiastic about it and to want to teach it.

Some comments regarding the role of coordinates in selected college geometry textbooks follow.

College Geometry, Lawrence A. Ringenberg. New York: John Wiley & Sons, 1968.

Chapters 2 through 10 of this textbook are devoted to a formal development of the foundations of Euclidean geometry with an approach that uses coordinates extensively.

Elementary Geometry from an Advanced Standpoint, Edwin E. Moise. Reading, Mass.: Addison-Wesley Publishing Co., 1963.

This textbook is an extensive, mathematically rigorous treatment of elementary geometry designed for prospective or in-service teachers. Coordinates on a line are introduced in chapter 3. Cartesian coordinate systems are the subject of chapter 18. Coordinates are used in several other places throughout the thirty-two chapters of this book.

Foundations of Geometry, C. Ray Wylie, Jr. New York: McGraw-Hill Book Co., 1964.

Designed for prospective teachers and teachers of secondary school mathematics, this textbook is a rigorous development of school geometry based on the postulates used in the SMSG-GW text, which Wylie helped to write. The book also includes a careful discussion of the axiomatic method, a treatment of a geometry of four dimensions, plane hyperbolic geometry, and an exposition of a Euclidean model of the hyperbolic plane.

Fundamentals of Geometry, Bruce E. Meserve and Joseph A. Izzo. Reading, Mass.: Addison-Wesley Publishing Co., 1969.

This textbook is designed to provide a broad geometric background that will help students see how Euclidean geometry is related to other geometries. It does not contain a complete formal development of coordinates on a line or in a plane or in space. Assuming the existence of such coordinate systems (as in elementary analytic geometry), nonhomogeneous coordinates are developed and used to extend and relate Euclidean geometry to other geometries.

Fundamentals of College Geometry, 2d ed., Edwin M. Hemmerling. New York: John Wiley & Sons, 1970.

The last chapter of this textbook is devoted to coordinate geometry.

REFERENCES

1. Ringenberg, Lawrence A. *College Geometry.* New York: John Wiley & Sons, 1968.
2. Ringenberg, Lawrence A., and Richard S. Presser. *Geometry.* Benziger-Wiley Mathematics Program series. Beverly Hills, Calif.: Benziger, 1971.
3. School Mathematics Study Group. *Geometry.* New Haven, Conn.: Yale University Press, 1960, 1961.
4. ———. *Geometry with Coordinates.* New Haven, Conn.: Yale University Press, 1962.
5. Wylie, C. Ray., Jr. *Foundations of Geometry.* New York: McGraw-Hill Book Co., 1964.

6

A Transformation Approach
to Geometry

ARTHUR F. COXFORD, JR.

THIS CHAPTER is the third in a series of five chapters in Part II of this
yearbook describing different approaches to school geometry. In each
of these chapters a case is made for the use of a particular approach to
school geometry. The approach described in this chapter is based on the
use of certain transformations of the plane. These transformations are
identified below and are described in detail later. For now, it is sufficient
to recall that a transformation of the plane is a one-to-one mapping of the
plane onto itself.

As indicated above, the purpose of this chapter is to describe one approach to school geometry. But how does one describe an approach? The
following summary of the organization and content of the chapter gives
the reader an idea of what the author means by "describing an approach."
It should also help the reader keep perspective as he moves through the
details of the description.

The first section of the chapter discusses the properties of the major
transformations used in the approach described herein. The first five transformations are *reflection, rotation, translation, glide reflection,* and *dilation*
(also called homothety, dilatation, enlargement, or central similarity).
Each of these transformations is a special case of a similarity transformation, which is discussed separately under *other similarity transformations.*

There are two reasons for including the material on the transformations and their properties. First, many who read this chapter will not have had in their formal education the opportunity to study these transformations. Thus the material will provide to these readers an introduction to transformation ideas and some of the fascinating associated ideas. The second reason, and related to the first, is that a reader cannot understand and appreciate this approach to geometry without a basic understanding of the transformation ideas used. Therefore, the material is included to make the probability of understanding the subsequent sections as great as possible. The reader who has studied transformations and their properties may either pass over this discussion or skim the material to refresh his memory.

The second and third sections argue that transformation ideas ought to be included in school geometry and that they ought to be central to the development of school geometry. These arguments set the stage for the fourth section, the major portion of the chapter. In this section the course objectives are given, and the scope and sequence of a course in school geometry based on transformation ideas is discussed in great detail. The development of standard topics such as parallelism, congruence, and similarity is discussed along with the less standard topics of transformations, matrices, and groups. The objective is to give the reader an accurate picture of such a course with detail enough to provide him with a sound basis for comparing this approach with others described in this yearbook.

The last two sections of the chapter are short ones. The first examines some issues that would arise if the described transformation approach were to be implemented. The final section includes a few personal observations of the author.

Transformations of the Plane

The transformations of the plane—reflection, rotation, translation, glide reflection, dilation, and similarity—are informally described in turn in this section. For each transformation one or more figures that illustrate the definition of the transformation or the properties are given. All examples and illustrations deal with transformations of the plane. The reader can easily construct analogous illustrations for the transformations of space.

Four of the transformations of the plane are called the *isometries*. They are reflection, rotation, translation, and glide reflection.

Reflection

Reflection in a line plays a fundamental role in the study of isometries. It may be used to generate the remaining three isometries. This is illustrated in the discussions of rotation, translation, and glide reflection.

Figures 6.1 and 6.2 illustrate the definition of the reflection of a point *A* over the line m. Notice that when A is not on m, as in figure 6.1, then the reflection of A over m is the point A' such that m is the perpendicular bisector of segment AA' $(\overline{AA'})$. When A is on m, as in figure 6.2, then the image of A is itself.

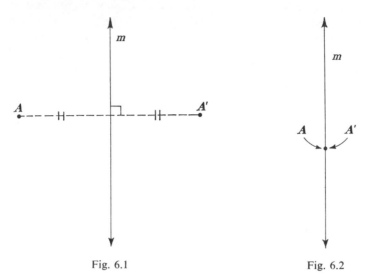

Fig. 6.1   Fig. 6.2

The line m is called the *mirror* (or *axis* or *reflecting line*) of the reflection. The reflection in the line m is denoted M_m. If the mirror is unspecified or unknown, a reflection is denoted M. With this symbolism, the statement "A' is the reflection of A in m" is written

$$M_m(A) = A'.$$

Throughout this discussion, primed letters will denote image points and unprimed letters will denote preimage points.

To construct the reflection image of a set of points, all one needs to do is to reflect each point in the set. The results of doing this for one geometric figure are illustrated in figure 6.3. For example, the set of points $\overline{AC}$ maps onto the set of points $\overline{A'C'}$.

Figure 6.3 also illustrates several properties of geometric figures and their images under reflection. These properties are summarized as follows:

1. If A is not on m, then A and A' are on opposite sides of m.

2. Each point of m is its own image under M_m; the points of m are *fixed points* of the transformation M_m. The line m is a *pointwise fixed line*. The line $\overleftrightarrow{AA'}$, where $\overleftrightarrow{AA'} \perp m$, is its own image, too. It is also a fixed line, but it is not pointwise fixed because each point is **not its own image**.

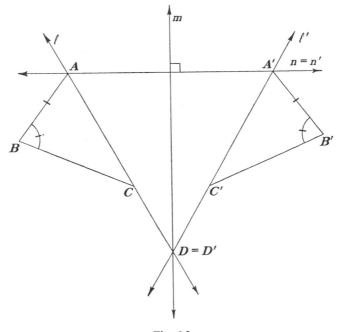

Fig. 6.3

3. The image of any straight line is again a straight line. In figure 6.3, $M_m(l) = l'$. Thus, reflection in a line preserves lines.

4. Reflection in a line preserves distance between points, maps angles onto angles, and preserves the measure of angles. In figure 6.3, $AB = A'B'$, $M_m(\angle ABC) = \angle A'B'C'$, and $m\angle ABC = m\angle A'B'C'$.

5. Reflection in a line reverses the orientation of three noncollinear points. The three points A, B, C (in that order) are counterclockwise oriented while their images A', B', C' (in that order) are clockwise oriented.

6. Given two arbitrary points A and A', there is a unique reflection M such that $M_m(A) = A'$. It is the reflection determined by the perpendicular bisector of $\overline{AA'}$.

7. Given two rays DA and DC', there is a unique reflection M_m interchanging $\overrightarrow{DA}$ and $\overrightarrow{DC'}$. The mirror m is the bisector of $\angle ADC'$.

Rotation

The second isometry discussed is the rotation. Figures 6.4 and 6.5 illustrate the most common way of thinking about a rotation. A point O is given as the center of the rotation, and a measure of an angle θ is specified.

Then a point $A \neq O$ (fig. 6.4) is rotated θ units about O to a new position A'. A' is the rotation image of A under a rotation with center O and magnitude θ. If A happens to equal O (fig. 6.5), then A is its own image. Notice that $OA = OA'$ and $m\angle AOA' = \theta$, properties that are characteristc of a rotation.

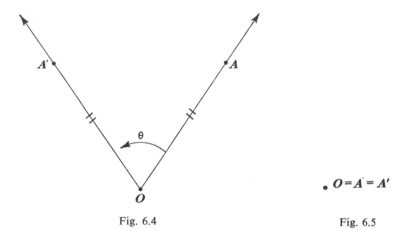

Fig. 6.4   Fig. 6.5

If you think of the rays OA and OA' as an ordered pair and think of θ as a measure of $\angle AOA'$ from $\overrightarrow{OA}$ to $\overrightarrow{OA'}$, then θ is considered positive. Similarly, when θ is negative, the angle is measured in a clockwise manner— from $\overrightarrow{OA'}$ to $\overrightarrow{OA}$ in figure 6.4. A positive magnitude θ corresponds to a counterclockwise rotation; negative θ implies a clockwise rotation. The letter R is used to denote a rotation.

Figure 6.6 illustrates several important properties of geometric figures and their images under a rotation R. They are summarized as follows:

1. Rotation is a one-to-one mapping of the plane onto itself.

2. Under a rotation the image of a line is a line. In figure 6.6, $R(l) = l'$.

3. Rotation is a distance-preserving and angle measure–preserving transformation. For example, $BC = B'C'$, $R(\angle BCA) = \angle B'C'A'$, and $m\angle BCA = m\angle B'C'A'$ in figure 6.6.

4. The angle between a line and its image under a rotation with magnitude θ is also θ. The angle between the lines is measured from the preimage line to its image. In figure 6.6, the angle between l and its image l' is θ. The exceptions to this occur when θ is an integral multiple of 180°. A rotation of 180° is a *half-turn*.

5. Generally a rotation R has exactly one fixed point, namely, the center O. Exceptions occur when θ is an integral multiple of 360°. In these

cases all points remain fixed. Thus any rotation R with magnitude an integral multiple of 360° is the *identity* transformation.

6. Rotation preserves the orientation of three noncollinear points. Compare, for example, points A, B, C and A', B', C' of figure 6.6.

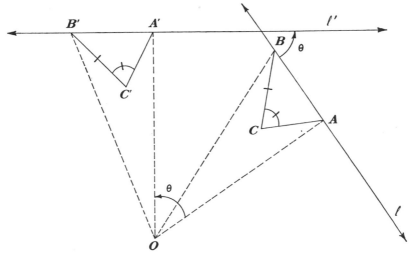

Fig. 6.6

The rotation transformation may also result when the operation *composition* is applied to two transformations. Composition of transformations is identical to composition of functions. If A is a point and G and H are transformations, composing G and H is done by first finding $G(A) = A'$, then finding $H(A') = A''$. Composing G and H results in a new transformation, $H \circ G$, which maps A onto A''. This transformation is called the *product* of H and G.

When G and H are reflections in lines intersecting at a point O such as M_l and M_m shown in figure 6.7, the product $M_m \circ M_l$ is a rotation R with center O. It is easy to see that $M_m \circ M_l$ is a rotation with center O. Certainly $AO = OA_l = OA'$, so $OA = OA'$. Moreover, if $M_m \circ M_l(A) = A'$, then $m \angle AOA'$ is independent of A. Thus $M_m \circ M_l$ is a rotation with center O and magnitude $m \angle AOA' = \theta$.

It is interesting to note the relationship between θ, the magnitude of the rotation, and ϕ, the measure of the acute or right angle determined by the lines l and m. As shown in figure 6.7, $\theta = 2\phi$. Thus the magnitude θ of a rotation is twice the measure of the acute or right angle determined by the intersecting mirrors.

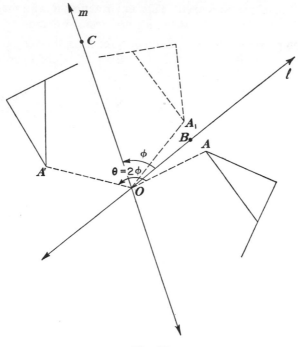

Fig. 6.7

Since the magnitude of the rotation is twice the angle between the mirrors, it follows that any pair of mirrors making this angle at the point O determines the same rotation R. Thus one pair of mirrors may be replaced by a second pair making the same angle at O. The reader is encouraged to verify this for himself by constructing the rotation image of a figure using several pairs of mirrors making the same angle at O. For example, in figure 6.8, $M_{m_1} \circ M_{l_1} = R = M_{m_2} \circ M_{l_2}$. Choose three noncollinear points such as A, B, C and verify that $M_{m_1} \circ M_{l_1}(\triangle ABC) = M_{m_2} \circ M_{l_2}(\triangle ABC)$.

The fact that any rotation can be defined in terms of the product of reflections suggests the fundamental role that reflection plays in transformation geometry. The definitions of the remaining two isometries are stated in terms of products of reflections to illustrate this role further.

Translation

A rotation about a point arises when the product of two reflections in intersecting lines is considered. But when the lines are parallel—the only other possible situation in the plane—then the product of reflections is a translation. Figure 6.9 illustrates the translation $T = M_m \circ M_l$ where $m \parallel l$.

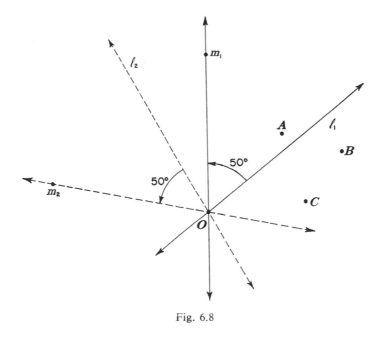

Fig. 6.8

We include in the definition of parallel lines the case where l and m are the same line; that is, two lines in a plane are parallel if and only if their intersection is empty or they are the same line.

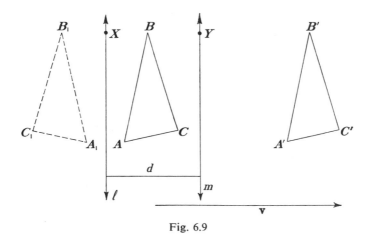

Fig. 6.9

Because each point A is mapped onto a point A' under the translation T by successive reflections in parallel lines, A and A' are on the line perpendicular to l and to m. Thus the lines determined by corresponding points under T are parallel. ($\overleftrightarrow{AA'} \parallel \overleftrightarrow{BB'} \parallel \overleftrightarrow{CC'}$ in fig. 6.9.)

By using directed distances, it can be shown that each point B in the plane is moved the same distance by a translation T. For example, figure 6.9 shows that

$$
\begin{aligned}
BB' &= BY + YB' \\
&= BB_l + B_lY + YB' \\
&= BB_l + 2B_lY \\
&= BX + XB_l + 2(B_lX + XY) \\
&= 2XY.
\end{aligned}
$$

Thus, not only is each point moved the same distance, but also that distance is twice the distance from l to m.

Since each point is moved the same distance and in the same direction under a translation T, only one pair of corresponding points needs to be known to determine T. A translation, then, can be represented by any arrow of the correct length and direction. The equivalence class of all such arrows is the *vector* of the translation T. In figure 6.9 the vector of T is denoted **v**.

Since T, a translation, is determined by the direction and distance any point moves, the mirrors l and m in the definition may be replaced by any pair of parallel mirrors the same directed distance apart. What property of rotations is similar to this?

Several properties of a translation are illustrated in figure 6.9:

1. Translation is a one-to-one mapping of the plane onto itself.

2. Under a translation, the image of a line is a line ($T(\overleftrightarrow{AB}) = \overleftrightarrow{A'B'}$ in figure 6.9). Moreover, a line and its image are parallel ($\overleftrightarrow{AB} \parallel \overleftrightarrow{A'B'}$).

3. Translation is a distance-preserving and angle measure–preserving transformation.

4. A translation has no fixed points.

5. Translation preserves the orientation of three noncollinear points. (Compare A, B, C and A', B', C' in fig. 6.9.)

Glide reflection

The fourth and final isometry of the plane is the glide reflection. A glide reflection is the product of three reflections in lines that are neither concurrent nor mutually parallel. Figure 6.10 illustrates the glide reflection $G = M_n \circ M_m \circ M_l$.

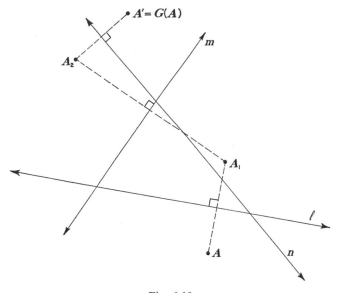

Fig. 6.10

Fortunately there are simpler and more easily applied ways to characterize a glide reflection G. Consider $G = M_n \circ M_m \circ M_l$ in figure 6.11.

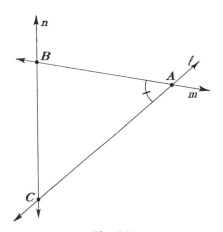

Fig. 6.11

Replace mirrors l and m by l' and m' such that they make the same directed
angle at A and such that $m' \perp n$ (fig. 6.12). Thus

$$G = M_n \circ M_m \circ M_l = M_n \circ M_{m'} \circ M_{l'}.$$

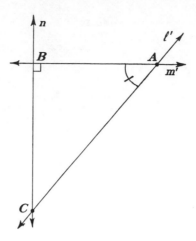

Fig. 6.12

Now replace mirrors n and m' by n' and m'' such that they are also perpendicular at B but such that $n' \perp l'$ (fig. 6.13). Thus

$$G = M_{n'} \circ M_{m''} \circ M_{l'}.$$

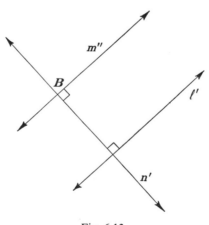

Fig. 6.13

Notice that $m'' \parallel l'$. Thus $M_{m''} \circ M_{l'}$ is a translation T. Finally, substituting T for $M_{m''} \circ M_{l'}$,

$$G = M_{n'} \circ T,$$

or in words:

A glide reflection G is the product of a reflection and a translation in the direction of the mirror.

The mirror is the *axis* of the glide reflection. (See fig. 6.14.) Is it also true that G is the product of a half-turn and a reflection?

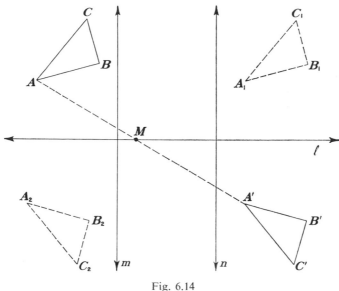

Fig. 6.14

Figure 6.14 also illustrates the fact that G may be obtained either by first translating then reflecting (A to A_1 to A') or by first reflecting then translating (A to A_2 to A'). It is also evident in figure 6.14 that the axis l bisects a segment determined by A and $G(A)$. In the figure, M is the midpoint of $\overline{AA'}$. Further properties of a glide reflection are similar to those listed for a reflection and are not repeated here.

Dilation

The first similarity to be illustrated that is not also an isometry is the dilation. (It is also called homothety, dilatation, and central similarity.) Given a point O in the plane and a positive real number k, the dilation image of a point $A \neq O$ is the point A' on $\overrightarrow{OA}$ such that $OA' = k \cdot OA$. If $A = O$, then the image and the preimage coincide. The letter D is used to denote a dilation. Figure 6.15 illustrates this definition for several points in the plane.

Some of the dilation properties that are illustrated in figure 6.15 follow:

1. The image of the center O is O. Thus a dilation is a one-to-one mapping of the plane onto itself. O is the only fixed point for all values of k.

2. Dilation is a line-preserving transformation. ($D(l) = l'$ in fig. 6.15.)

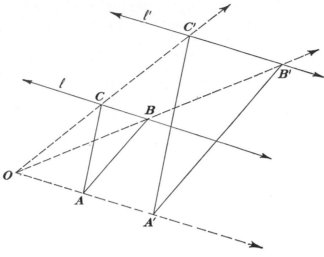

Fig. 6.15

3. Dilations preserve angles and angle measure. $(D(\angle ACB) = \angle A'C'B'$ and $m\angle ACB = m\angle A'C'B'$.)

4. All lines containing the center O are their own images; thus they are fixed lines. Are they pointwise fixed lines?

5. Under a dilation with magnitude k, the distance $A'B'$ between the images of A and B is given by $A'B' = k \cdot AB$.

6. A dilation is an orientation-preserving transformation. (Compare A, B, C and A', B', C' in fig. 6.15.)

Other similarity transformations

The five transformations discussed and illustrated in this section are examples of similarity transformations. Yet there are similarity transformations that are neither isometries nor dilations. These other similarity transformations are products of isometries and dilations. An example of such a similarity transformation is shown in figure 6.16. The similarity transformation mapping $\triangle ABC$ onto $\triangle A'B'C'$ is the product of a reflection, a translation, and a dilation.

The properties of these similarity transformations, which are referred to henceforth as simply similarity transformations, are those properties that are common both to dilations and to isometries. Some of these properties are outlined as follows:

1. A similarity transformation is a one-to-one mapping of the plane onto itself.

2. Under a similarity transformation, the image of a line is a line.

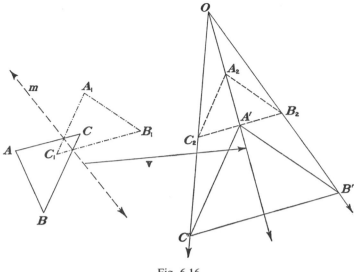

Fig. 6.16

3. Similarity transformations preserve measures of angles. (In fig. 6.16, $m\angle ABC = m\angle A'B'C'$.)

4. Under a similarity transformation, the ratio of distances is preserved. $\left(\text{In fig. 6.16, } \dfrac{AB}{BC} = \dfrac{A'B'}{B'C'}.\right)$

Just as the number of isometries is finite, so too is the number of similarity transformations of the plane. The minimal set of similarity transformations is discussed in the section on the groups of transformations.

These six line-preserving transformations and their products form a basis for the study of Euclidean geometry. The central relations between figures in Euclidean geometry such as parallelism, congruence, similarity, perpendicularity, or intersection are relations that are preserved under these transformations. Thus one could, by studying the invariants of the transformations, study Euclidean geometry. The remainder of this chapter is a description of a means by which these transformations can be employed in school geometry.

Why Transformations Should Be Included in Geometry

Increasing criticism has been aimed at the school geometry that has been taught for the last half century. One of the major areas for criticism is the mathematical content taught. The content and organization of school geometry courses were patterned after the reorganization of Euclid due to Legendre, whose work dates from 1794. Some topics and ideas that were important a half century ago have less importance today; other ideas retain their importance. Still other ideas, even though long recognized as important, needed present-day mathematical advances to highlight their fundamental importance in mathematics.

One such idea that has been conspicuous by its absence from geometry texts is function. The importance of function is attested to by its use in the mathematical sciences and by the 1959 *Report of the Commission on Mathematics,* published by the College Entrance Examination Board. Isometries and dilations are functions, and thus they can provide the learner with immediately useful applications of the function concept. Other areas in school mathematics have more difficulty in showing significant (to the average learner) uses of functions. In geometry the function can become truly useful. (These uses are described fully in later sections.)

In addition, employing isometries and dilations in geometry provides "readiness" experiences that will stand the learner in good stead if he studies affine geometry, projective geometry, or topology. For example, the notion of invariance, so easily brought to the fore while using isometries and dilations in school geometry, is likewise important in these other geometries. Perhaps the learner will better appreciate the significance of invariant properties of sets of points under a transformation if he begins his study with transformations that preserve nearly all properties of the sets, as do the isometries, and studies later other transformations that preserve fewer and fewer properties. This sequence appears to be a major goal of the English School Mathematics Project materials, *Books 1–5* (8; 9; 10; 11; 12).

If for no other reason, transformations should be introduced in geometry in order to provide another tool useful in solving geometric problems. A favorite example follows:

> Given a triangle *ABC,* construct a square with two vertices on $\overline{AB}$ and one vertex on each other side.

This problem is most easily solved using dilations. Many other similar problems are in the literature. The two-volume set *Geometric Transformations I and II* by I. M. Yaglom (17; 18) and *Geometry Revisited* by Coxeter and Greitzer (2) represent sources for such problems. Another useful source is *Transformation Geometry* by Max Jeger (5).

Not all the uses of transformations as tools are elementary. Advanced topics in geometry employ transformation ideas. For example, in his book *Plane Geometry and Its Groups,* Guggenheimer (4) uses transformations in his development of hyperbolic geometry. True, the transformation he uses is inversion in a circle, but this turns out to be interpretable as reflection in a line in the hyperbolic plane. The point is, these valuable tools can and should be introduced to school-age youngsters not only for their immediate use but also for their long-range use.

Structure in mathematics has been established as a central notion in contemporary school mathematics. Courses in algebra emphasize the field of real numbers as the basis for algebraic manipulation and deduction. Geometry, too, has its set of axioms that give it its structure. Transformations provide the basis for an axiom system different from present examples.

The introduction of transformations provides, additionally, opportunities to examine structures that are common both to algebraic and arithmetic objects and to geometric objects. For example, the set of isometries is a group under composition; subgroups also exist and are easily identified. In addition, the finite symmetry groups of geometric figures provide further examples that have as their objects geometric entities rather than numerical ones. These examples should help the learner to fathom the importance of common structures for all mathematics and to begin to see the power of abstraction and generalization.

Transformations provide an additional way to emphasize the interplay between algebra and geometry. Initially, transformations may be introduced in completely geometric terms, as was done in the first section of this chapter. Such a description can be thought of as a geometric representation of the mathematical ideas. Subsequently, transformations may also be represented algebraically with matrices. Studying two representations of the same mathematical idea, one geometric and one algebraic, should help the student gain confidence in his ability to do mathematics by making him aware of various approaches to the same idea. It should also help him in his ability to interpret different mathematical representations.

The preceding paragraphs include some more or less mathematical reasons for introducing transformations into the school geometry program. Essentially, the main point is that mathematical transformations, their properties, and their representations are important both to mathematics and to a student of mathematics. A final argument for transformations is more pedagogical in nature. Its validity is partially based on the author's personal experience in teaching a transformation approach to geometry.

When trying to teach a youngster a mathematical concept, we seek a development that is closely associated with his previous intuitive notions. We try to give him an intuitive peg on which to anchor the ideas to be learned. For example, in teaching congruence, we usually say it means the "same size and shape." The learner is able to determine whether two figures are congruent by deciding physically or mentally whether they "fit" one onto the other. Often when the formalization of the concept occurs in geometry classes, little attempt is made to use the intuitive peg of the learner (because superposition is not accepted). Instead, a correspondence is defined. What could be more natural to the learner than to use isometries that map one figure onto another in the formalization of the concept of congruence?

The argument, then, is that transformations can be used to organize instruction so that it is more closely related to youngsters' intuitive ideas. There are other pedagogical and mathematical benefits also. For one, transformations provide a means to present geometry in a more unified fashion than is usually the case. Their use also allows greater student flexibility and creativity in constructing proofs. These latter two points will be more fully amplified in the remainder of this chapter.

The entire argument for the incorporation of transformations in geometry can be succinctly summarized as follows: Transformations provide a unified and mathematically contemporary orientation to geometry that is extremely sound pedagogically.

A Fundamental Question

Having presented the case for including transformations in the study of geometry, a fundamental question yet remains; namely, in what manner and form should transformations be incorporated in the school geometry work?

This question has many answers, but there are two types of approaches that include extended work with transformations. These approaches are attempts to do more with transformations than simply add a chapter to a standard text. One approach makes the transformations and their properties the objects of study in geometry. Such an approach is well represented by the works of Yaglom (17; 18) and Jeger (5). Each of these works assumes that the student has a knowledge of much traditional Euclidean geometry. Euclidean results are used to derive properties of the transformations and relationships among transformations. Each work has a large collection of problems whose solutions follow from applications of transformations. These works are a good source of ideas and problems, but as they stand, they are beyond the reach of many school students.

An English series that takes a similar point of view, yet written expressly for British students of ages eleven to sixteen, is the School Mathematics Project's (SMP) *Books 1–5* (8; 9; 10; 11; 12). In these materials, transformations are introduced intuitively in the first two books. In the same books, topics such as angle, polygons, area, volume, and the Pythagorean theorem are also developed intuitively. A similar pattern is evident throughout the series of texts, namely, the parallel development of some standard Euclidean topics with the transformation ideas. But as the student progresses through the series, the geometric emphasis centers more and more on the transformations themselves, as the following remarks suggest:

> In the S.M.P. texts, Transformation Geometry makes an early appearance and one consequence of this is that the treatment is biased towards the intuitive. . . . it is in Book 2 that one begins to study the transformations in earnest. [9, p. v]

> In this chapter [chapter 2] the emphasis shifts from the geometrical figure to the transformation, and the results obtained and the experience gained are more relevant to later work on matrices, functions and groups than to work on pure geometry. [10, p. v]

This last quotation suggests another feature of the School Mathematics Project materials:

> The emphasis now begins to be placed on the interrelations that exist between the various topics. For example, matrices, which were first encountered in Book 2, are now used to illumine the study of topology, transformation geometry, relations and inverse functions. [10, p. v]

Another feature of these materials is their lack of clearly delineated assumptions. The reader of these materials will be hard put to discover a basic set of assumptions for geometry or any other mathematical topic. This should not be construed as a criticism of the materials, for it is consistent with SMP's policy to encourage student discovery and with its further policy not to encourage geometric proofs for children in the age range of twelve to sixteen years (8, p. vii). Furthermore, *Book 5* includes a summary of each topic studied. In some sense these summaries could be taken as the assumptions of the mathematical topics in the course. It can be argued convincingly that the assumptions made regarding a mathematical topic ought to be the end product of the study rather than the beginning point, for it is then that the important ideas are clearly recognized, and it is then that these basic ideas can truly organize and structure the topic. In any case, undefined terms and assumptions are not evident in the SMP texts.

In the author's opinion, the SMP *Books 1–5* are excellent materials. They satisfy most of the reasons given for including transformations in geometry. Unfortunately, the development of geometry is done over a five-year period. Presently this fact precludes their use in most secondary schools in the United States. However, if the organization of mathematical instruction in the United States were to be modified, these materials would constitute an excellent source of ideas for developing new sequences for mathematics courses.

The materials briefly described above should be classified as transformation geometries in that the major emphasis is the study of the transformations themselves. For example, the table of contents of Jeger's *Transformation Geometry* reads in part (5, p. 11): (1) Mappings, (2) Reflection in a Line, (3) Translations, (4) Rotations, (5) The Group of Isometries, (7) Enlargements, (8) Similarities, and (9) Affine Transformations. Similarly, chapter titles in the SMP *Books 1–5* include Symmetry (8, chap. 13), Topology (9, chap. 1), Similarity and Enlargement (9, chap. 3), Reflections and Rotations (9, chap. 5), Translations and Vectors (9, chap. 7), Isometries (10, chap. 2), Shearing (10, chap. 12), Matrices in Transformations (11, chap. 1), Isometries (11, chap. 5), Vector Geometry (11, chap. 12), and Invariants in Geometry (12, chap. 6). These titles clearly suggest that the emphasis is on the transformations and the study of their properties.

The second type of approach to the inclusion of transformations in school geometry fits more readily into the present American curriculum sequence. It assumes that Euclidean geometry is a valid area of study, or that at least certain portions of Euclidean geometry are appropriate to study. This type of approach manifests itself in two forms.

The first form is represented by the text by Paul J. Kelly and Norman E. Ladd (6). In the first six chapters of this text, topics such as the foundations of geometry, congruence, parallelism, similarity, and convexity are discussed. Much of the content in these chapters is similar to that found in contemporary geometry texts. For example, in the chapter on congruence, segments are defined to be congruent if and only if they have the same length (6, p. 101). Similarly, angles are congruent if and only if they have the same measure. Congruence of triangles is a special one-to-one correspondence between triangles, one in which the corresponding sides and angles are congruent (6, p. 104). In order to have a basis for demonstrating triangles congruent, the side-angle-side (SAS) congruence axiom is stated. Other content is not nearly as familiar—for example, the work on the convexity of plane and solid figures (6, pp. 318–70).

In the seventh chapter, geometric mappings are introduced. Two ideas are initially emphasized. The first is the generality of this idea of correspondence or mapping or transformation of one set onto a second. The second emphasis is that some property or relation of one set may carry over to a second set under a transformation. This is the idea of invariance, or preservation.

The point of view of the work with transformations is well stated by the authors:

> We will be particularly interested, for example, in certain mappings of space onto itself, such as a rotation of space, in which congruence is pre-

served. Such a mapping is called a *motion* of space. When you apply a motion to space, you automatically know that *every* figure maps onto a congruent figure. Such mappings will give us a way of defining congruence for general sets, and they will extend our method of congruent triangles by providing us with an infinite number of congruences simultaneously. [6, p. 372]

Thus mappings will be used to extend the idea of congruence (and also similarity, as we shall see later).

The particular isometries are introduced as examples of distance-preserving mappings. They are introduced only after extensive work is done with distance-preserving mappings in general. The sequence of major definitions and theorems for this development follows. (Prior to this seqeunce, invariance of measure and relation on sets was defined and shown to hold for the product of mappings.)

DEFINITION. A mapping in which the measure of distance between pairs of points is invariant is called a *distance-preserving mapping* (6, p. 386).

THEOREM. If R and S are distance-preserving mappings and the product RS exists, then the product is also distance preserving (6, p. 386).

THEOREM. If R is a distance-preserving mapping, then its inverse is a distance-preserving mapping (6, p. 386).

THEOREM. If R is a distance-preserving mapping of T onto S, then point C is between points A and B in T if and only if its image point $R(C)$ is between the image points $R(A)$ and $R(B)$ in S (6, p. 388).

LEMMA. If P, Q, and R are three collinear points and if A and B are two points such that $AB = PQ$, then there is exactly one point C in $\overleftrightarrow{AB}$ such that $AC = PR$ and $BC = QR$ (6, p. 388).

THEOREM. In a distance-preserving mapping, the image of a line is a line (6, p. 389).

THEOREM. In a distance-preserving mapping, the image of a plane is a plane (6, p. 390).

THEOREM. In a distance-preserving mapping, the image of a triangle is a congruent triangle, and the image of an angle is a congruent angle (6, p. 391).

DEFINITION. If set S is the image of set T in a distance-preserving mapping, then T and S are said to be *congruent sets* (6, p. 392).

The remainder of the chapter deals with a study of reflections, rotations, and translations as examples of distance-preserving mappings and with similarity mappings, which are defined as follows:

DEFINITION. A mapping R is a *similarity mapping* if there exists a positive number k such that, for every pair of points P and Q in the domain of R, $R(P)R(Q) = k \cdot PQ$ (6, p. 413).

The major points of the work on similarity mappings include these:

THEOREM. In a similarity mapping, the image of a triangle is a similar triangle, and the image of an angle is a congruent angle (6, p. 417).

DEFINITION. If a set S is the image of a set T in a similarity mapping, then T and S are *similar sets* (6, p. 418).

Special instances of similarity mappings are expansion and contraction (6, p. 418). Thus the notions of congruent and similar sets—notions that began with the notions of segments, angles, and triangles—are extended to any set of points.

An alternative to the Kelly-Ladd treatment is one in which transformations—specifically the isometries, dilations, and their products—are introduced first, then assumptions are made about them, and finally they are used to help develop and unify topics that are typical of geometry in American schools today. This approach is less of a departure from the American tradition than is the study of transformation geometry described earlier. Because it can be incorporated into the sequence of mathematics courses that exist today and because it employs transformations in a fundamental manner from the beginning of the course, it is carefully and fully developed in the subsequent sections of this chapter. The reader should note, however, that this choice does not imply that the particular development described is the best one possible, nor does it imply that the study of transformation geometry is inappropriate for American schools. Rather, a decision was made to describe one particular point of view. No judgments were made regarding other approaches except those relating to the possibility of implementation in the present American system. Thus the central message of this chapter is that transformations can, and probably should, be incorporated into school geometry in some manner. The manner to do so is not specified. The description following delineates one possible organization that has been verified as feasible by teachers and learners in regular classrooms.

Course Objectives for Geometry Using Transformations

At present the author is aware of only two textbooks that use transformations in the way indicated above to develop the ideas of Euclidean geometry. The first of these is a Dutch series written by Troelstra, Habermann, de Groot, and Bulens (13; 14; 15). This series is written in Dutch. The second is a book by Coxford and Usiskin called *Geometry: A Transformation Approach* (3). The second book, being more familiar and written in English, forms the basis of the subsequent detailed discussion.

The major objectives of the geometry course using transformations follow. The geometry course is designed—

1. to transmit the factual aspects of Euclidean geometry that are commonly included in American high school geometry courses—aspects including properties of polygons, relations between lines, congruence, similarity, intuitive geometry of space, circles, and parallelograms;

2. to introduce the concept of transformation and its particular instances, isometry and similarity;

3. to use transformations and their elementary properties as a basis for an axiomatic treatment of geometry;

4. to extend student understanding of the nature of proof by employing deductive reasoning to verify geometric propositions in text and in exercises;

5. to introduce matrices and matrix representations of transformations and geometric figures in order to highlight the relation between algebra and geometry;

6. to introduce the algebraic structure of a group and to examine the group structure of sets of transformations under composition;

7. to provide a pedagogically and mathematically defensible and unified development of congruence and similarity.

Scope and Content of Geometry Using Transformations

Prerequisite and review materials

The course employs postulates similar to the ruler and protractor postulates of the School Mathematics Study Group. These make it necessary for students to review certain algebraic topics. Students are expected to have studied such ideas as real numbers, equality and its properties, operations on the reals, properties of these operations, and inequality relations and properties thereof. The properties of the operations and inequalities mentioned here are standard, with the exception of one that states, "If a and b are positive, then $a + b > a$ and $a + b > b$" (3, p. 6). This is called the *positive number-sum property* and is used in the study of the geometric inequalities of the triangle.

Some knowledge of absolute value is also assumed, as well as familiarity with sets and the relations between sets. Two sets are defined to be equal if and only if each is a subset of the other. Familiarity with operations on sets, namely, intersection and union, is also a prerequisite expectation. As is the case in most contemporary American school geometries, geometric figures are thought of as sets of points that may intersect or that may be defined as the union of two or more sets of points.

Even though, strictly speaking, no geometric knowledge is assumed, practically speaking the student is expected to be familiar with many fundamental geometric concepts and their properties. In particular, the undefined ideas of point, line, and plane and their relationships are assumed to be familiar to students. The first postulate in this development summarizes these relationships (3, pp. 18–20).

POSTULATE 1 (point-line-plane postulate).
   a) *A line is a set of points and contains at least two points.*
   b) *Two different lines intersect in at most one point.*
   c) *Every pair of distinct points lies on at least one line.*
   d) *A plane is a set of points and contains at least three points that are not collinear.*
   e) *There is exactly one plane that contains three noncollinear points.*
   f) *If two points lie in a plane, the line determined by these points is a subset of the plane.*
   g) *If two different planes intersect, their intersection is a line.*

As with postulate 1, postulate 2 is thought of as a summary of ideas already intuitively accepted and reasonably well understood by students. The student's previous experience that forms the basis for this understanding is his work with the number line in elementary and junior high school.

POSTULATE 2 (ruler postulate). *The points of a line can be placed in correspondence with the real numbers so that—*
   a) *to every point of the line there corresponds exactly one real number called its* coordinate;
   b) *to every real number there corresponds exactly one point of the line;*
   c) *to each pair of points there corresponds a unique number called the* distance between the points;
   d) *given any two distinct points A and B of a line, the line can be coordinatized in such a way that A corresponds to zero and B corresponds to a positive number.* [3, p. 22]

Students are expected to have had some experience with deduction in previous mathematics work. Throughout the initial chapters of these materials informal deduction is often used to support the truth of geometric assertions. The point of view taken is that the deductive chain forms an argument that informs the reader that the proposition can be verified on the basis of previously accepted notions.

Several theorems are informally deduced from postulate 2. Some of these follow (3, pp. 22–34):

1. *A line is an infinite set of points.*
2. *If point B is between points A and C, then AB + BC = AC.*
3. *On a ray there is exactly one point at a given distance from the end-point.*
4. *A segment has exactly one midpoint, and if M is the midpoint of $\overline{AB}$, then $AM = \frac{1}{2} AB$.*

The third postulate deals with separation (3, p. 36):

POSTULATE 3 (plane-separation postulate). *Every line separates the plane into two convex sets.*

The development of this postulate assumes some student acquaintance with separation and convexity, but clearly not as much familiarity as is expected for other topics. These ideas are used to review the ideas of the interior and exterior of a polygon and of a convex polygon.

The final area in which students are assumed to have had quite extensive previous experience deals with the angle, its measure, and its special properties. An angle is defined as the union of two rays with a common endpoint for which the following assumptions are made (3, p. 46):

POSTULATE 4 (protractor postulate). *To each $\angle AOB$ there corresponds a unique real number greater than or equal to 0 and less than or equal to 180 called the measure of the angle, written m $\angle AOB$, so that—*
 a) the angle formed by two identical rays has measure 0;
 b) the angle formed by two opposite rays has measure 180;
 c) if $\overrightarrow{OC}$ (except for point O) is in the interior of $\angle AOB$, $m\angle AOC + m\angle COB = m\angle AOB$;
 d) if $\overrightarrow{OA}$ is one ray of an angle, there is exactly one ray in each half-plane of the line $\overleftrightarrow{OA}$ with a given measure between 0 and 180.

Notice that parts *a* and *b* of postulate 4 deal with angles with measures of 0 and 180. This assumption is not made explicitly in many popular American geometry texts. The major reason for including these ideas is that it is important to have rotations through angles that measure 0 (the identity) and 180 (a half-turn). The only apparent difficulties arising from angles with measure 180 are that either half-plane may be chosen as the interior of the angle and that an angle may not be a subset of a unique plane. These difficulties are less constraining than would be the case if half-turns had to be deleted. (In the School Mathematics Project's *Book 1*, *angle* is defined as the amount of turning [8, p. 55]. The term *turn* is not defined. Moreover, in addition to acute and obtuse angles, the SMP text defines a reflex angle as greater than a half-turn. This suggests that there is little consensus on how *angle* should be treated.)

The assumptions expressed in postulate 4 are used informally, yet deductively, to verify standard properties of supplementary, complementary, and vertical angles. For example, the following theorems are presented (3, pp. 50–54):

1. *All supplements (complements) to an angle have the same measure.*

2. *If the noncommon sides of two adjacent angles are opposite rays, the angles are supplementary.*

3. *Vertical angles have the same measure.*

Two lines are defined to be perpendicular if and only if they form a right angle. Part *d* of postulate 4 is employed to emphasize the fact that at a point on a line there is a unique perpendicular to the line. This fact, together with the uniqueness of the midpoint of a segment, permits the deduction that a segment has a unique perpendicular bisector, or mediator. Similarly, the uniqueness of the angle bisector is pointed out.

Even though students may not be equally familiar with each topic discussed above, it is evident from the materials that some familiarity is assumed. Postulates are stated as summaries of "known" facts and relations. Theorems are informally deduced, and generally their content is found in junior high school texts. Finally, in the suggestions for teaching, the teacher is encouraged to move his class through this material quite rapidly. These factors justify, at least partially, considering this material as prerequisite and review material. The material described in the next section cannot be similarly categorized.

Transformations

As decided on earlier, the approach to geometry being discussed here is designed to use transformations in a fundamental manner to develop other Euclidean topics. Thus it is mandatory in this approach to develop student comprehension of transformation ideas and to present an axiomatic basis for the topics to be developed later. In the Coxford and Usiskin materials, as well as in the Dutch materials by Troelstra et al., transformations are introduced by means of the specific transformation called reflection in a line. Intuitive work is done prior to the formalization of the idea. The definitions of reflection given in each work are essentially the same; each emphasizes the relation between a point and its image (3, p. 70).

DEFINITION. *If A is not on line m, the* reflection image *of A over line m is the point A' if and only if m is the perpendicular bisector of $\overline{AA'}$. If A is on m, then the reflection image of A is A itself.*

Notice that, strictly speaking, the transformation *reflection* has not been defined. Rather, the *reflection image* of a point in the plane is defined. The reflection itself is the mapping of the plane onto itself such that each point is associated with its image by means of the definition given above. In the Troelstra et al. and in the Coxford-Usiskin works, the emphasis on transformations is on mapping one geometric figure onto another rather than on thinking of a mapping of the plane onto itself. Only when the group characteristics of certain sets of transformations are discussed is the transformation itself (not the transformation of figures) the object of study. The reasons for this are at least twofold: (1) Pedagogically it is easier to examine a figure and its image than it is to examine the plane and its image, and (2) the materials are designed to include the study of figures and their properties and relations between figures. Thus figures and their images are of prime importance. It is posited that the image of a figure is the set of images of the points making up the figure.

An important result deduced from the definition of reflection over a line is that if B is the reflection image of A over m, then A is the reflection image of B over m. By the use of the notation introduced in the first section, this can be stated as follows:

$$\text{If } M_m(A) = B, \quad \text{then } M_m(B) = A.$$

In the Troelstra materials, only four axioms are given in the entire work. Many other assumptions are tacitly made. Three of these axioms deal with reflections:

AXIOM 1. *The mirror image of a straight line is again a straight line* (13, p. 44).

AXIOM 2. *A line segment and the image line segment are the same length* (13, p. 44).

AXIOM 3. *An angle and the image angle have the same measure* (13, p. 44).

The fourth axiom is equivalent to the parallel postulate:

AXIOM 4. *If a quadrilateral has three right angles, then the fourth angle is also right* (13, p. 61).

The assumptions concerning reflections made in the American materials follow. Each part of these assumptions is intuitively introduced and discussed. The learners' intuitions are summarized and stated as assumed (accepted) properties of the transformation (3, p. 73–85).

POSTULATE 5.
   a) *Given a line of reflection, every point has exactly one image.*
   b) *The reflection image of a line over any reflecting line m is a line.*

c) *The reflection image of P is between the images of A and B if and only if P is between A and B.*

d) *Let m be a reflecting line. Let A' be the image of A over m and B' the image of B. Then AB = A'B'.*

e) *The reflection image of an angle is an angle of the same measure.*

POSTULATE 6.

a) *Let A_1, A_2, . . . , A_n be the n vertices of a convex polygon. Then the path from A_1 to A_2 to A_3 to A_n is either clockwise oriented or counterclockwise oriented and not both.*

b) *Let polygon $A_1'A_2'A_3'$ . . . A_n' be the reflection image of polygon $A_1A_2A_3$ . . . A_n. If A_1 to A_2 to A_n is clockwise oriented, then A_1' to A_2' to A_n' is counterclockwise oriented, and vice versa.*

Each of these assumptions can be convincingly illustrated by simple constructions of the images of special sets of points and corresponding measurements or other observations of the resulting image sets. Logically, the assumptions made are not independent. For example, preservation of betweenness can be deduced from the assumptions and theorems stated previously. However, from a pedagogical point of view, the effort needed to prove this result far overshadows its instructional benefits.

The image of a polygon is the polygon determined by the images of the vertices of the polygon. Conversely, when two polygons are images of each other, then the vertices are images of each other. When a polygon is its own image under reflection in a line, the reflecting line is a line of symmetry for the polygon. The polygon is said to be symmetric with respect to the line. Thus the idea of reflection allows a clear and concise definition of *line symmetry*.

The definition of transformation given by Coxford and Usiskin is, *A* transformation *is an operation such that each point in the preimage set has a unique image, and each point in the image set is the image of exactly one point* (3, p. 100). *Operation* is essentially an undefined term. It could be substituted for by *rule, mapping,* or perhaps *assignment*. Under this definition, reflection is a transformation.

The composition of transformations is introduced as the operation of applying one transformation to a figure, then applying to the resulting image a second transformation. As a result of composing transformations, new transformations are generated—they are called composites, or products. In particular, reflections are composed to get composites of reflections. When the reflecting lines number two and lie in a plane, the product of reflection is one of two transformations. The new transformations are the translation and the rotation. Even though translations and rotations are new, they are familiar to students because their physical representations, the slide and the turn, are so common in experience. The definitions follow (3, pp. 105–8):

1. *A transformation T is a* translation *if and only if* $T = M_m \circ M_l$ *where* $m \parallel l$ *or* $m = l$.

2. *A transformation R is a* rotation *if and only if* $R = M_m \circ M_l$ *where m and l have a point in common.*

Notice that the possibility of an identity translation and rotation can be handled by these definitions. For translation, the identity (or zero translation) occurs when $m = l$. The situation is similar for rotation; however, it is not explicitly mentioned that when "m and l have a point in common," they may coincide.

The introduction to these transformations is quite informal. Translations and rotations are examined mainly as examples of transformations. Their properties, which depend only on the fact that they are products (composites) of reflections, are discussed. Discussing the relations between the angle of rotation or the length of translation to the reflecting lines is delayed until parallelism and congruence have been studied in some detail. Then these properties are derived as applications of congruence and parallelism.

A major result deduced at this time is that every product of reflections preserves angle measure, betweenness, collinearity (lines), and distance. Thus translations and rotations preserve these properties.

The discussion of products of reflections lays the foundation for the definition of congruent figures. First, an *isometry* is defined as a reflection or a product of reflections. Then the following definition for congruent figures is stated (3, p. 116):

Two figures α and β are congruent *if and only if there is an isometry such that the image of α is β.*

Two immediate results of this definition are that (1) congruent segments have the same length, and (2) congruent angles have the same measure. These two results follow because isometries, as products of reflections, preserve the length of segments and the measure of angles.

The concept of congruent figures as defined here is a completely general one. The definition is not restricted to segments, to angles, or to triangles. Any pair of geometric figures that are "alike in size and shape" or that are images of one another under an isometry are congruent.

The definition also provides a straightforward strategy for determining whether two figures are congruent. All one needs to do is to find an isometry that maps one figure onto the other. Even though the strategy is clear, the details may vary from one student to another. Thus there is room for creativity also. This point is amplified in the discussion of congruent triangles later in this chapter.

Proof

Formal proof is an integral part of geometry in the United States. One reason for this is that geometry provides a good medium in which to practice discovering and writing proofs. It is a good medium because proof can be stimulated by the abundance of plausible propositions available in geometry. Moreover, the figures used in geometry provide a heuristic to proof.

In any treatment of geometry, written proofs usually deal initially with reasonably well understood concepts. The reason for this is that we wish to emphasize the mechanics of organizing and writing a proof rather than the discovery of the proposition to be proved. Later, the discovery of propositions and the writing of proofs can be combined. In the transformation approach to geometry, the student repertoire of geometric concepts available at the time proof writing is introduced includes fundamental properties of lines, planes and angles, reflections, and the properties of reflections. The latter concepts are used to introduce the learner to proof. Thus after a discussion of "if-then" statements, axiomatics, and deductive chains, written proofs are introduced through the content of reflections and their fundamental properties.

The first examples are very simple and involve short deductive chains. (See fig. 6.17.) For example:

$$\text{If } M_l(A) = B \text{ and } M_l(C) = D, \text{ then } BC = AD.$$

The proof comes directly from fundamental ideas:

1. $M_l(A) = B$, $M_l(C) = D$ (given)

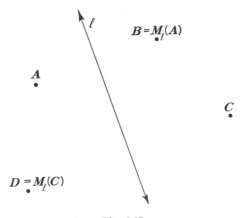

Fig. 6.17

2. $M_l(B) = A$ (why?)
3. $BC = AD$ (reflection preserves distance)

The practice material is also centered on reflections and their properties. One example follows (see fig. 6.18):

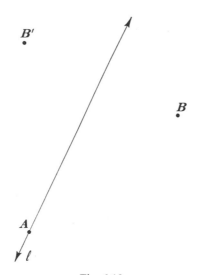

Fig. 6.18

Given: $M_l(B) = B'$ and figure 6.18.
Prove:

a) $M_l(\overline{BA}) = \overline{B'A}$
b) $\overline{AB} = \overline{AB'}$
c) $AB = AB'$

Part *a* follows because of the segment reflection theorem, which states that the reflection image of a segment $\overline{AB}$ is the segment determined by the images of the endpoints of $\overline{AB}$. Part *b* follows from the definition of congruence and a ray reflection theorem similar to the above given segment reflection theorem. Part *c* is a direct consequence of postulate 5 (*d*), which asserts that reflection preserves distance.

The introduction to formal proof through propositions dealing with reflections seems natural. The students use directly in proofs ideas they have just studied in a less formal manner. Moreover, using such a means allows for the maintenance and consolidation of essential notions that are not as familiar as many other ideas in geometry. Proof is continually emphasized throughout the course. However, this is not unique to a transformation approach.

Symmetry

Two activities are central to geometry: the first is the study of individual figures and their properties, and the second is the study of relationships between figures. Symmetry with respect to a line, as defined earlier, is useful in each of these activities, but it is vital in the first.

One of the simplest geometric figures with a line of symmetry is the segment. In fact, the segment has two lines of symmetry, *m* and *n* as shown in figure 6.19. The useful symmetry line is *m,* the unique perpendicular bisector (mediator) of the segment.

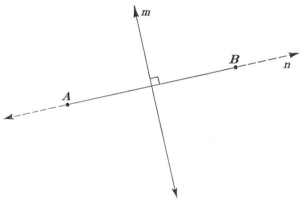

Fig. 6.19

A second fundamental geometric figure with line symmetry is the angle. The protractor postulate can be used to prove that the image of each side of an angle over the bisector of the angle is the other side of the angle. Thus reflection of an angle over its bisector switches the sides of the angle. It is clear, then, that the bisector of an angle is a symmetry line for the angle.

The line symmetry of the segment and the angle are fundamental in examining the symmetry of other important geometric figures. The isosceles triangle has a line of symmetry: the bisector of the vertex angle. In figure 6.20, *m* is the bisector of the vertex angle, *B,* of isosceles triangle *ABC.* A proof goes as follows. Because an angle is symmetric with respect to its bisector, the images of $\overrightarrow{BA}$ and $\overrightarrow{BC}$ are $\overrightarrow{BC}$ and $\overrightarrow{BA}$, respectively. Moreover, the image of *A* is *C* because the image of *A* is on $\overrightarrow{BC}$ and the same distance from *B* as is *C.* Similarly, the image of *C* is *A.* *B* is its own image. Thus the image of triangle *ABC* over *m* is triangle *CBA.* These are two names for the same triangle. So *m* is a line of symmetry.

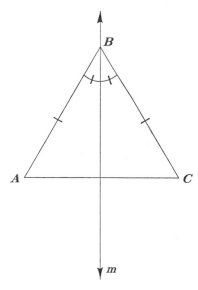

Fig. 6.20

The symmetry of the isosceles triangle is used to derive important properties (3, pp. 154–55):

THEOREM. *Base angles of an isosceles triangle are congruent.*

One proof uses the fact that the image of $\angle BAC$ is $\angle BCA$. Thus by definition of congruence, $\angle BAC \cong \angle BCA$.

THEOREM. *The bisector of the vertex angle of an isosceles triangle is the same line as the mediator of the base of the triangle.*

This follows because the unique angle bisector of $\angle B$ perpendicularly bisects $\overline{AC}$. Since the mediator of $\overline{AC}$ is unique, the two lines are one.

The symmetries of the three figures—segment, angle, and isosceles triangle—are used to examine symmetries of other figures and the properties of these figures. Some examples follow. The proofs are left for the reader (3, pp. 174–89).

DEFINITION. *A kite is a quadrilateral with two distinct pairs of adjacent sides congruent.*

THEOREM. *A kite has a symmetry line, the line determined by the endpoints of the pairs of congruent sides.* (Proof hint: Think of the mediator of $\overline{AC}$ in fig. 6.21.)

What properties of a kite can you deduce, knowing that it has line symmetry?

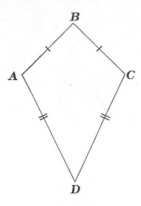

Fig. 6.21

THEOREM. *A rhombus has two symmetry lines, the lines containing the diagonals.*

What are some consequences of the symmetry of a rhombus?

THEOREM. *A circle has many symmetry lines, the lines containing the center.*

THEOREM. *A line has many symmetry lines: (a) itself or (b) any line perpendicular to it.*

The symmetry of figures is further developed when parallelograms are studied. (In that material the line symmetries of the rectangle and the square are discussed.) These topics follow a study of parallelism and the introduction of a parallel postulate.

An interesting outcome of the work on symmetric figures is the development of necessary and sufficient conditions for the congruence of segments and angles. It has already been demonstrated that if two segments or angles are congruent, then the segments or angles have the same measure. The sufficiency of these conditions is easily established. Only the case of segments is presented in detail here.

THEOREM. *If $AB = CD$, then $\overline{AB} \cong \overline{CD}$.*

Proof. Let m be the mediator of $\overline{AC}$. Then $M_m(A) = C$ and $M_m(B) = B'$.

1. If $B' = D$, then $M_m(\overline{AB}) = \overline{CD}$ and $\overline{AB} \cong \overline{BD}$ by definition (see fig. 6.22).

2. If $B' \neq D$, then $M_m(\overline{AB}) = \overline{CB'}$ and $AB = CB' = CD$. $\triangle DCB'$ is isosceles and reflection over n, the bisector of $\angle C$ maps $\overline{CB'}$ onto $\overline{CD}$. Thus $M_n \circ M_m(\overline{AB}) = \overline{CD}$ and $\overline{AB} \cong \overline{CD}$. (See fig. 6.23.)

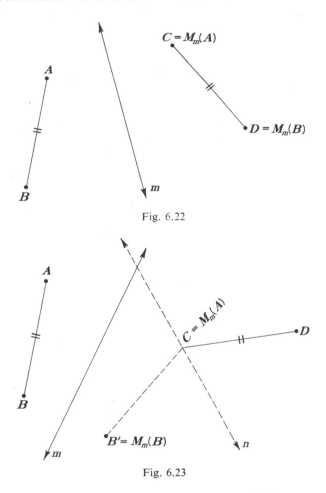

Fig. 6.22

Fig. 6.23

A similar argument can be used for angles (3, p. 161):

THEOREM. *If* $m\angle ABC = m\angle DEF$, *then* $\angle ABC \cong \angle DEF$.

It is interesting to note that these theorems involving the necessary and sufficient conditions for the congruence of segments and angles are often taken as definitions in nontransformation approaches to geometry. The generality of the definition of congruent figures in terms of isometries is apparent.

Parallelism

The treatment of parallelism in a transformation approach does not have as many unique features as are apparent for many other topics. The assumption made is well known (3, p. 203).

POSTULATE 7 (parallel postulate). *Through a point not on a line there is exactly one line parallel to a given line.*

Transformations are used in the proofs of a few of the standard theorems on parallel lines. The use of transformations is not extensive. A few examples of theorems follow (3, pp. 196–204):

1. *Two lines perpendicular to the same line are parallel.*
2. *If m is a line and P a point not on m, then there is at least one line containing P and parallel to m.*
3. *If a line is perpendicular to one of two parallel lines, it is perpendicular to the other.*

The theorems on the symmetry of a rectangle and a square were mentioned previously. If a rectangle has two symmetry lines—the mediators of the opposite sides—what properties of a rectangle can be deduced? If a square is a quadrilateral, which is a rhombus and a rectangle, what symmetry lines does it have? What are its properties?

Perhaps the most distinct feature of the transformation approach to parallelism is exhibited in the development of the alternate interior angles theorem. The key theorem in that development and its proof are given below (3, p. 214):

If two parallel lines are cut by a transversal, then the acute alternate interior angles are congruent.

In figure 6.24, parallel lines m and n are cut by transversal t at P and Q, and $\angle P_1$ and $\angle Q_1$ are acute. Construct perpendiculars $\overleftrightarrow{QR}$ and $\overleftrightarrow{PS}$ to m

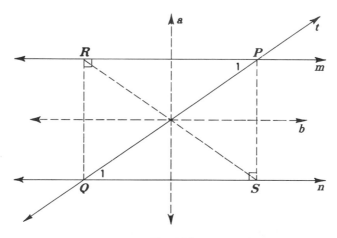

Fig. 6.24

and n, respectively. Then $\overleftrightarrow{RQ} \perp n$, $\overleftrightarrow{SP} \perp m$, and $RQSP$ is a rectangle. Let a and b be the symmetry lines for the rectangle. Then

$$M_a(\angle P_1) = \angle PRS$$
$$M_b(\angle PRS) = \angle Q_1.$$

Thus

$$M_b \circ M_a(\angle P_1) = \angle Q_1.$$

Finally

$$\angle P_1 \cong \angle Q_1.$$

Since the acute angles are congruent, so are their supplements. If $t \perp m$ and $t \perp n$, then all angles are right angles and thus congruent. This takes care of all cases; so the alternate interior angles theorem is demonstrated.

The remaining work with parallelism does not make specific use of transformations. It includes other theorems (and their converses) on the angles formed by parallels and a transversal, the theorem about the sum of the interior angles of a triangle, and the relations between interior and exterior angles of a triangle.

Congruent triangles

Whatever the figures, they are congruent if and only if one maps onto the other under an isometry. This is the unifying idea whenever congruence is studied. Just as with segments and angles, certain conditions on pairs of figures are sufficient for the congruence of those figures. In the case of triangles, three pairs of conditions are needed to insure that the triangles are congruent: these sufficient conditions for the congruence of triangles are commonly abbreviated SAS, ASA (or AAS), and SSS.

In the transformation approach to geometry, the sufficient conditions SAS, ASA, and SSS are demonstrable. In other more or less axiomatic treatments, one or more of them are postulated. Pedagogically, it is more satisfying to derive these results than to assume one or more of them because none of the conditions is readily accepted by students without extensive intuitive work. Also, transformations provide a common unifying point of view toward these conditions and other similar ones for right triangles. Students can observe how each specified condition on the triangles is used in showing that one may be mapped onto the other by an isometry.

Even though there is a clear strategy for proving SAS, ASA, and SSS, there are many variations possible in the proofs themselves. The variations allow student creativity in proof construction. This is exemplified by the three proofs of the SAS congruence theorem given below.

Given: $\triangle ABC$ and $\triangle XYZ$ (see fig. 6.25)
    $AC \cong XZ$
    $BC \cong YZ$
    $\angle C \cong \angle Z$

Prove: $\triangle ABC \cong \triangle XYZ$

Proof 1. Translate $\triangle ABC$ so that C maps onto Z: $\triangle ABC \rightarrow \triangle A'B'Z$. Let m be the bisector of $\angle A'ZX$; m is also the bisector of $\angle B'ZY$. Reflect $\triangle A'B'Z$ over m. Then $M_m(Z) = Z$. $M_m(B') = Y$ and $M_m(A') = X$ because $\triangle A'ZX$ and $\triangle B'ZY$ are isosceles triangles. Thus $\triangle ABC$ is mapped onto $\triangle XYZ$ under the product of a translation and a reflection—an isometry. Therefore $\triangle ABC \cong \triangle XYZ$.

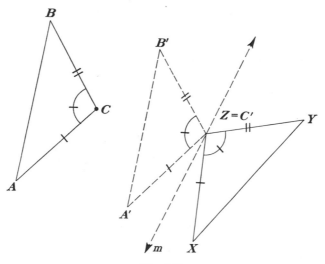

Fig. 6.25

Proof 2. Reflect $\triangle ABC$ over m, the mediator of BY. (See fig. 6.26.)
$$M_m(\triangle ABC) = \triangle A'YC'.$$

Reflect $\triangle A'YC'$ over n, the bisector of $\angle C'YZ$. (See fig. 6.27.)
$$M_n(\triangle A'YC') = \triangle A''YZ.$$

Reflect $\triangle A''YZ$ over $\overleftrightarrow{YZ}$. The image is $\triangle XYZ$ (why?). Thus $\triangle ABC$ maps onto $\triangle XYZ$ under an isometry, and $\triangle ABC \cong \triangle XYZ$.

Proof 3. Since $\overline{AC} \cong \overline{XZ}$, we can map $\overline{AC}$ onto $\overline{XZ}$ by an isometry. (See fig. 6.28.) Under this isometry, B maps onto B', and B' is in one of the half-planes determined by $\overleftrightarrow{XZ}$. If B' and Y are in opposite half-planes, then reflection over $\overleftrightarrow{XZ}$ maps B' onto Y and $\triangle XB'Z$ onto XYZ (why?).

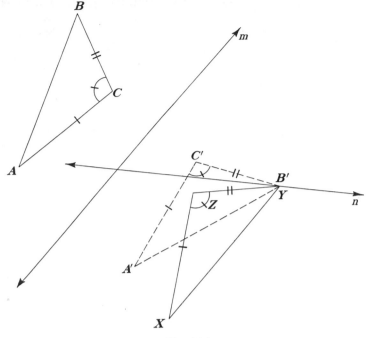

Fig. 6.26

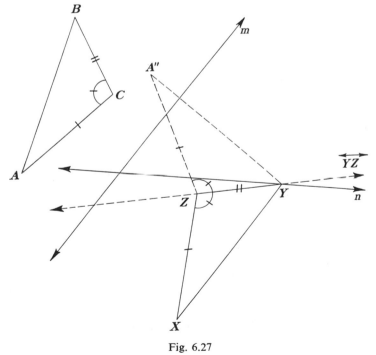

Fig. 6.27

Thus $\triangle ABC \cong \triangle XYZ$. If B' and Y are in the same half-plane of $\overleftrightarrow{XZ}$, reflect $\triangle XB'Z$ over $\overleftrightarrow{XZ}$ to $\triangle XB''Z$. Then B'' and Y are in opposite half-planes. This is the situation discussed above, and thus $\triangle ABC \cong \triangle XYZ$.

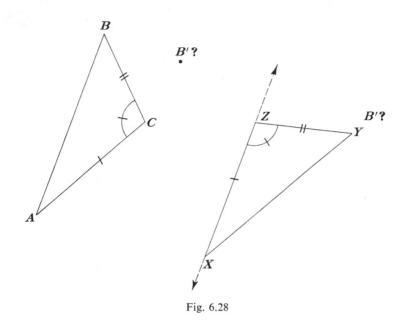

Fig. 6.28

The three proofs shown are representative of the types of proofs available. The first allows the use of any isometry desired; the second uses only reflections; the third uses more abstract reasoning. Most students can construct at least one of these types of proofs.

Proofs of ASA and SSS are left for the reader to construct. He should construct proofs when the triangles have opposite orientation and when they have the same orientation. What is the maximum number of reflections needed to map one congruent triangle onto another when the orientations are the same? The opposite?

When SAS, ASA, and SSS are demonstrated, they are used as tools to obtain other familiar results. These results include the conditions on right triangles sufficient for congruence, the AAS theorem, the converse of the base angles of an isosceles triangle theorem, the properties of isosceles trapezoids, and the properties of parallelograms (3, pp. 244–75).

The results indicated in the previous paragraph could be demonstrated by means of transformation arguments. However, students should be given opportunities to apply the triangle congruence theorems because they form the basis of an important strategy for geometric proof. Many proofs based on congruent triangles are also quite easy to organize. Thus students get the opportunity not only to gain familiarity with a new mode of proof but also to gain that familiarity in problem situations that promote successful proof writing.

Finally, congruent triangles are used to help demonstrate properties of translations and rotations. These properties include the following theorems (3, pp. 277–82):

1. *Under a translation every line is parallel to its image.*

2. *Under a translation segments determined by points and their images are parallel and congruent.*

3. *If $m \parallel n$ and $P' = r_m \circ r_n(P)$, then $\overleftrightarrow{PP'} \perp m$ (and n), and PP' is twice the distance between m and n.*

4. *If a point is on both m and n, then it coincides with its image under $R = r_m \circ r_n$. (The point is a fixed point under the rotation R.)*

5. *Let P' and Q' be the images of P and Q under a rotation with center C. Then $m\angle P'CP = m\angle Q'CQ$.*

6. *The magnitude of the rotation $R = r_m \circ r_n$ is twice the measure of the angle between m and n, measured in a direction from n to m.*

The theorem stating that *the segment connecting the midpoints of two sides of a triangle is parallel to the third side and is half the length of the third side* is used in the demonstration showing that the midpoint of a segment and its glide reflection image is on the axis of the glide reflection (3, p. 273).

The summary of congruence work done from a transformation approach suggests the importance of two tools: (1) the transformations themselves in deriving sufficient conditions for the congruence of triangles, and (2) the sufficient conditions in demonstrating properties of common figures and of less familiar transformations. Both aspects are valuable in geometry.

Similarity

The development of similarity, according to the course objectives, is to be related to congruence. This is done in the transformation approach to geometry by defining similar figures as those that are images under a similarity mapping. A special case of a similarity mapping is an isometry, which will be discussed later. In order to get the most general similarity mappings, one new transformation is introduced. It is the dilation. The dilation is called a *size transformation* by Coxford and Usiskin. The definition they give follows (3, p. 293):

DEFINITION. *Let C be a point and k be a positive real number. For any* *point A, let S(A) be the point A' on $\overrightarrow{CA}$ whose distance from C is k times the distance of A from C, A'C = k · AC. Then S is the size transformation with magnitude k and center C.*

Notice that the given definition specifies that k is greater than zero. Other definitions allow k to be any nonzero real number. The magnitude k was chosen to be positive for pedagogical reasons; using negative real numbers would be more difficult for the learner both conceptually and manipulatively and might interfere with understanding the transformation itself. In addition, if one wanted to introduce negative magnitudes, he could do so easily by defining a size transformation with a negative magnitude k as the product (composite) of a size transformation with magnitude k and a half-turn about the center of the size transformation.

The definition of a size transformation, in a manner similar to the definition of reflection, defines the image of one point. The image of a set of points is the set of image points. As with the other transformations, the initial emphasis in work with size transformations is on figures and their images. Thus the mapping of the plane onto itself is justifiably de-emphasized.

One final note on the definition of a size transformation. The image point A' is defined to be the point k times as far from C as is A, that is, $A'C = k · AC$. This statement is an attempt to bypass the well-documented difficulties students have with ratios. The emphasis is on the multiplicative nature of the relationship rather than on the constant quotient, that is, $\frac{A'C}{AC} = k$.

Properties of a figure and its image under a size transformation are intuitively discovered by constructing images of sets of points. The essential properties are assumed in postulate 8 (3, pp. 298–302):

POSTULATE 8 (size transformation postulate). *Each size transformation preserves (a) angles and angle measure, (b) betweenness, and (c) collinearity (lines); (d) under a size transformation of magnitude k, the distance between image points is k times the distance between their preimages.*

The following theorems are consequences of postulate 8 (3, pp. 299–310):

1. *Under a size transformation, a line is parallel to its image.*

2. *Under a size transformation, the image of polygon ABC . . . is polygon A'B'C' . . . where A' is the image of A, and so on.*

3. *If a line parallel to one side $\overline{BC}$ of $\triangle ABC$ intersects the other two sides in distinct points P and Q, then there is a size transformation S with $S(\triangle APQ) = \triangle ABC$.*

4. *A line parallel to a side of a triangle intersecting the other two sides in distinct points splits these sides into four proportional segments (side-splitting theorem).*

5. *If a line intersects two sides of a triangle and cuts off segments proportional to these sides, then the line is parallel to the third side.*

The proof of the last theorem stated above goes as follows. (See fig. 6.29.)

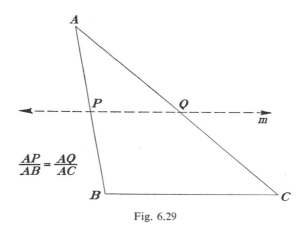

Fig. 6.29

Proof. Consider the size transformation S with center A and magnitude $k = \dfrac{AP}{AB} = \dfrac{AQ}{AC}$. Then $S(B) = P$, and $S(C) = Q$. (Why?) Thus $S(\overleftrightarrow{BC}) = \overleftrightarrow{PQ}$ and $\overleftrightarrow{BC} \parallel \overleftrightarrow{PQ}$. (Why?)

In order to introduce the similarity transformation in all its forms, the concepts of the identity transformation and the inverse transformations are needed. The identity transformation, I, is the transformation that maps each point onto itself, that is, $I(P) = P$ for all P. For example, a translation or rotation with magnitude zero and a size transformation with magnitude 1 are each an example of the identity transformation.

Two transformations, T and T', are inverse transformations if the product $T \circ T' = I$, that is, $T \circ T'(P) = P$. Examples of inverse transformations include the following:

1. T: Rotation about C, magnitude θ T': Rotation about C, magnitude $-\theta$

2. T: Reflection over m T': Reflection over m

3. T: Size transformation, center C, magnitude $k \neq 0$ T': Size transformation, center C, magnitude $\dfrac{1}{k}$

Dealing with identity and inverse transformations is preparatory to introducing groups of transformations. These ideas are also used immediately in the work on similarity transformations.

Whereas an isometry is a reflection or a product of reflections, a similarity transformation is more than a product of size transformations (3, p. 319).

DEFINITION. *A similarity transformation is a product of isometries and size transformations.*

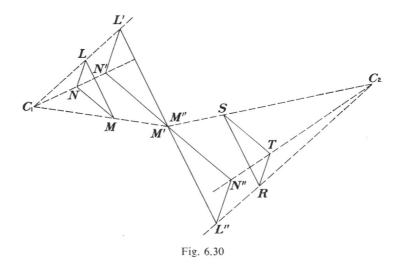

Fig. 6.30

The diagram in figure 6.30 depicts a similarity transformation, S, that maps $\triangle LMN$ onto $\triangle RST$. S is the product of a size transformation (center C_1), a half-turn with center M' and a second size transformation (center C_2).

Similarity transformations, being products of isometries and size transformations, preserve those properties of sets of points that are preserved by both isometries and size transformations. Thus images and preimages under a similarity transformation have their corresponding angles congruent and their lengths of corresponding segments proportional.

DEFINITION. *Two figures α and β are similar if and only if there is a similarity transformation S with $S(\alpha) = \beta$.* [3, p. 321]

Since similar figures are related by a similarity transformation, it follows that—

1. corresponding angles of similar figures are congruent;
2. corresponding segments of similar figures are proportional.

Any isometry can be thought of as the product of the identity size transformation and the isometry. Thus two figures related by an isometry are related by a similarity transformation. Congruent figures are, therefore, similar figures. This relationship adds unity to the study of congruence and similarity.

Sufficient conditions for the similarity of two triangles are stated in the AA, SAS, and SSS similarity theorems (3, p. 326–28). These theorems are easily proved using the definition of similar figures. As with congruent triangle work, the strategy is to find a product of size transformations and isometries that map one figure onto the other. The AA similarity theorem is proved below. The proof is due to a ninth-grade student in a class taught by the author. (The girl made the proof while being observed by a methods class of twenty-five college students. See fig. 6.31.)

Given. $\triangle ABC$ and $\triangle XYZ$ with $\angle A \cong \angle X$, $\angle B \cong \angle Y$.

Prove. $\triangle ABC \sim \triangle XYZ$.

Proof. Choose a point P as the center of a size transformation H with magnitude $k = \dfrac{AB}{XY}$.

$$H(\triangle XYZ) = \triangle X'Y'Z'$$
$$\angle Y' \cong \angle Y \cong \angle B$$
$$\angle X' \cong \angle X \cong \angle A$$
$$X'Y' = k \cdot XY = \frac{AB}{XY} \cdot XY = AB$$

Thus $\triangle X'Y'Z' \cong \triangle ABC$ by ASA.

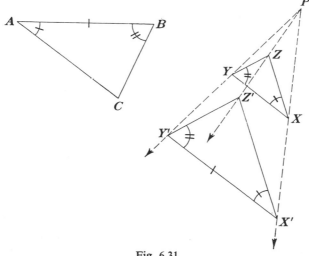

Fig. 6.31

By definition of congruence there is an isometry T such that $T(\triangle X'Y'Z')$ $= \triangle ABC$. Thus $T \circ H(\triangle XYZ) = \triangle ABC$, and $\triangle XYZ \sim \triangle ABC$.

The proofs of SAS and SSS are similarly simple and are left for the reader. After demonstrating the sufficient conditions for triangle similarity, several of the standard applications can be investigated. These include right-triangle similarities, the theorem of Pythagoras and its converse, trigonometric ratios, the slope of a line, parallel and perpendicular lines, and certain theorems on circles (3, pp. 331–80).

Matrices

Mathematical entities are abstractions. They have no physical existence. However, they may be represented in various ways—for example, a point may be represented geometrically by a "dot." The same point may be represented algebraically by an ordered pair of numbers or by a pair of distinct linear equations satisfied by that ordered pair. In a similar manner, a line may be represented by a mark on paper, by a linear equation, or by two planes intersecting.

Geometric figures and transformations, being mathematical entities, may have various representations. If coordinates are available, a useful means of representing geometric figures and transformations is the matrix.

The study of different representations of ideas, especially algebraic representations of geometric concepts, is important in school mathematics for several reasons. In much of advanced mathematics algebraic representations predominate, but geometric terminology is widely used in the descriptions. Moreover, the ability to use different representations of the same mathematical idea does not appear to be a strength of most secondary students. This could hinder the development of problem-solving skills. Finally, studying the same ideas in various representations cannot help but suggest to the student that mathematics is a unified whole and need not be thought of as algebra, geometry, trigonometry, and so on.

As represented by the Coxford and Usiskin materials, the work on matrices and their uses is not extensive (3, pp. 459–510). However, there is the possibility of extensive and meaningful work with matrices. In the descriptions that follow, we shall not feel restricted by what is actually included in the texts; we shall, however, be guided by the actual content found. (For an example of the pervasive use of matrices in school mathematics, the reader is referred to the School Mathematics Project, *Books 1–5* (8; 9; 10; 11; 12). Matrices are used consistently to represent many different mathematical ideas, not just those ideas related to geometry. In fact, in the School Mathematics Project materials, matrices serve as unifiers of many mathematical topics.)

In the initial stages of instruction, matrices are defined as rectangular arrays of numbers, and equality is discussed. Matrices may be applied immediately; points in a coordinate plane can be represented by a column matrix. For example:

$$(x, y) = \begin{bmatrix} x \\ y \end{bmatrix}$$

A polygon, being determined by its vertices taken in order, may also be represented by a matrix. In figure 6.32, quadrilateral $ABCD$ is represented in matrix form by

$$\begin{bmatrix} -3 & 1 & 2 & -1 \\ 1 & 2 & -1 & -3 \end{bmatrix} = ABCD.$$

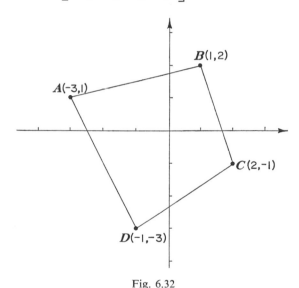

Fig. 6.32

In anticipation of representing certain transformations of the plane by 2×2 matrices, the multiplication of 2×2 matrices and the properties of 2×2 matrices under matrix multiplication may be developed. Important results are listed as follows:

1. The product of two 2×2 matrices is a 2×2 matrix.

2. Multiplication of 2×2 matrices is not commutative.

3. Multiplication of 2×2 matrices is associative.

4. $\begin{bmatrix} 1 & 0 \\ 0 & 1 \end{bmatrix}$ is the identity for multiplication of 2×2 matrices.

5. A matrix $\begin{bmatrix} a & b \\ c & d \end{bmatrix}$ is invertible if and only if $ad - bc \neq 0$.

When the determinant $D = ad - bc$ is not zero, the inverse of $\begin{bmatrix} a & b \\ c & d \end{bmatrix}$ is

$$\begin{bmatrix} d/D & -b/D \\ -c/D & a/D \end{bmatrix}.$$

By the manipulation of 2×2 matrices, each of these results is easily demonstrated in detail. Property 5 is important, for only invertible matrices can be used to represent one-to-one "onto" transformations of the plane.

How can a matrix be used to represent a transformation? Recall that a transformation is an operation (1) that assigns the points in one set to those in another such that each point has a unique image, and (2) such that each point in the image set has exactly one preimage.

Let any point P be represented by the column matrix $\begin{bmatrix} x \\ y \end{bmatrix}$. Also let

$$A = \begin{bmatrix} a & b \\ c & d \end{bmatrix}$$

be an invertible 2×2 matrix. Define the image of P to be the point

$$P' = AP = \begin{bmatrix} ax + by \\ cx + dy \end{bmatrix}.$$

It follows from the definition of transformation that the matrix A represents a transformation. First, P' is a unique point. Second, the point P' has exactly one preimage because A is invertible and thus has a unique inverse (see property 5) that when multiplied by P' yields the point P.

We know that the transformations studied so far map lines onto lines. Do the transformations represented by 2×2 matrices do the same? Fortunately, the answer is yes. This fact may be demonstrated by showing that betweenness is preserved under a transformation represented by a 2×2 invertible matrix.

As a very special case, we show in figure 6.33 that the midpoint of the segment with endpoints $A = (0,2)$ and $B = (2,4)$ maps into the midpoint of the image segment under the transformation represented by the matrix $\begin{bmatrix} 5 & 3 \\ 4 & 2 \end{bmatrix}$.

Proof. The midpoint of $\overline{AB}$ is $M = (1,3)$.

$$A' = \begin{bmatrix} 5 & 3 \\ 4 & 2 \end{bmatrix} \begin{bmatrix} 0 \\ 2 \end{bmatrix} = \begin{bmatrix} 6 \\ 4 \end{bmatrix}$$

$$B' = \begin{bmatrix} 5 & 3 \\ 4 & 2 \end{bmatrix} \begin{bmatrix} 2 \\ 4 \end{bmatrix} = \begin{bmatrix} 22 \\ 16 \end{bmatrix}$$

The midpoint of $\overline{A'B'}$ is

$$M' = \left(\frac{6 + 22}{2}, \frac{4 + 16}{2} \right) = (14,10).$$

Under the transformation, the image of $M(1,3)$ is

$$\begin{bmatrix} 5 & 3 \\ 4 & 2 \end{bmatrix} \begin{bmatrix} 1 \\ 3 \end{bmatrix} = \begin{bmatrix} 14 \\ 10 \end{bmatrix} = M'.$$

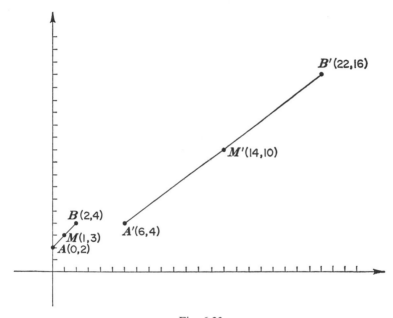

Fig. 6.33

Since lines map onto lines, the image of a polygon is a polygon. An example is shown in figure 6.34. The image is found by multiplying the matrix representing the polygon by the transformation matrix:

$$\begin{bmatrix} 1 & -2 \\ 2 & 1 \end{bmatrix} \begin{bmatrix} -2 & 2 & 0 \\ -1 & 1 & 1 \end{bmatrix} = \begin{bmatrix} 0 & 0 & -2 \\ -5 & 5 & 1 \end{bmatrix}$$

The image of a polygon represented by a matrix B (a $2 \times n$ matrix for an n-gon) is the polygon represented by the matrix $A \cdot B$ (again $2 \times n$).

Only certain similarity transformations are representable by 2×2 matrices, namely, those for which the origin is its own image. $\left(\text{For all } A = \begin{bmatrix} a & b \\ c & d \end{bmatrix}, A \cdot \begin{bmatrix} 0 \\ 0 \end{bmatrix} = \begin{bmatrix} 0 \\ 0 \end{bmatrix}. \right)$ These include the following transformations:

$$M_{x\text{-axis}} = \begin{bmatrix} 1 & 0 \\ 0 & -1 \end{bmatrix} \qquad M_{y\text{-axis}} = \begin{bmatrix} -1 & 0 \\ 0 & 1 \end{bmatrix}$$

$$M_{x=y} = \begin{bmatrix} 0 & 1 \\ 1 & 0 \end{bmatrix} \qquad M_{x=-y} = \begin{bmatrix} 0 & -1 \\ -1 & 0 \end{bmatrix}$$

$$R_{90°} = \begin{bmatrix} 0 & -1 \\ 1 & 0 \end{bmatrix} \qquad R_{180°} = \begin{bmatrix} -1 & 0 \\ 0 & -1 \end{bmatrix}$$

$$R_{270°} = \begin{bmatrix} 0 & 1 \\ -1 & 0 \end{bmatrix} \qquad I = R_{360°} = \begin{bmatrix} 1 & 0 \\ 0 & 1 \end{bmatrix}$$

Size transformation, center 0, magnitude $k = \begin{bmatrix} k & 0 \\ 0 & k \end{bmatrix}$

Each of these matrices may be obtained by examining the effect of the transformation on the two points $\begin{bmatrix} 1 \\ 0 \end{bmatrix}$ and $\begin{bmatrix} 0 \\ 1 \end{bmatrix}$. For example, under a reflection in the x-axis, $\begin{bmatrix} 1 \\ 0 \end{bmatrix}$ has image $\begin{bmatrix} 1 \\ 0 \end{bmatrix}$, and $\begin{bmatrix} 0 \\ 1 \end{bmatrix}$ has image $\begin{bmatrix} 0 \\ -1 \end{bmatrix}$.

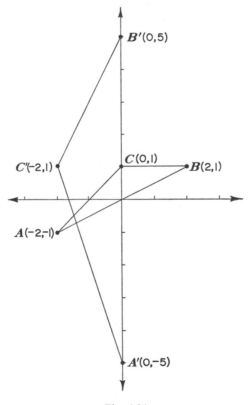

Fig. 6.34

Thus the matrix for $M_{x\text{-axis}}$ is $\begin{bmatrix} 1 & 0 \\ 0 & -1 \end{bmatrix}$. In general, if a transformation maps $\begin{bmatrix} 1 \\ 0 \end{bmatrix}$ onto $\begin{bmatrix} a \\ c \end{bmatrix}$ and $\begin{bmatrix} 0 \\ 1 \end{bmatrix}$ onto $\begin{bmatrix} b \\ d \end{bmatrix}$, then the transformation has matrix $\begin{bmatrix} a & b \\ c & d \end{bmatrix}$.

The composition of transformations is accomplished by the multiplication of matrices. That is, if two transformations T_1 and T_2 are represented by matrices A_1 and A_2, then the product transformation $T = T_2 \circ T_1$ is represented by the product matrix $A = A_2 \cdot A_1$. This idea can be used to derive the matrix representations of the rotations of 90°, 180°, and 270° about the origin.

As an example, the product of $M_{x\text{-axis}}$ with $M_{x\,=\,y}$ is a rotation of 90° about the origin:

$$M_{x\text{-axis}} = \begin{bmatrix} 1 & 0 \\ 0 & -1 \end{bmatrix} \qquad M_{x\,=\,y} = \begin{bmatrix} 0 & 1 \\ 1 & 0 \end{bmatrix}$$

$$R_{90°} = M_{x\,=\,y} \circ M_{x\text{-axis}} = \begin{bmatrix} 0 & 1 \\ 1 & 0 \end{bmatrix} \begin{bmatrix} 1 & 0 \\ 0 & -1 \end{bmatrix} = \begin{bmatrix} 0 & -1 \\ 1 & 0 \end{bmatrix}. \text{ (Why not }$$

use the opposite order?)

Similarly, $R_{90°} \circ R_{90°} = R_{180°}$; so

$$R_{180°} = \begin{bmatrix} 0 & -1 \\ 1 & 0 \end{bmatrix} \begin{bmatrix} 0 & -1 \\ 1 & 0 \end{bmatrix} = \begin{bmatrix} -1 & 0 \\ 0 & -1 \end{bmatrix}.$$

Simple facts about the trigonometric ratios for 30°, 45°, and 60° angles could yield rotation matrices for these angles. Further grasp of trigonometry would allow the development of the general matrix for rotation about the origin through an angle with measure θ:

$$\begin{bmatrix} \cos \theta & -\sin \theta \\ \sin \theta & \cos \theta \end{bmatrix}$$

The major emphasis of the work on matrix representations of transformations should probably be on finding the representation for familiar transformations. However, there are easy examples of the opposite emphasis; for example, the transformation called the *shear* with x-axis fixed can be defined to be the transformation represented by the matrix $\begin{bmatrix} 1 & k \\ 0 & 1 \end{bmatrix}$. Its effect on figures can be investigated algebraically. For example, in figure 6.35 the rectangle $\begin{bmatrix} 0 & 0 & 3 & 3 \\ 0 & 1 & 1 & 0 \end{bmatrix}$ under the shear $\begin{bmatrix} 1 & 4 \\ 0 & 1 \end{bmatrix}$ is mapped onto the parallelogram as follows:

$$\begin{bmatrix} 1 & 4 \\ 0 & 1 \end{bmatrix} \begin{bmatrix} 0 & 0 & 3 & 3 \\ 0 & 2 & 2 & 0 \end{bmatrix} = \begin{bmatrix} 0 & 4 & 7 & 3 \\ 0 & 2 & 2 & 0 \end{bmatrix}$$

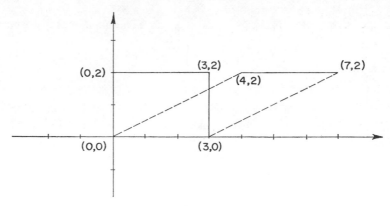

Fig. 6.35

The properties of the shear are summarized in the next theorem. The proof of the theorem is left for the reader (3, pp. 499–501).

Let H be a shear represented by $\begin{bmatrix} 1 & k \\ 0 & 1 \end{bmatrix}$. *Then—*

(a) $H\begin{bmatrix} x \\ y \end{bmatrix} = \begin{bmatrix} x + ky \\ y \end{bmatrix}$;

(b) *A point and its image are on the same horizontal line;*

(c) *H preserves collinearity and betweenness;*

(d) *H preserves distance between points on horizontal lines;*

(e) *H does not preserve angle measure or distance, in general.*

The reader should also convince himself that the areas of a figure and its image under a shear are identical. This idea is used extensively in the School Mathematics Project materials.

Translations, other than the identity translation, do not leave the origin fixed. They are not representable by 2 × 2 matrices. They may be represented by 2 × 1 matrices, however. In this case, a translation that moves every point a units on the horizontal and b units on the vertical is given by $\begin{bmatrix} a \\ b \end{bmatrix}$. To apply this translation to a point, $\begin{bmatrix} x \\ y \end{bmatrix}$, matrix addition is used; that is, the image is given by $\begin{bmatrix} x \\ y \end{bmatrix} + \begin{bmatrix} a \\ b \end{bmatrix} = \begin{bmatrix} x + a \\ y + b \end{bmatrix}$. This type of work is included in the School Mathematics Project materials but is not extensively developed in American materials. Time permitting, the topic would fit nicely into the other work on matrix representations of transformations.

Two other transformations not found in American materials are the one-way stretch, $\begin{bmatrix} k & 0 \\ 0 & 1 \end{bmatrix}$ or $\begin{bmatrix} 1 & 0 \\ 0 & k \end{bmatrix}$, and the two-way stretch, $\begin{bmatrix} m & 0 \\ 0 & n \end{bmatrix}$.

The reader may wish to investigate the properties of these transformations. They are discussed in the School Mathematics Project materials (11, p. 11; 12, chap. 13).

Groups of transformations

Using matrices to represent geometric objects and to represent certain transformations provides one link between geometry and algebra. A second link is provided by examining the algebraic structure of sets of transformations under the operation composition. The structure found is the group. A set S with an operation is a group if and only if—

1. there is an identity for the operation in $S;$

2. the inverse of any element is in $S;$

3. the operation is closed in $S;$

4. the operation is associative.

With the composition operation many sets of transformations form groups. For example, translations, rotations and translations (direct isometries), rotations about a point, all isometries (congruence group), size transformations with a given center, and all similarity transformations are groups with composition. Additionally, the symmetries of a geometric figure form a group—for example, the symmetries of an equilateral triangle, of a square, of a rectangle, or of a parallelogram (see fig. 6.36). These groups are finite-order groups.

In the Troelstra and in the Coxford-Usiskin materials only a selection of these groups is discussed. In the Troelstra materials each of the groups of isometries mentioned above is discussed along with one symmetry group, that of the equilateral triangle (15, chap. 7).

After introducing the idea of a group and some familiar examples, the first group of transformations studied in the American materials is the symmetry group of the equilateral triangle. This example is used to motivate students to discover the following general result (3, p. 576):

With composition, the set of all isometries which map a figure onto itself forms a group.

A consequence is that every figure has an associated symmetry group. For "irregular" figures the group has only one element, the identity.

The most important group discussed that is not of finite order is the congruence group, the group of all isometries. The mathematical development of this result is interesting. The first theorem is left for the reader to prove (3, p. 580):

THEOREM. *With composition, the set of distance-preserving transformations forms a group.*

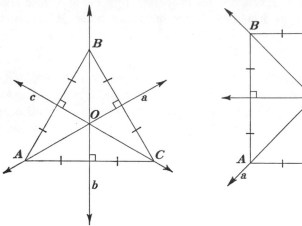

Members of group: I, M_a, M_b,
M_c, $R_{120°}$, $R_{240°}$ (both about O)

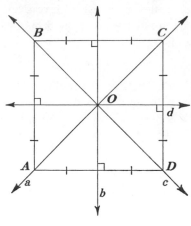

Members of group: I, M_a, M_b,
M_c, M_d, $R_{90°}$, $R_{180°}$, $R_{270°}$ (all about O)

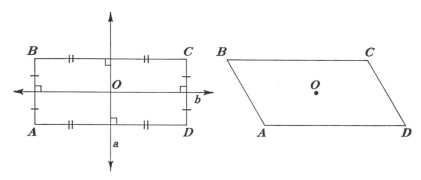

Members of group: I, M_a,
M_b, $R_{180°}$ (about O)

Members of group: I,
$R_{180°}$ (about O)

Fig. 6.36

Now the task is to show that the set of isometries is also the set of distance-preserving transformations. Clearly, since reflections preserve distance, all isometries do also. What about the converse (3, p. 581)?

THEOREM. *Every distance-preserving transformation is an isometry.*

Proof—part 1. Suppose T is distance preserving, and furthermore, that T fixes a point P. Let A be another point and $T(A) = A'$. Since T preserves distance, $PA = PA'$. Let B be another point with $T(B) = B'$. Thus $\triangle PAB \cong \triangle PA'B'$. The two possible configurations are shown in figures 6.37 and 6.38.

In figure 6.37, T is a rotation with center P and magnitude $m\angle APA'$

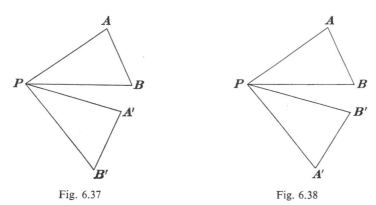

Fig. 6.37   Fig. 6.38

(prove this). In figure 6.38, T is a reflection over the mediator of $\overline{AA'}$ (prove this). Thus if a distance-preserving transformation has a fixed point, it is a reflection or a rotation—an isometry.

Proof—part 2. Suppose T has no fixed point. Then $T(P) = P'$. Let M be the reflection over the mediator of $\overline{PP'}$. Then $M(P') = P$ and $M[T(P)] = P$. That is, P is a fixed point for the distance-preserving transformation $M \circ T$. By part 1, $M \circ T$ is an isometry. Certainly then, $M \circ (M \circ T)$ is an isometry. But $M \circ M = I;$ so $I \circ T = T$ is an isometry.

The set of distance-preserving transformations is the set of isometries. The group of isometries (distance-preserving transformations) is the congruence group. The set of orientation-preserving isometries is also a group that is a subgroup of the congruence group.

The isometries form a group, but the elements of that group are not explicitly known. That is, one does not know how many isometries there are or what they are. Several theorems are needed to deduce that there are exactly four isometries.

THEOREM 1. *Suppose l and m are lines intersecting at P. Let l' and m' be the images of l and m under any rotation with center P. Then*

$$M_m \circ M_l = M_{m'} \circ M_{l'}.$$

Proof. The magnitude of the rotation $M_m \circ M_l$ is twice the measure of the angle from l to m. This measure is unchanged by rotating about P. Therefore $M_m \circ M_l$ and $M_{m'} \circ M_{l'}$ have the same magnitude. They also have the same center P and are thus the same rotation.

THEOREM 2. *Suppose $l \parallel m$. Let l' and m' be the images of l and m under any translation. Then*

$$M_m \circ M_l = M_{m'} \circ M_{l'}.$$

The proof is similar to that for the last theorem.

The essence of these two theorems is this: The reflecting lines used to define a rotation or a translation may be replaced by another pair with the same characteristics.

THEOREM 3. *The product of three reflections in parallel lines is a reflection* (see fig. 6.39).

Proof. Let $T = M_n \circ M_m \circ M_k$. Apply theorem 2 to replace k and m by lines k' and $m' = n$. (That is, translate k and m to k' and n.) Thus

$$T = M_n \circ M_m \circ M_k$$
$$= M_n \circ M_n \circ M_{k'}$$
$$= I \circ M_{k'}$$
$$= M_{k'}.$$

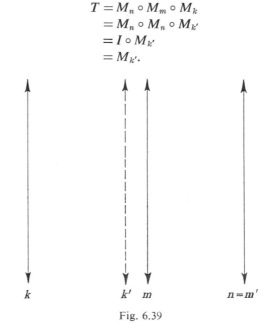

$$k \qquad\qquad\qquad\qquad k' \quad m \qquad\qquad\qquad n = m'$$

Fig. 6.39

THEOREM 4. *A translation is the product of two half-turns* (see fig. 6.40).

Proof. Let translation $T = M_n \circ M_m$. Let l be perpendicular to m and n and let $l \cap m = Q$, $l \cap n = R$. Since $M_l \circ M_l = I$, $T = M_n \circ M_l \circ M_l \circ M_m = H_R \circ H_Q$ where H_R and H_Q are half-turns about R and Q respectively.

THEOREM 5. *The product of three half-turns is half-turn* (see fig. 6.41).

Proof. Let E, F, and G be the centers of the half-turns. Let $\overleftrightarrow{EF} = l$ and let k, m, n be perpendicular to l through E, F, and G, respectively. Let o be perpendicular to n at G.

$$H_G \circ H_F \circ H_E = M_o \circ M_n \circ M_m \circ M_l \circ M_l \circ M_k$$
$$= M_o \circ M_n \circ M_m \circ M_k$$
$$= M_o \circ M_{k'} \text{ (by theorem 3)}$$
$$= H_p$$

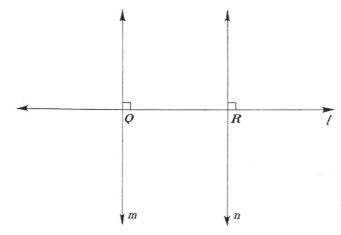

Fig. 6.40

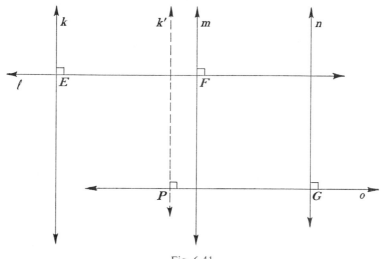

Fig. 6.41

THEOREM 6. *The product of two translations is a translation.*

Proof. Let the translations be T_1 and T_2. By theorem 4,

$$T_1 = H_A \circ H_B \qquad \text{and} \qquad T_2 = H_C \circ H_D.$$
$$T_2 \circ T_1 = H_C \circ H_D \circ H_A \circ H_B$$
$$= H_C \circ H_P \text{ (by theorem 5)}$$
$$= T \text{ (a translation by theorem 4)}$$

The remaining theorems are left for the reader to prove.

THEOREM 7. *The product of three reflections in concurrent lines is a reflection.*

THEOREM 8. *The product of two rotations of magnitudes a and b is a rotation with magnitude a + b. If a + b = 180, then the product is a translation.*

THEOREM 9. *The product of a translation and a rotation is a rotation.*

THEOREM 10. *The product of 2n reflections is a rotation or a translation.*

THEOREM 11. *The product of 2n + 1 reflections is the product of at most three reflections.*

THEOREM 12. *The product of three reflections is either a reflection or a glide reflection.* (The proof of this theorem is given in the proofs of theorems 3 and 7 and in the initial section of this chapter.)

Theorem 13 should now be clear:

THEOREM 13. *The product of any number of isometries is a reflection, a rotation, a translation, or a gilde reflection.*

A development as detailed as that given in theorems 1–13 is not found in either the Dutch or the American texts for high school students. It is included here for its intrinsic interest and because these theorems are appropriate for inclusion in geometry classes studying transformations as a topic. Other theorems that may be proved are the following:

THEOREM 14. *With composition, translations form a group.*

THEOREM 15. *The product of a half-turn with itself is the identity transformation.*

THEOREM 16. $H_E \circ H_F \circ H_G = H_G \circ H_F \circ H_E$ *where* H_E, H_F, *and* H_G *are any half-turns.*

THEOREM 17. $T_2 \circ T_1 = T_1 \circ T_2$ *where* T_1 *and* T_2 *are translations.*

There is another more elegant but perhaps less instructive proof of the fact that there are only four isometries of the plane. In the proof that every distance-preserving transformation of the plane is an isometry, it can be recalled that if an isometry T had a fixed point, it was either a reflection or a rotation and the product of *at most two reflections.*

Let T be any isometry that fixes no point. Then there is a reflection M such that $M \circ T$ fixes a point. Thus

$$M \circ T = \text{reflection, or } M \circ T = \text{rotation.}$$

Consequently,

$$M \circ (M \circ T) = M \circ (\text{reflection}), \text{ or } M \circ (M \circ T) = M \circ (\text{rotation}).$$

But since $M \circ M = I$,

$$T = M \circ \text{(reflection)}, \text{ or } T = M \circ \text{(rotation)}.$$

Thus T, an isometry with no fixed point, is the product of *at most three reflections*. We have proved, then, the following theorem:

Every isometry T is the product of at most three reflections.

It is an easy exercise to deduce that the isometries with no fixed points are translation (two reflections) and glide reflection (three reflections). The proof of the latter fact follows by using theorems 1 and 2 to prove theorem 12. It is suggested that the reader construct the proof.

The only other group explicitly discussed in the Coxford-Usiskin materials is the similarity group. This group is the set of similarity transformations with the composition operation. The discussion provides a bare introduction to the similarity group (3, pp. 588–89).

If a teacher wishes to expand the treatment of the similarity group, he can identify the minimal set of transformation types that make up the elements of the group. This minimal set includes the isometries translation and glide reflection (where a reflection is a special case of a glide reflection). In addition to these transformations, there are only two others needed to form a closed set of transformations under the operation composition. They are the *spiral similarity* and the *dilative reflection*.

DEFINITION. *A spiral similarity with center C, rotation angle θ and similarity coefficient k is the product of a dilation with center C and similarity coefficient k and a rotation about C through the angle θ, taken in either order. (See fig. 6.42.)*

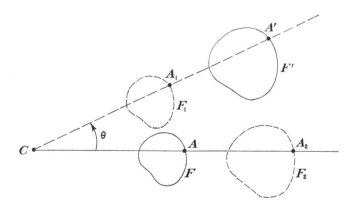

Fig. 6.42

In figure 6.42, the figure F has as its image under the spiral similarity the figure F'. Corresponding points are A and A'. To transform F to F', we may use either F_1 or F_2 as the intermediate image. In the one case, we rotate first, then dilate ($F \to F_1 \to F'$). In the other, we dilate first, then rotate ($F \to F_2 \to F'$).

DEFINITION. *A dilative reflection is the product of a dilation with center C and similarity coefficient k and a reflection in a line m containing C, taken in either order.* (See fig. 6.43.)

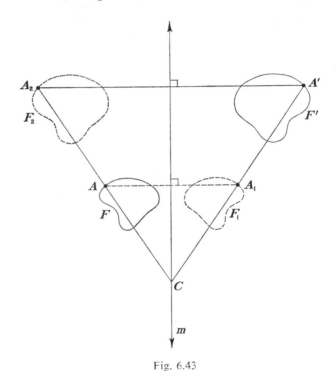

Fig. 6.43

Figure 6.43 depicts a dilative reflection of figure F to figure F'. The order of operation is optional here; one may reflect in m first, then dilate ($F \to F_1 \to F'$), or one may dilate, then reflect in m ($F \to F_2 \to F'$).

For a complete discussion of spiral similarities and dilative reflections, the reader is referred to Yaglom's *Geometric Transformations II* (18, pp. 9–62).

A most impressive conclusion to be drawn from an examination of elements of the similarity group is that two similar plane figures that are not congruent are related by either a spiral similarity or a dilative reflection. Moreover, if the figures are directly oriented, then they are related by exactly one transformation—the spiral similarity. Similarly, the dilative reflection is the only transformation needed to map one of two similar figures into a second that is oppositely oriented. Constructions for these transformations are found in Yaglom. Verification of the constructions is not beyond the high school level.

This concludes the discussion on the scope and sequence of the content of the transformation approaches to geometry found in print at the present. Other sequences are possible, and no claim is made that the sequence described in this chapter is the best mathematically or pedagogically.

Often other content was described in addition to what was actually included in any one set of materials. The reader may think of these comments as describing variations of, or additions to, the main sequence. The place where the greatest opportunity for variation in content exists is in the extent of the work on the transformations themselves. Time permitting, a great deal of additional work can be done with matrices and with transformation groups. Assuming a certain amount of geometric knowledge, the study of geometry using transformations can easily turn into the study of transformation geometry. As youngsters come out of junior high schools with a better understanding of Euclidean geometry, more time will be available to study the transformations themselves. As this occurs, the scope and sequence, as described here, must change.

Pedagogical Questions

In the scope and sequence section of this chapter, the content of a geometry course based on transformations was described. Associated with this content, as is the case with any mathematical content suggested for secondary school consumption, are many pedagogical questions. Such questions include, but are not limited to, the following:

1. What evidence exists concerning the "teachability" of the material?
2. What ability must a youngster have in order to succeed with the material?
3. What effect will incorporating this material into the curriculum have on other work in mathematics?
4. Is teacher training necessary in order to teach such material?

These questions are considered individually below.

Teachability

Several sources on the evidence of the teachability of the material are available. The major source is a report of an investigation carried out by

Usiskin during the 1968/69 school year (16). Thirteen schools were involved in the study; six used the Coxford-Usiskin materials, and seven used standard geometry texts. There were 425 youngsters and eight teachers using the transformation approach, and 475 youngsters and nine teachers using contemporary approaches.

The geometric knowledge of both groups of students was measured in the fall and again at the end of the school year. The instruments used were alternate forms of the Cooperative Geometry Test. These tests measured traditional geometric knowledge; they included no questions on transformations.

In addition to the measure of standard geometric knowledge, the youngsters using the transformation approach completed a test on transformation ideas at the end of the year. All teachers were surveyed in regard to their opinions concerning the transformation materials.

The students using the transformation approach spent approximately five or six weeks studying topics unique to the transformation approach. The remainder of the time, thirty or thirty-one weeks, was spent on relatively standard geometric topics. In contrast to this division of time, the control group spent thirty-six weeks on standard geometric content.

On the measures of the achievement of standard geometric concepts, the group using the transformation approach scored slightly lower than the control group: the transformation group answered correctly approximately 95 percent as many questions as did the control group. On the test of transformation concepts administered at the end of the year, the youngsters who had studied transformation ideas performed as well as they did on the standard content.

One school that used the transformation materials again in 1969/70 and gave the same Cooperative Plane Geometry Test at the end of the year reported no differences in achievement for those studying the standard and transformation approaches.[1] These data suggest that the transformation approach is teachable. The author's experience with teaching the approach to ordinary tenth graders confirms this. Additionally, the schools that used the approach in the experimental setting continue to use it, and teacher comments continue to be favorable.

Ability

The evidence concerning achievement for different ability levels of students is similar to that discussed above for geometry students in general. The lower-ability youngsters using the transformation approach answered correctly nearly 41 percent of the items on the final test of transformation

---

1. Zalman Usiskin, 1970: personal communication.

ideas and also approximately 41 percent of the items on the test of standard geometric ideas. The control students answered correctly approximately 45 percent of the items on the test of standard geometric ideas. Neither of these percents is large. The transformation approach appears to be as appropriate for lower-ability youngsters as other geometry courses, but none are appropriate as they exist today. In order to adapt a geometry course to the rate of learning exhibited by slower students, topics would have to be deleted and emphases changed, whether the course be transformation, vector, affine, or standard. In particular, the emphasis on proof would have to be diminished, for it was this area that caused most slower learners the greatest difficulty.

The rate of correct response on both the final test covering transformation ideas and the final test of standard geometric ideas was approximately 64 percent for the higher-ability students in the transformation approach. The higher-ability students studying the standard treatment answered correctly approximately 67 percent of the items on the final test of standard geometric content. These rates are quite similar. In each approach additional concepts could easily be incorporated into the course. Certainly the problem of differences among individuals is not solved by using a transformation approach to geometry. Greater changes would have to be made. But perhaps a transformation approach offers accessible alternatives for both the slow and the fast learner in that the transformations lend themselves either to a manipulative and constructive approach or to a more formal treatment that culminates in the study of abstract groups.

Effects on other mathematics

One may consider this issue in two parts: (1) prior mathematics and (2) subsequent mathematics.

As far as effects on prior mathematics are concerned, there appear to be none. That is, no apparent changes in prior mathematics work are necessary in order to institute a transformation approach to geometry. As evidence, one may cite the fact that such materials have been successfuly taught in a number of high schools throughout the nation with no changes in the mathematical work done in the elementary or junior high school.

The evidence concerning effects on subsequent mathematics is sparse and must be considered as tentative because the approach has not yet been used extensively in the United States. The one bit of evidence available is due to Kort (7). Kort investigated the effects of studying a transformation approach to geometry in the tenth grade on eleventh-grade students' retention of geometric concepts and on their achievement in function and inverse function concepts. Kort found that there was no difference in overall retention of geometric concepts but that the transformation-studying group was significantly superior in the retention of congruence, similarity, and symmetry. In addition, the transformation group was significantly superior in the achievement of functions and relations and inverses of functions and relations. No differences were found in the achievement of circular functions and inverse circular functions.

The only other evidence available on this question is subjective. Several teachers who have used the transformation approach have also taught subsequent courses. They reported no uncommon difficulties or successes in that work. These results and those reported by Kort are not unexpected because the transformation approach that was used emphasized Euclidean geometry. Only in the work on congruence, similarity, and symmetry were transformations emphasized. If Kort's results can be replicated, then the transformations approach to congruence, similarity, and symmetry may lead to a better understanding and retention of these ideas. The greater achievement in functions and relations is, if replicable, an added benefit, for functions and relations are often difficult to teach effectively.

Teacher training

Training teachers in the concepts, problems, and procedures of transformation geometry would enhance a school geometry course based on transformations. With such training, a teacher would feel more confident, would allow students to be more creative in proof and problem solving, and would be able to provide more stimulating enrichment topics. If transformation geometry per se is to form a significant portion of the course taught, then teacher training in transformations is an absolute must because such work is not at present a part of the standard collegiate mathematical preparation of teachers.

The evidence presently available to the author strongly suggests that specific teacher training in transformation geometry is not necessary (even though it is desirable) in order to implement the sequence outlined in this chapter. The major evidence comes from the research of Usiskin (16). In his project he offered several training services to the teachers. These included a question-answering service—by telephone and by mail—and the opportunity for regular in-service meetings of all the teachers using the transformation approach. No telephone or mail request for assistance was received. The in-service meetings were dropped because only one or two teachers regularly attended. Additional evidence comes from the fact that many schools have taught the transformation approach successfully using only the text materials and a brief teacher's commentary.

The argument here should not be taken to imply that teacher training in transformations is not desirable. It is! Most teachers will do a better job of teaching when they have had training directly related to the material to be taught. The suggestion here is simply that training in geometric transformations does not appear to be necessary to instituting the course described in this chapter.

Concluding Remarks

The course described in this chapter is obviously only one of many possible courses using transformations. A basic underlying assumption of that course is that transformations have a place in school geometry and that they ought to be employed in a pedagogically and mathematically sound manner. It is hoped that the course described in this chapter satisfies this assumption. If it does not, new and distinct alternatives will be forthcoming. But this is exactly as it should be—the only way that the mathematics taught in the schools can remain fresh, alive, and relevant is for people with ideas to present those ideas to teachers and to students, to try them in the testing ground of the classroom, to modify or discard them, and to present materials reflecting those good ideas to the public. With such a continual evaluation of the school mathematics taught, there is hope that the most appropriate mathematics for the schools will continue to evolve.

REFERENCES

1. Coxeter, H. S. M. *Introduction to Geometry*. New York: John Wiley & Sons, 1961.
2. Coxeter, H. S. M., and Samuel L. Greitzer. *Geometry Revisited*. New Mathematical Library, no. 19. New York: Random House, 1967.
3. Coxford, Arthur F., and Zalman P. Usiskin. *Geometry: A Transformation Approach*. River Forest, Ill.: Doubleday & Co., Laidlaw Brothers, 1971.
4. Guggenheimer, Heinrich W. *Plane Geometry and Its Groups*. San Francisco: Holden-Day, 1967.
5. Jeger, Max. *Transformation Geometry*. Translated by A. W. Deicke and A. G. Howson from the 3d German ed. New York: John Wiley & Sons, 1966.
6. Kelly, Paul J., and Norman E. Ladd. *Geometry*. Chicago: Scott, Foresman & Co., 1965.
7. Kort, Anthone P. "Transformation vs. Non-Transformation Tenth-Grade Geometry: Effects on Retention of Geometry and on Transfer in Eleventh-Grade Mathematics." Ph.D. dissertation, Northwestern University, 1970.
8. School Mathematics Project. *Book 1 (Metric)*. London: Cambridge University Press, 1969.
9. ———. *Book 2*. London: Cambridge University Press, 1966.
10. ———. *Book 3*. London: Cambridge University Press, 1967.
11. ———. *Book 4 (Metric)*. London: Cambridge University Press, 1969.
12. ———. *Book 5 (Metric)*. London: Cambridge University Press, 1969.

13. Troelstra, R., A. N. Habermann, A. J. de Groot, and J. Bulens. *Transformatie-meetkunde 1*. Groningen, The Netherlands: J. B. Wolters, 1965.

14. ———. *Transformatiemeetkunde 2*. Groningen, The Netherlands: J. B. Wolters, 1965.

15. ———. *Transformatiemeetkunde 3*. Groningen, The Netherlands: J. B. Wolters, 1966.

16. Usiskin, Zalman P. "The Effects of Teaching Euclidean Geometry via Transformations on Student Achievement and Attitudes in Tenth-Grade Geometry." Ph.D. dissertation, The University of Michigan, 1969.

17. Yaglom, I. M. *Geometric Transformations*. Translated by Allen Shields. New Mathematical Library, no. 8. New York: Random House, 1962.

18. ———. *Geometric Transformations II*. Translated by Allen Shields. New Mathematical Library, no. 21. New York: Random House, 1968.

An Affine Approach
to Euclidean Geometry

C. RAY WYLIE

AFFINE GEOMETRY can be loosely described as Euclidean geometry stripped of its capacity to measure lengths and angles. Conversely, Euclidean geometry can be approached through affine geometry, as we shall point out in this chapter, by introducing into the affine plane suitable definitions of distance and angle measure. This raises two important questions:

1. What is an affine plane?
2. How can an affine plane be obtained?

Affine planes arise in various ways. For instance, it is possible to construct an affine plane, or an affine space of any number of dimensions for that matter, by defining the affine space to be a set of elements, called *points,* on which the additive group of a linear vector space of the appropriate dimension acts as a group of translations (3, pp. 1–10). However, since this process requires a substantial background in the theory of vector spaces, it is probably inappropriate as the starting point of a development intended for high school students.

A more familiar plan is to regard the affine plane as a specialization of the classical projective plane, Π_2, in which one line, λ, is singled out to be the so-called ideal line, or "line at infinity" (1, pp. 206–15; 5, pp. 230–34). Lines of Π_2 that intersect on λ are then defined to be parallel in the reduced system $\Pi_2 - \lambda$, where properties that depend on parallelism but not on measurement can now be investigated. Analytically, this amounts to the study of the properties of Π_2 that are invariant under those projective transformations that leave λ fixed. These linear transformations form a subgroup of the projective group known as the affine group. This approach, presupposing a knowledge of projective geometry, also seems to be too sophisticated for high school students.

Finally, in addition to the approach through linear algebra and the approach through projective geometry, affine geometry can be developed independently from the appropriate set of axioms. This has the advantages of presupposing no sophisticated background material and of providing substantial, though not excessive, experience with the axiomatic method— two features that recommend it as a possible program for high school students. The bulk of this chapter will be devoted to a discussion of the appropriateness of this approach and an exposition of the material itself, beginning with the necessary axioms and culminating in the Euclidean plane.

However, before embarking on this venture, we shall attempt to answer the first question we raised above, namely, *What is an affine plane?*, by describing in some detail two affine planes, one finite, one infinite.

Two Particular Affine Planes

First example

Our first example of an affine plane is a very simple one suggested by our earlier observation that affine geometry can be regarded as a specialization of projective geometry.

It is well known, and easily verified, that the system PG_4 (the symbol PG_k is commonly used to denote a finite projective geometry in which every line contains exactly k points), consisting of 13 points, $P_1, P_2, \ldots, P_{13}$, and 13 lines, $l_1, l_2, \ldots, l_{13}$, connected by the incidence relations shown in table 7.1, satisfies the usual incidence and connection axioms of projective geometry, namely:

1. For any two points there is one and only one line that contains both points.
2. For any two lines there is at least one point that lies on both lines.
3. Every line contains at least three points.
4. All points do not lie on the same line.
5. There exists at least one line.

If some line of PG_4, say l_{13}, is singled out to play the role of the ideal line, or line at infinity, then in the reduced system $PG_4 - l_{13}$, described by the incidence table shown in table 7.2, lines that in PG_4 intersected in a point of l_{13} no longer have an intersection. In other words, they are parallel.

TABLE 7.1

l_1	l_2	l_3	l_4	l_5	l_6	l_7	l_8	l_9	l_{10}	l_{11}	l_{12}	l_{13}
P_1	P_3	P_2	P_3	P_2	P_1	P_1	P_2	P_3	P_1	P_4	P_7	P_{10}
P_6	P_5	P_4	P_4	P_6	P_5	P_4	P_5	P_6	P_2	P_5	P_8	P_{11}
P_7	P_8	P_9	P_7	P_8	P_9	P_8	P_7	P_9	P_3	P_6	P_9	P_{12}
P_{10}	P_{10}	P_{10}	P_{11}	P_{11}	P_{11}	P_{12}	P_{12}	P_{12}	P_{13}	P_{13}	P_{13}	P_{13}

In fact, the system $AG_3 = PG_4 - l_{13}$ is a finite affine geometry in which axioms 1, 3, 4, 5 are still valid but in which axiom 2 is replaced by the parallel axiom:

2′. If l is a line and if P is a point that is not on l, then there is one and only one line that contains P and has no point in common with l.

TABLE 7.2

l_1	l_2	l_3	l_4	l_5	l_6	l_7	l_8	l_9	l_{10}	l_{11}	l_{12}
P_1	P_3	P_2	P_3	P_2	P_1	P_1	P_2	P_3	P_1	P_4	P_7
P_6	P_5	P_4	P_4	P_6	P_5	P_4	P_5	P_6	P_2	P_5	P_8
P_7	P_8	P_9	P_7	P_8	P_9	P_8	P_7	P_9	P_3	P_6	P_9

For example, in $PG_4 - l_{13}$ the unique parallel to the line $l_4 = \{P_3, P_4, P_7, (P_{11})\}$ through the point P_5 is the line $l_6 = \{P_1, P_5, P_9, (P_{11})\}$, since the point of intersection of these lines in PG_4, namely, the point P_{11}, is not a point of $PG_4 - l_{13}$. In contrast, if another line of PG_4, say l_{10}, is identified as the ideal line, then in the new system $AG_3{}^* = PG_4 - l_{10}$, described by the incidence table shown in table 7.3, the lines $l_4 = \{(P_3), P_4, P_7, P_{11}\}$ and $l_6 = \{(P_1), P_5, P_9, P_{11}\}$ are no longer parallel, since their intersection in PG_4, namely, P_{11}, not being a point of l_{10}, is still a point of $PG_4 - l_{10}$. In $AG_3{}^*$ the unique parallel to l_4 through P_5 is the line $l_2 = \{(P_3), P_5, P_8, P_{10}\}$, since the intersection of l_2 and l_4 in PG_4, namely, P_3, being a point of l_{10}, is not a point of $AG_3{}^* = PG_4 - l_{10}$.

TABLE 7.3

l_1	l_2	l_3	l_4	l_5	l_6	l_7	l_8	l_9	l_{11}	l_{12}	l_{13}
P_6	P_5	P_4	P_4	P_6	P_5	P_4	P_5	P_6	P_4	P_7	P_{10}
P_7	P_8	P_9	P_7	P_8	P_9	P_8	P_7	P_9	P_5	P_8	P_{11}
P_{10}	P_{10}	P_{10}	P_{11}	P_{11}	P_{11}	P_{12}	P_{12}	P_{12}	P_6	P_9	P_{12}

Conversely, beginning with AG_3, say, it is possible to recover PG_4 by adjoining to AG_3 a line at infinity, that is, $l_{13} = \{P_{10}, P_{11}, P_{12}, P_{13}\}$, whose points are in one-to-one correspondence with the families of parallel lines in AG_3,

$$l_1, l_2, l_3 \longleftrightarrow P_{10}$$
$$l_4, l_5, l_6 \longleftrightarrow P_{11}$$
$$l_7, l_8, l_9 \longleftrightarrow P_{12}$$
$$l_{10}, l_{11}, l_{12} \longleftrightarrow P_{13}$$

and serve as the intersections of any two lines of the parallel families to which they correspond. Similarly, of course, PG_4 can be recovered by adjoining the necessary line of points to AG_3*.

Among other things, these observations suggest, by extrapolation to the familiar Euclidean plane, that there is a sense in which it is meaningful to say that "parallel lines meet at infinity."

The preceding example is especially appropriate as a first illustration because it begins with an extended system in which the points and lines that appear as ideal elements are present from the outset and are clearly no different from any other points and lines. Moreover, it illustrates in an exceedingly simple setting how points that are ideal elements for one affine specialization become ordinary points in another and, conversely, how ordinary points in one affine specialization become ideal points for another.

Second example

Our second example is more sophisticated because, in the first place, it involves an infinite rather than a finite number of points and lines and, secondly, it begins with a system in which the ideal elements are not apparent and have to be postulated rather than perceived.

Our starting point is the coordinatized Euclidean plane, that is, the familiar Cartesian plane, which we shall henceforth denote by the symbol E. Our first step is to create a conceptual extension of this plane in which without exception every two lines will have a point in common. This we do by postulating the existence of additional points, called ideal points, with the property that—

1. each line contains exactly one ideal point;
2. all lines of the same parallel family contain the same ideal point;
3. the locus of ideal points is a line.

Clearly, the ideal points that we have thus added to the original Euclidean plane, E, are in one-to-one correspondence with the directions in E, and lines that are parallel in E can be described in the extended plane as lines that intersect on, or are concurrent with, the ideal line.

A more detailed discussion of the introduction of ideal elements can be found in *Introduction to Projective Geometry* (5, pp. 28–42). However, for our expository purposes whatever mystery may seem to surround these new objects can best be dispelled by a careful reconsideration of the reversible relation between PG_4 and AG_3 (or between PG_4 and $AG_3{}^*$), which we described in our first example. There these ideas can be traced out and understood without any of the perceptual difficulties that arise from our long experience with the Euclidean plane and our unfamiliarity with any larger, more inclusive plane.

We now raise the following question: *Is it possible in the extended system we have just created to single out another line to play the role of the ideal line and to say that lines that are concurrent with it are parallel, even though to our eyes, prejudiced in favor of the Euclidean plane with which we started, they still appear to intersect?* This can, in fact, be done, and as preparation for our subsequent discussion of the affine approach to Euclidean geometry, we shall explore this matter in some detail and describe how perpendicularity can be defined and distances and angles measured in a way consistent with our new definition of parallelism.

To be specific, in E let us select the line $y = 2$ to be the new ideal line, and let us select the lines $x = 0$ and $y = x$ as the axes of what we hope, after a suitable new definition of perpendicularity, will be a new rectangular coordinate system. Furthermore, let I be the point with coordinates $(1,1)$ in E, let J be the point with E-coordinates $(0,1)$, let U be the point with E-coordinates $(2,2)$, and let V be the point with E-coordinates $(0,2)$, as seen in figure 7.1.

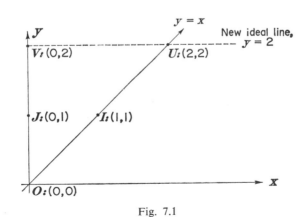

Fig. 7.1

Our first step is to introduce new coordinate systems on the proposed new axes, $\overleftrightarrow{OU}$ and $\overleftrightarrow{OV}$. In doing this we must remember that the ideal points originally postulated on $\overleftrightarrow{OU}$ and $\overleftrightarrow{OV}$ are now perfectly "respectable" points and must have coordinates, while U and V themselves, being now ideal points, do not belong to the system we are constructing and, therefore, cannot have coordinates in the new plane. A further requirement that we intend to impose is that on $\overleftrightarrow{OU}$ the coordinates of O and I are to be 0 and 1, respectively, and that on $\overleftrightarrow{OV}$ the coordinates of O and J are to be 0 and 1, respectively.

To set up the required coordinate system on $\overleftrightarrow{OU}$, we begin by considering the set Λ_V consisting of all the lines on V with the exception of the new ideal line $\overleftrightarrow{VU}$. Since these lines are all concurrent with the ideal line $\overleftrightarrow{VU}$ at the one ideal point V, they are, by definition, parallel in the new plane. In terms of the rectangular coordinate system in E, the equation of any line of this family can be written in the form

$$\lambda(y - 2) + x = 0.$$

The lines of Λ_V are thus in one-to-one correspondence with the values of the parameter λ. Moreover, since the lines of Λ_V are clearly in one-to-one correspondence with the points of the set $\overleftrightarrow{OU} - U$, it follows that there is a one-to-one correspondence between the values of λ and the points of $\overleftrightarrow{OU} - U$. Thus if P is the arbitrary point of $\overleftrightarrow{OU} - U$, the value of λ corresponding to the line $\overleftrightarrow{VP}$ can be thought of as the coordinate of P, and we shall make this identification. In particular, since the equation of the line $\overleftrightarrow{VO}$ is $x = 0$, that is $0(y - 2) + x = 0$, and since the equation of $\overleftrightarrow{VI}$ is $(y - 2) + x = 0$, it follows that the coordinates of O and I are 0 and 1, respectively. Two further properties of the new coordinate system are worthy of note. Since the original ideal point on $\overleftrightarrow{OU}$ is, by definition, common to $\overleftrightarrow{OU}$ and the line on V that is parallel to $\overleftrightarrow{OU}$, namely, the line $-(y - 2) + x = 0$, it follows that the new coordinate of this point is -1. Also, points on $\overleftrightarrow{OU}$ that in E are arbitrarily close to U have new coordinates that are arbitrarily large in absolute value. Specifically, consider the general point P: (u,u) on $\overleftrightarrow{OU}$. The line $\lambda(y - 2) + x = 0$ will contain this point if and only if

$$\lambda(u - 2) + u = 0 \qquad \text{or} \qquad \lambda = \frac{u}{2 - u},$$

and clearly as u approaches 2, and therefore P approaches U, the coordinate, λ, of P becomes infinite (see fig. 7.2).

Similarly, to establish the required new coordinate system on $\overleftrightarrow{OV}$, we

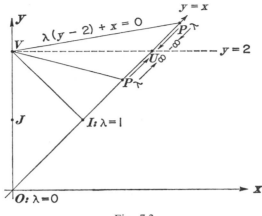

Fig. 7.2

consider the set Λ_U consisting of all the lines on U with the exception of the ideal line $\overleftrightarrow{UV}$. The equation of any line of Λ_U can be written in the form $\mu(y-2)+(y-x)=0$, and we choose as the new coordinate of an arbitrary point P of the set $\overleftrightarrow{OV}-V$ the value of μ in the equation of the line $\overleftrightarrow{VP}$. In particular, it is easy to verify that the coordinates thus assigned to O and J are, respectively, 0 and 1, that the coordinate of the original ideal point on $\overleftrightarrow{OV}$ is -1, and that points arbitrarily close to V have coordinates that are arbitrarily large in absolute value (see fig. 7.3).

An arbitrary point in the new plane we are evolving has coordinates that are defined in terms of the new coordinate systems on $\overleftrightarrow{OI}$ and $\overleftrightarrow{OJ}$ in the

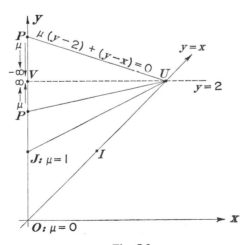

Fig. 7.3

following way. Let P be an arbitrary point and consider the unique line through P that is parallel to $\overleftrightarrow{OJ}$. In terms of the new definition of parallelism, this line, of course, is the line $\overleftrightarrow{PV}$. This line necessarily intersects the axis $\overleftrightarrow{OI}$ in some point, and the new coordinate, a, of this intersection is defined to be the abscissa, X, of P. Similarly, the ordinate, Y, of P is defined to be the new coordinate, b, of the intersection of $\overleftrightarrow{OJ}$ and the line through P that is parallel to $\overleftrightarrow{OI}$, that is, the line $\overleftrightarrow{PU}$. Clearly, under this scheme the ordinate of any point on $\overleftrightarrow{OI}$ is 0, and the abscissa of any point on $\overleftrightarrow{OJ}$ is 0 (see fig. 7.4).

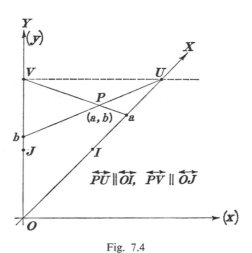

Fig. 7.4

Given an arbitrary point, P, it is now necessary to obtain the equations relating its coordinates in the old and the new coordinate systems. To do this, let (x_0, y_0) be the coordinates of P in the coordinate system in E. Then the line $\lambda(y - 2) + x = 0$ on V will contain P if and only if $\lambda(y_0 - 2) + x_0 = 0$, that is, if and only if $\lambda = \dfrac{x_0}{2 - y_0}$. By definition, this is the abscissa, X, of P in the coordinate system in the new plane. Likewise, the line $\mu(y - 2) + (y - x) = 0$ on U will contain P if and only if $\mu(y_0 - 2) + (y_0 - x_0) = 0$, that is, if and only if $\mu = \dfrac{(y_0 - x_0)}{(2 - y_0)}$, and this, by definition, is the ordinate, Y, of P in the new coordinate system. Thus we have

(1) $$X = \frac{x}{2 - y} \quad \text{and} \quad Y = \frac{y - x}{2 - y}$$

as the equations expressing the new coordinates (X,Y) in terms of the (x,y) coordinates in E. Conversely, solving these equations for x and y, we obtain

(2) $$x = \frac{2X}{X + Y + 1} \quad \text{and} \quad y = \frac{2(X + Y)}{X + Y + 1}$$

as the equations that express the (x,y) coordinates in E in terms of the new coordinates, X,Y. From these two sets of equations it is clear that points for which $y = 2$, that is, points on the new ideal line, have no coordinates in the new plane, since in fact they do not belong to this plane. Similarly, points for which $X + Y + 1 = 0$ have no (x,y) coordinates. They, of course, are the ideal points for the original plane, E, and $X + Y + 1 = 0$ is the equation of this line in the new plane, where it is just an ordinary line.

Consider now the set of all lines except those that pass through the point V, that is, the set of all lines except those that in the new plane are parallel to the Y-axis. In terms of the (x,y) coordinates, the equation of any line of this family is of the form

$$ax + by + c = 0, \quad c \neq -2b.$$

If this equation is expressed in terms of the (X,Y) coordinates by means of equations (2), we obtain

$$Y = -\left(\frac{2a + 2b + c}{2b + c}\right) X - \frac{c}{2b + c}.$$

Under the assumption that $c \neq -2b$, this equation is always meaningful and in fact has the familiar form $Y = MX + B$. By analogy with the situation in E, we shall define

(3) $$M = -\frac{2a + 2b + c}{2b + c}$$

to be the slope of the given line in the new plane, even though as yet we have no definition of angle measures and, therefore, cannot interpret M as the tangent of any angle. Clearly, if $c = -2b$, the slope M is undefined. In this case, as we have already observed, the line is parallel to the Y-axis, and the fact that it has no slope is just what we should expect.

It is now interesting and important to note that in our new plane two lines are parallel if and only if neither has a slope or they have the same slope. The first part of this assertion is obvious, since a line has no slope if and only if it is parallel to the Y-axis, and two such lines are necessarily parallel to each other.

To check the second assertion, consider two lines

$$Y = M_1 X + B_1 \quad \text{and} \quad Y = M_2 X + B_2$$

such that $M_1 = M_2$. From equation (3) this means that

$$-\frac{2a_1 + 2b_1 + c_1}{2b_1 + c_1} = -\frac{2a_2 + 2b_2 + c_2}{2b_2 + c_2},$$

or simplifying,

$$\frac{a_2c_1 - a_1c_2}{a_1b_2 - a_2b_1} = 2,$$

which is easily verified to be the necessary and sufficient condition that the intersection of the lines

$$Y = M_1X + B_1 \equiv a_1x + b_1y + c_1 = 0$$

and

$$Y = M_2X + B_2 \equiv a_2x + b_2y + c_2 = 0$$

should lie on the line $y = 2$. Thus $M_1 = M_2$ if and only if the lines intersect in an ideal point, that is, are parallel.

Once our new plane is coordinatized, an arbitrary line, l, that is not parallel to either the X-axis or the Y-axis bears two coordinate systems, one defined by the X-coordinates of the points of l and the other defined by the Y-coordinates of the points of l. These two systems are related by the equation of l, say $Y = MX + B$. On a line parallel to the X-axis all points have the same ordinate, hence the Y-coordinates do not form a coordinate system although, clearly, the X-coordinates do. Similarly, on a line parallel to the Y-axis, all points have the same abscissa, hence the X-coordinates do not form a coordinate system although the Y-coordinates do.

At this stage, the distance between two points is not defined in our new plane, but the *ratios* of the distances between points *on the same line* are defined to be the ratios of the absolute values of the differences of the coordinates of the points in either of the coordinate systems on the line. Although similar triangles are not available for use in the proof, by using the equation of the line, $Y = MX + B$, it is easy to show that the ratio of two distances on a general line is the same whether the X-coordinates or the Y-coordinates are used in the computation.

We have now developed a coordinatized affine plane. From it we shall construct, as an illustration of the central theme of this chapter, a new Euclidean plane, E^*, by introducing suitable definitions for the measurement of distances and angles. Until we have these definitions, however, we are restricted to the comparison of distances on the same line and, roughly speaking, affine geometry is concerned with theorems that involve, in addition to parallelism, only this limited type of measurement. As an example of such a theorem, let us consider the following result: *A line parallel to one side of a triangle divides the other two sides in the same ratio.* Although commonly taught and accepted as a theorem of Euclidean geometry, this is really an affine theorem that is true under weaker assumptions than those that characterize the Euclidean plane.

Let the vertices of $\triangle PQR$ be, respectively, the points whose coordinates in E are $\left(1, \dfrac{3}{2}\right), \left(\dfrac{5}{3}, \dfrac{5}{3}\right)$, $(4,1)$ and whose X,Y-coordinates are, therefore, $(2,1)$, $(5,0)$, $(4,-3)$ by equation (1). (See fig. 7.5.) Easy calculations, identical to those of elementary analytic geometry, give us the X,Y-equations of the lines that contain the sides of $\triangle PQR$:

$$\overleftrightarrow{PQ}: X + 3Y = 5$$
$$\overleftrightarrow{QR}: 3X - Y = 15$$
$$\overleftrightarrow{RP}: 2X + Y = 5$$

Now consider a general line that is parallel to $\overleftrightarrow{PQ}$. Since we have just observed that its slope must be the same as the slope of $\overleftrightarrow{PQ}$, it must have an equation of the form $X + 3Y = B$. This line intersects $\overleftrightarrow{PR}$ in the point S: $\left(\dfrac{-B + 15}{5}, \dfrac{2B - 5}{5}\right)$ and intersects $\overleftrightarrow{QR}$ in the point T: $\left(\dfrac{B + 45}{10}, \dfrac{3B - 15}{10}\right)$. Hence, using the X-coordinates on $\overleftrightarrow{PR}$ and $\overleftrightarrow{QR}$, we easily find that

$$\frac{PS}{SR} = \frac{|-B + 5|}{|B + 5|} = \frac{QT}{TR}.$$

Thus for all values of B except $B = -5$, that is, for all lines parallel to PQ except the one that passes through R, we have verified that

$$\frac{PS}{SR} = \frac{QT}{TR}.$$

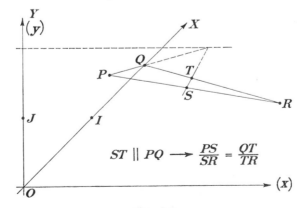

Fig. 7.5

Other affine theorems that are instructive to verify are (1) *the diagonals of a parallelogram bisect each other,* and (2) *the medians of a triangle are concurrent.*

Ordinarily the notion of perpendicularity is subordinate to the idea of angle measurement. It is possible, however, to make perpendicularity a primary concept and use it in the definition of distance and angle measurement. To implement this point of view, we observe that perpendicularity is a relation between lines which possesses the following properties:

1. No line is perpendicular to itself.
2. If a line l is perpendicular to a line m, then m is perpendicular to l.
3. For any line l and any point P, there is one and only one line that contains P and is perpendicular to l.

The obvious way to define a relation with these properties is to construct an equation $f(M_1,M_2) = 0$ such that—

1. $f(M_1,M_1) = 0$ has no (real) roots;
2. $f(M_1,M_2) = 0 \longleftrightarrow f(M_2,M_1) = 0$;
3. $f(M_1,M_2) = 0$ has a unique solution for M_i given M_j;

and then say that two lines are perpendicular if and only if their slopes, M_1 and M_2, satisfy this equation. The simplest equation with these properties is $a(M_1M_2) + b(M_1 + M_2) + c = 0$, or

$$(4) \qquad\qquad M_2 = -\frac{bM_1 + c}{aM_1 + b}, \qquad b^2 - ac < 0.$$

While equation (4) does define a perpendicularity relation with the required properties , it is too general for our purposes. In particular, we want the new axes, $\overleftrightarrow{OI}$ and $\overleftrightarrow{OJ}$, to be perpendicular. Since the slope *of* $\overleftrightarrow{OI}$ is $M_1 = 0$ and since $\overleftrightarrow{OJ}$ has no slope, this requires that in equation (4) the coefficient b must be zero (and therefore $-ac < 0$), giving us the condition

$$(5) \qquad M_2 = -\frac{c}{aM_1} \qquad \text{or} \qquad M_1M_2 = -\frac{c}{a} = K, \quad K < 0.$$

For all negative values of K, equation (5) provides a definition of perpendicularity according to which the new axes, $\overleftrightarrow{OI}$ and $\overleftrightarrow{OJ}$, are perpendicular, as desired. However, even this is too general to be consistent with the familiar definition of distance that we hope to introduce as we continue on from E via our affine plane to a new Euclidean plane, E^*. Naturally, we want both the Pythagorean theorem and the usual distance formula to hold; that is, if $\overleftrightarrow{PQ} \perp \overleftrightarrow{RQ}$, we want the formula $(PR)^2 = (PQ)^2 + (RQ)^2$ to hold, and we want the distance between the points P_1: (X_1, Y_1) and P_2: (X_2, Y_2) to be given by the formula

$$(P_1P_2)^2 = (X_2 - X_1)^2 + (Y_2 - Y_1)^2.$$

To see how this restricts the perpendicularity relation (5), let us apply the distance formula to two points P_1 and P_2 such that $\overleftrightarrow{OP_1} \perp \overleftrightarrow{OP_2}$ according to equation (5). The equation of $\overleftrightarrow{OP_1}$ is $Y = M_1X$ and the equation of $\overleftrightarrow{OP_2}$ is $Y = M_2X$, where $M_1M_2 = K$. The coordinates of P_1 and P_2 are then, respectively, (p_1, M_1p_1) and (p_2, M_2p_2). The distance formula, which we accept as a matter of definition, thus gives us $(OP_1)^2 = p_1^2 + p_1^2M_1^2$, $(OP_2)^2 = p_2^2 + p_2^2M_2^2$, and $(P_1P_2)^2 = (p_2 - p_1)^2 + (p_2M_2 - p_1M_1)^2$. Using these, we observe that the Pythagorean theorem will hold if and only if $(P_1P_2)^2 = (OP_1)^2 + (OP_2)^2$, or

$$(p_2 - p_1)^2 + (p_2M_2 - p_1M_1)^2 = (p_1^2 + p_1^2M_1^2) + (p_2^2 + p_2^2M_2^2),$$

which simplifies immediately to

$$M_1M_2 = -1,$$

which, of course, is just the familiar perpendicularity condition of elementary analytic geometry. Thus in (5) we must take $K = -1$.

Many students, even those with a sufficient flair for abstraction to grasp the idea of extending the Euclidean plane into a new system in which, without exception, every two lines intersect, find it difficult to understand how two lines that intersect (even outside the system in which they were originally parallel) can still have the property of equidistance. To explore this question, let us consider two lines that are parallel in E^* (and therefore intersect on $y = 2$ in E), say

$$Y = MX + B_1 \qquad \text{and} \qquad Y = MX + B_2, \quad B_1 \neq B_2.$$

An arbitrary line perpendicular to these two lines is defined by the equation $Y = -\dfrac{1}{M}X + B$, and its intersections with the two parallel lines are easily found to be

$$P_1: \left(\frac{(B - B_1)M}{1 + M^2}, \frac{B_1 + M^2B}{1 + M^2}\right) \qquad \text{and} \qquad P_2: \left(\frac{(B - B_2)M}{1 + M^2}, \frac{B_2 + M^2B}{1 + M^2}\right).$$

The distance between these two points of intersection, that is, the perpendicular distance between the two parallel lines, we find to be

$$(P_1P_2)^2 = \left(\frac{(B_1 - B_2)M}{1 + M^2}\right)^2 + \left(\frac{B_2 - B_1}{1 + M^2}\right)^2.$$

Since this is clearly independent of B, that is, independent of which common perpendicular of the two lines is used, the equidistance property is verified for this particular pair of parallel lines in E^*, in spite of whatever subjective reservations to the contrary we may have. (See fig. 7.6.)

Given two intersecting rays, $\overrightarrow{QP}$ and $\overrightarrow{QR}$, it is now possible to show that the quantity Z defined by the equation

(6) $\qquad (PR)^2 = (QP)^2 + (QR)^2 - 2(QP)(QR)Z$

has the following properties:

1. It is independent of the points P and Q on the respective rays.
2. It satisfies the inequality $-1 \le Z \le 1$.
3. It is 1 if and only if $\overrightarrow{QP}$ and $\overrightarrow{QR}$ are the same ray; it is 0 if and only if $\overrightarrow{QP} \perp \overrightarrow{QR}$; and it is -1 if and only if $\overrightarrow{QP}$ and $\overrightarrow{QR}$ are opposite rays.

Thus Z has the necessary properties to permit it to be taken as the definition of $\cos \angle PQR$. With this definition, equation (6) becomes the familiar law of cosines, and with it angles can now be measured and the law of sines and the angle-sum theorem verified. In particular, for the triangle shown in figure 7.5 an easy calculation shows that

$$\cos \angle PQR = 0, \quad \cos \angle QRP = \frac{1}{\sqrt{2}}, \quad \cos \angle RPQ = \frac{1}{\sqrt{2}}$$
$$m\angle PQR = 90°, \quad m\angle QRP = 45°, \quad m\angle RPQ = 45°.$$

Hence $\triangle PQR$, despite its appearance and its actual dimensions in E, is an isosceles right triangle in E^*.

Many other interesting observations may be made about configurations considered simultaneously in E and E^*. For instance, the unit circle, $X^2 + Y^2 = 1$, in E^*, is in E the curve whose equation, by (1), is

$$\left(\frac{x}{2-y}\right)^2 + \left(\frac{y-x}{2-y}\right)^2 = 1 \qquad \text{or} \qquad x^2 - xy + 2y = 2.$$

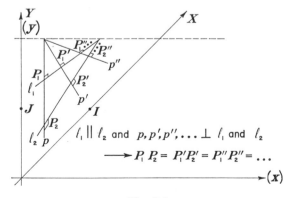

Fig. 7.6

This curve, plotted in E, is the hyperbola shown in figure 7.7, yet it is the locus in E^* of points at a unit distance from the origin.

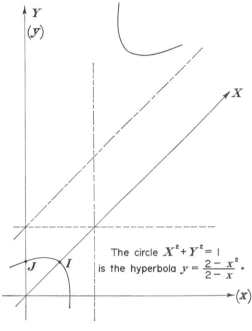

The circle $X^2 + Y^2 = 1$ is the hyperbola $y = \dfrac{2 - x^2}{2 - x}$.

Fig. 7.7

Similarly, the circle whose equation in E is $x^2 + y^2 = r^2$ is in E^* the conic whose equation is, by (2),

$$(8 - r^2)X^2 + (8 - 2r^2)XY + (4 - r^2)Y^2 - 2r^2X - 2r^2Y - r^2 = 0.$$

The discriminant of this conic is

$$(8 - 2r^2)^2 - 4(8 - r^2)(4 - r^2) = 16(r^2 - 4).$$

Hence if $r^2 < 4$, the curve in E^* is an ellipse; if $r^2 = 4$, the curve in E^* is a parabola; and if $r^2 > 4$, the curve in E^* is a hyperbola. It is instructive to note that $r^2 < 4$ is the condition that in E the given circle does not intersect the line $y = 2$, which is the ideal line, or line at infinity, in E^*. Likewise $r^2 = 4$ is the condition that in E the given circle is tangent to the ideal line of E^*, and $r^2 > 4$ is the condition that in E the given circle intersects the ideal line of E^* in distinct (real) points. This, perhaps, sheds new light on the observation sometimes made in elementary analytic geometry that ellipses are conics that do not intersect the line at infinity, parabolas are conics that are tangent to the line at infinity, and hyperbolas are conics that intersect the line at infinity in two points, the tangents at these points being the asymptotes of the hyperbola.

We have now explored in some detail the nature of a particular affine plane and have traced the evolution of a Euclidean plane from such a plane. What we have done should be helpful as an illustration of the more abstract approach we shall begin to describe in the next section. However, an affine approach to Euclidean geometry obviously cannot begin with a preexistent Euclidean plane, as did our example. Instead, as we suggested earlier, what we shall describe in the rest of this chapter is, first, the axiomatic development of affine geometry, and second, the evolution of the Euclidean plane from the affine plane.

The affine approach to Euclidean geometry is probably to be regarded as still experimental, and although it has been well tested, a variety of suitable texts is not available. Accordingly, the presentation we shall now make follows closely the original work of the Wesleyan Coordinate Geometry Group as it appears in the text that resulted from that work, *Modern Coordinate Geometry* (4).

Pedagogical Considerations

One of the major characteristics of mathematics today is its emphasis on synthesis. Working mathematicians are continually striving to discover relations between fields that previously seemed unrelated and to create structures capable of subsuming what were previously disjoint topics. At the same time, the rigid barriers that once separated the various courses in the mathematical curriculum are being broken down, and unified courses are appearing at all levels. Separate courses in plane and solid geometry are largely a thing of the past. Analytic geometry and calculus are now normally taught as an integrated subject, and at least some linear algebra is being incorporated into a variety of courses.

The affine approach to Euclidean geometry is an example at the precalculus level of this tendency toward unification. It leads to all the results obtained in traditional courses in geometry, and at the same time, as an essential part of the development, it leads to substantially all the analytic geometry of the straight line and the circle. It thus provides an excellent background for a modern integrated course in analytic geometry and calculus. Furthermore, it introduces and makes effective use of the cosine function and thus prepares the student for a more detailed study of the trigonometric functions in a subsequent course in the elementary functions.

Since the development proceeds in a rigorous fashion from a set of carefully stated axioms, the student obtains considerable experience in the axiomatic method, so important in contemporary mathematics. But since the number of axioms is small (only five, compared to twenty or twenty-five in the usual development of Euclidean geometry), deductive reasoning from them does not dominate the work. At the same time, this approach is much more algebraic than is the usual course in geometry. Hence the student, instead of having a year's "vacation" from algebra, is given ample opportunity to consolidate and extend his previous work in algebra.

Minimum subject matter prerequisites for this approach are probably no more than a good one-year course in algebra together with a reasonable background in intuitive geometry. However, because of the serious use of such relatively sophisticated ideas as mappings and their composition and restrictions, affinities, similitudes, isometries, and oblique coordinate systems, as well as the general level of precision and rigor to be maintained, an intangible prerequisite of maturity is highly desirable. The need for maturity and a certain flair for abstraction is underscored by the fact that in the nature of the affine approach the early work *is* affine and not conventionally Euclidean. As an example, the introduction of coordinate systems before the notion of perpendicularity is defined requires, both for clarity and for honesty, that they be drawn as though they were oblique. The great advantage of such systems, stemming from the ease with which they can be adapted to particular configurations, is not always sufficient to outweigh their seeming strangeness for some students.

The experience of the several test centers that tried out this approach using the materials prepared by the Wesleyan Coordinate Geometry Group in the summer of 1964 seemed to indicate that it is suitable primarily for students capable of college work and that for the upper 30 or 40 percent of these, it is a genuinely exciting intellectual adventure.

It is doubtful that a course based on the affine approach should be considered a replacement for the courses in geometry normally offered for the average student. For a very select group of students it might be offered in the tenth grade and then followed by courses in the elementary functions, advanced algebra, and analytic geometry and calculus. More realistically, it probably should be offered in the eleventh grade to students with a background of two years of algebra. A subsequent year devoted to the elementary functions and an introduction to analytic geometry and calculus would then provide an excellent preparation for college-level mathematics.

The Affine Line

Fundamental to the affine approach is the notion of an *affine function,* that is, a function, f, defined by the linear rule

$$f(x) = ax + b, \ a \neq 0.$$

At the outset, a number of properties of affine functions must be established. Among these, the following are the most important:

1. Every affine function is a one-to-one mapping from the set of real numbers to the set of real numbers.
2. The inverse of an affine function is an affine function.
3. The composition of two affine functions is an affine function.
4. If p and q are distinct real numbers and if r and s are distinct real numbers, then there is one and only one affine function, f, such that

$$f(p) = r \qquad \text{and} \qquad f(q) = s.$$

5. If p, q, r, s are real numbers such that $r \neq s$, then f is an affine function if and only if

$$\frac{f(p) - f(q)}{f(r) - f(s)} = \frac{p - q}{r - s}$$

6. If f is an affine function, then $f(q)$ is between $f(p)$ and $f(r)$ if and only if q is between p and r.

Careful proofs of these results should be well within the grasp of any student who has successfully completed a one-year course in high school algebra and should provide desirable reinforcement of his previous experience in algebra before it is put to work in geometry.

With these algebraic preliminaries out of the way, the concept of a line can be introduced. However, unlike the usual treatment in which a line is initially a very simple object on which a coordinate system is gradually created, here a line appears at once with a full-blown coordinate structure. Specifically, consider a set, L, of unspecified objects and a nonempty collection, C, of one-to-one correspondences between L and the set, R, of all real numbers. Then the pair (L, C) is said to be a line if and only if the following two axioms hold:

A. 1. If a and b are distinct elements of the set L, then there is one and only one correspondence in the set C that pairs a with the number 0 and b with the number 1.

A. 2. If c and c' are any correspondences in the set C and if r and r' are, respectively, the numbers that c and c' pair with the element x of L, then there exists an affine function, f, such that $r' = f(r)$ for all x in L.

From these axioms it is clear that a line is a complicated thing that cannot be visualized simply as a set of unspecified objects, called *points*, but must be understood to involve both a set of elements, L, and a set of coordinate systems, C. In spite of this, however, we shall often refer to the line (L,C) simply as the line L.

Because these axioms are quite different from those encountered in the usual course in geometry, it may be helpful to illustrate their significance by a specific example. In the coordinatized Euclidean plane, let L be the set of parabolas that pass through the origin and the point $(1,1)$ and have their axes vertical. The equation of any such parabola is of the form $y = ax^2 + (1 - a)x$, and if the degenerate parabola $y = x$, corresponding to the value $a = 0$, is included, this equation establishes a one-to-one correspondence between the set of parabolas, L, and the set of real numbers, R. Because of this, we are free to speak of "the parabola a," meaning, of course, the parabola whose equation is determined by the value a.

A set, C, of one-to-one correspondences between L and R can now be established in a variety of ways. For instance, the equation $r = pa^3 + q$ defines such a correspondence, provided only that $p \neq 0$. To verify axiom 1, let a_0 and a_1 be two particular parabolas that are to correspond, respectively, to the real number 0 and 1. Then p and q must satisfy the equations

$$0 = pa_0{}^3 + q \qquad \text{and} \qquad 1 = pa_1{}^3 + q,$$

and these equations have the unique solution

$$p = \frac{1}{a_1{}^3 - a_0{}^3}, \qquad q = -\frac{a_0{}^3}{a_1{}^3 - a_0{}^3}.$$

Hence there is one and only one correspondence in the set defined by the equation $r = pa^3 + q$, namely,

$$r = \left(\frac{1}{a_1{}^3 - a_0{}^3}\right) a^3 - \frac{a_0{}^3}{a_1{}^3 - a_0{}^3},$$

which pairs the parabola a_0 with the number 0 and the parabola a_1 with the number 1.

To verify axiom 2, let

$$c: r = pa^3 + q \qquad \text{and} \qquad c': r' = p'a^3 + q', \qquad p,p' \neq 0,$$

be any two correspondences of the set C. Then the relation between r and r', obtained by eliminating a^3 between these equations, is

$$r' = \left(\frac{p'}{p}\right) r + \frac{pq' - p'q}{p},$$

which clearly is the rule of an affine function. Thus both axiom 1 and axiom 2 are satisfied, and hence the set of parabolas, L, and the set of one-to-one correspondences, C, together constitute a line, (L,C).

Other lines could have been created from L by using other sets of correspondences. For instance, if C' is the set of correspondences defined by the equation $r = pa^5 + q$, $p \neq 0$, then the pair (L,C') is also a line. The pair $(L, C \cup C')$ is not a line, however, since now there are two members of the set $C \cup C'$, one in C and one in C', which pair a given parabola a_0 with the number 0 and a second parabola a_1 with the number 1.

The elements of any set L satisfying axioms 1 and 2 are naturally called *points,* and the correspondences in the set C are referred to as *coordinate systems on L.* For any particular coordinate system, c, on L, the points that C pairs with the numbers 0 and 1 are called, respectively, the *origin* and the *unit point of* c. Axiom 1 thus asserts that given any two points of L, there is a unique coordinate system in the set C for which these points are respectively the origin and the unit point. Axiom 2 further describes the structure of L by asserting that given any two coordinate systems on L there exists an affine function, f, such that for each point of L its coordinates, r and r' in the two systems, are related by the formula $r' = f(r)$.

On the basis of the coordinate systems identified by axioms 1 and 2, it is now a straightforward matter to define the relation of betweenness for the points of a line L in terms of the arithmetic order relations of the coordinates of the points and to prove that it is independent of the particular coordinate system used to identify the points. Segments and rays can now be defined in terms of the concept of betweenness, and the usual theorems can be proved. Finally, the distance between two points of L is defined as the absolute value of the difference between their coordinates in a particular system, and such questions as the relation between distances measured in different coordinate systems can be explored.

Affinities

Before taking the full step from the affine line to the affine plane, it is convenient to consider relations between two lines not thought of as embedded in any larger system. Specifically, since the points of any line are put in one-to-one correspondence with the real numbers by any coordinate system on the line, it is obvious that a one-to-one correspondence between the points of two lines can be established by setting up a one-to-one correspondence between their coordinates in any two particular coordinate systems. When the relation between the coordinates on the two lines is an affine function, the one-to-one correspondence induced between the sets of points on the two lines is known as an *affinity.* Affinities are crucial in the development of the affine plane. The roles of the two coordinate systems and the affine function in defining an affinity are shown graphically in the diagram on page 221.

It is important to distinguish clearly between an affine function and an affinity. An affine function is a one-to-one mapping from the set of reals to the set of reals defined by a rule of the form $r' = ar + b$, $a \neq 0$. By contrast, an affinity is the one-to-one mapping between the points of a line

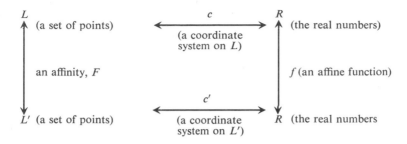

L and the points of a line L' that is induced when the coordinates of the points of L are related to the coordinates of the points of L' by an affine function. Three things are thus required for the definition of an affinity: a coordinate system on each of two lines and an affine function relating the coordinates in these two coordinate systems.

Although an affinity, F, is defined in terms of coordinates relative to two particular systems, it is actually just a one-to-one correspondence between two sets of points. It is therefore natural to ask if it does in fact depend on the particular coordinate systems employed in its definition. The answer, which is contained in the following theorem, shows that a given affinity can be defined equally well in terms of any coordinate systems on the respective lines.

THEOREM. *Let F be an affinity from a line L to a line L' defined in terms of the coordinate systems c and c' and the affine function f. Then if c_1 is any coordinate system on L and if c_1' is any coordinate system on L', there exists an affine function, f_1, such that F is also defined by c_1, c_1', and f_1.*

Various properties of affinities can now be established. In particular, the following properties are important for subsequent use in the study of the affine plane:

1. The inverse of an affinity is an affinity.
2. If P and Q are distinct points of a line L and if P' and Q' are distinct points of a line L', there is a unique affinity that maps P onto P' and Q onto Q'.
3. Under any affinity the image of a ray is a ray and the image of a segment is a segment.
4. If P', Q', R' are the images of the points P, Q, R under an affinity F, then Q' is between P' and R' if and only if Q is between P and R.
5. If P', Q', R' are the images of the points P, Q, R under an affinity F, then Q' divides the segment $\overline{P'R'}$ in the same ratio that Q divides the segment $\overline{PR}$.

6. If F is an affinity from a line L to a line L' and if G is an affinity from L' to a line L'', then the composition of G with F is an affinity from L to L''.

7. If an affinity F between a line L and a line L' is defined by the respective coordinate systems c and c' and the affine function, f, whose rule is $f(x) = ax + b$, and if p' and q' are the coordinates in c' of the images of the points whose coordinates in c are p and q, then

$$a = \frac{p' - q'}{p - q}.$$

8. If F is an affinity between a line L and itself and if the two coordinate systems, c and c', used in the definition of F are the same, then the value of a in the rule of the affine function $f(x) = ax + b$ is the same for all choices of the coordinate system on L.

The Affine Plane

Axioms 1 and 2 were essentially criteria for determining whether or not a set of elements, called *points,* could, when associated with certain correspondences, be considered to be a line. Given the points of a set of lines, we now require three more axioms to tell us whether or not such a set of points, α, can be considered to be a plane. The first two of these are completely familiar and completely expected:

A. 3. If P and Q are distinct points of α, there is one and only one line of α that contains them both.

A. 4. If L is a line of α and if P is a point of α, there is one and only one line of α that contains P and is parallel to L.

With axioms 3 and 4, it is now possible to construct a two-dimensional coordinate system in α. This is done by first choosing three noncollinear points, O, I, J, as base points. Axiom 1 guarantees us a unique coordinate system on the line $\overleftrightarrow{OI}$ with base points O and I and a unique coordinate system on the line $\overleftrightarrow{OJ}$ with base points O and J. The coordinatization of α is begun by assigning to an arbitrary point, A, on $\overleftrightarrow{OI}$ the ordered pair of coordinates $(x,0)$ where x is the coordinate of A in the one-dimensional coordinate system that we already have on $\overleftrightarrow{OI}$. Similarly, an arbitrary point, B, on $\overleftrightarrow{OJ}$ is assigned the ordered pair of coordinates $(0,y)$ where y is the coordinate of B in the one-dimensional coordinate system on $\overleftrightarrow{OJ}$. Finally, to an arbitrary point, P, which is neither on $\overleftrightarrow{OI}$ nor $\overleftrightarrow{OJ}$, coordinates are assigned in the following fashion: Through P there is, by axiom 4, a unique line L that is parallel to $\overleftrightarrow{OJ}$ and a unique line L' that is parallel to OI. It follows immediately by axiom 4 that L must intersect $\overleftrightarrow{OI}$, say in the point A, and that L' must intersect $\overleftrightarrow{OJ}$, say in the point B. If x is the coordinate of A in the one-dimensional coordinate system on $\overleftrightarrow{OI}$ and if y is the coordinate of B in the one-dimensional coordinate system on $\overleftrightarrow{OJ}$, the two-dimensional coordinates of P are defined to be (x,y).

Although the foregoing procedure is reminiscent of the process of assigning rectangular coordinates in the Euclidean plane, it differs from it in two significant respects. In the first place, in the Euclidean plane the axes $\overleftrightarrow{OI}$ and $\overleftrightarrow{OJ}$ are chosen to be perpendicular, whereas in the affine plane, since the notion of perpendicularity has not yet been defined, all that can be assumed is that $\overleftrightarrow{OI}$ and $\overleftrightarrow{OJ}$ are distinct, intersecting lines. Secondly, in the Euclidean plane the points A and B, whose one-dimensional coordinates determine the coordinates of a general point P, are located by considering perpendiculars from P to the axes, whereas in the affine plane, since the notion of perpendicularity is not available, A and B are determined by considering the lines that pass through P and are parallel to the axes.

It is tempting to say that in the affine plane one uses an oblique rather than a rectangular coordinate system, and the figures one draws make this seem reasonable. This is not a correct observation, however, for *oblique* in this context means *not perpendicular,* and until the notion of perpendicularity is defined it is equally meaningless to say that two lines are or are not perpendicular.

Not only is a two-dimensional coordinate system defined in α by two one-dimensional coordinate systems on a pair of intersecting lines, but, conversely, any two-dimensional coordinate system induces at least one one-dimensional coordinate system on each line of α. More specifically, we have the following theorem:

THEOREM. *Let L be a line of α and let $\mathcal{C}$ be a two-dimensional coordinate system in α ; then—*

1. *if L is not parallel to the y-axis of $\mathcal{C}$, there is a one-dimensional coordinate system on L in which the coordinate of each point of L is its x-coordinate in $\mathcal{C}$;*

2. *if L is not parallel to the x-axis of $\mathcal{C}$, there is a one-dimensional coordinate system on L in which the coordinate of each point of L is its y-coordinate in $\mathcal{C}$.*

The final axiom required to define the affine plane is motivated by the familiar notion of a parallel projection between a line L and a line L'. Such a projection clearly establishes a one-to-one correspondence between the points of L and the points of L', and axiom 5 asserts that this correspondence is an affinity:

A. 5.   If L and L' are lines of α , then any parallel projection from L to L' is an affinity.

To illustrate the meaning of axiom 5, let us return to the affine plane we constructed in the second section of this chapter and consider a parallel projection between any two of its lines, say

$$L_1: \ Y = M_1X + B_1 \qquad \text{and} \qquad L_2: \ Y = M_2X + B_2.$$

A parallel projection between these lines is defined by their intersections with the lines of a parallel family, say the family $y = MX + B$, B variable. Now the general line $Y = MX + B$ intersects L_1 and L_2, respectively, in the points

$$\left(\frac{B_1 - B}{M - M_1}, \ \frac{MB_1 - MB}{M - M_1} \right) \qquad \text{and} \qquad \left(\frac{B_2 - B}{M - M_2}, \ \frac{MB_2 - M_2B}{M - M_2} \right).$$

If we now relate the X-coordinates of corresponding points in this parallel projection by eliminating B between the equations

$$X_1 = \frac{B_1 - B}{M - M_1} \qquad \text{and} \qquad X_2 = \frac{B_2 - B}{M - M_2},$$

we obtain

$$X_2 = \left(\frac{M - M_1}{M - M_2} \right) X_1 + \frac{B_2 - B_1}{M - M_2},$$

which is the rule of an affine function. Since the correspondence defined by the parallel projection between L_1 and L_2 is thus equivalent to an affine function relating the X-coordinates of corresponding points, the parallel projection is an affinity, as required by axiom 5. Had the y-coordinates of corresponding points been employed, we would have found the relations

$$Y_1 = \frac{MB_1 - M_1B}{M - M_1}, \qquad Y_2 = \frac{MB_2 - M_2B}{M - M_2}$$

and, again eliminating B,

$$Y_2 = \left(\frac{M_2 \ (M - M_1)}{M_1 \ (M - M_2)} \right) Y_1 - \frac{M \ (M_2B_1 - M_1B_2)}{M_1 \ (M - M_2)}.$$

If neither M_1 nor M_2 is zero, that is, if neither of the given lines is parallel to the X-axis, then this equation is also the rule of an affine function, which confirms the fact that the correspondence defined by the parallel projection is an affinity. Incidentally, it also illustrates the fact that the same affinity may be defined by various affine functions, depending on the coordinate systems used on the respective lines.

It is now an easy matter to show that relative to any coordinate system in the affine plane a line, L, has an equation of the form $ax + by + c = 0$ in which a and b are not both zero. Although this result is completely familiar, the proof in the affine context is quite different from the usual Euclidean proofs and is perhaps worth presenting as an illustration of the differences in the two approaches. For simplicity, we shall consider only the general case in which the line is parallel to neither coordinate axis.

Let L be a line that is parallel to neither $\overleftrightarrow{OI}$ nor $\overleftrightarrow{OJ}$. Then, by a previously quoted theorem, both the x-coordinates and the y-coordinates of the points of L form a one-dimensional coordinate system on L. Moreover, for any two coordinate systems on L it follows from axiom 2 that there is an affine function relating the coordinates in one system to those in the other. Hence the x- and y-coordinates of any point on L satisfy an equation of the form

$$y = ax + b, \qquad a \neq 0.$$

Conversely, if (u,v) is a pair of numbers satisfying the equation $y = ax + b$, then there is a point, P, on L whose x-coordinate is u, since the x-coordinates on L form a one-dimensional coordinate system. Moreover, as we have just seen, the coordinates of any point, P, on L satisfy the equation $y = ax + b$. Therefore the y-coordinate of P is $au + b$, which by hypothesis is v. Hence u and v are the coordinates of P, which is known to be a point of L. Thus a point P is on the line L if and only if its coordinates satisfy the equation $y = ax + b$, which is of the asserted form.

Although there is as yet no basis for measuring angles in $\mathcal{A}$ and hence no basis for speaking of the trigonometric functions of an angle, it is natural to define the coefficient a in the equation $y = ax + b$ of a line L to be the *slope of L* and to say that a line with an equation of the form $x + b = 0$ has no slope. It follows that if (x_1,y_1) and (x_2,y_2) are two distinct points on a line L of slope a, then

$$a = \frac{y_2 - y_1}{x_2 - x_1}.$$

Furthermore, it is easy to prove the familiar theorem (which we illustrated in the discussion in the second section of this chapter) that two lines are parallel if and only if they have the same slope or neither has a slope. As usual, the equation $y = ax + b$ is said to be in the *slope-intercept form*. Similarly, the *point-slope form*

$$y - y_1 = a(x - x_1),$$

the *two-point form*

$$y - y_1 = \left(\frac{y_2 - y_1}{x_2 - x_1}\right)(x - x_1),$$

and the two-intercept form

$$\frac{x}{c} + \frac{y}{b} = 1$$

can be obtained.

Up to this point, not only the methods but also the results achieved by the affine approach have been more algebraic than geometric. It is now possible, however, to prove many of the purely geometric theorems encountered in the usual high school geometry course as well as some not ordinarily included in such a course. In fact, any results that do not depend on the measurement of angles, the comparison of lengths on different lines, or such related properties as perpendicularity, similarity, and congruence, are now accessible. In addition to many familiar theorems on triangles and parallelograms, these include such things as the following theorems:

1. THEOREM OF CEVA. *If $\triangle ABC$ is an arbitrary triangle, and if Q_a, Q_b, and Q_c are points on $\overleftrightarrow{BC}$, $\overleftrightarrow{CA}$, and $\overleftrightarrow{AB}$, respectively, then $\overleftrightarrow{AQ_a}$, $\overleftrightarrow{BQ_b}$, and $\overleftrightarrow{CQ_c}$ are concurrent if and only if*

$$\frac{AQ_c}{Q_cB} \cdot \frac{BQ_a}{Q_aC} \cdot \frac{CQ_b}{Q_bA} = 1,$$

the distances involved being considered directed distances.

2. THEOREM OF MENELAUS. *If $\triangle ABC$ is an arbitrary triangle, and if Q_a, Q_b, and Q_c are points on $\overleftrightarrow{BC}$, $\overleftrightarrow{CA}$, and $\overleftrightarrow{AB}$, respectively, then Q_a, Q_b, and Q_c are collinear if and only if*

$$\frac{AQ_c}{Q_cB} \cdot \frac{BQ_a}{Q_aC} \cdot \frac{CQ_b}{Q_bA} = -1,$$

the distances involved being considered directed distances.

3. THEOREM OF DESARGUES. *If the lines determined by corresponding vertices of two triangles are concurrent, then the intersections of the lines that contain corresponding sides of the two triangles are collinear.*

Moreover, because in the affine plane any two nonparallel lines can be chosen as the axes of a coordinate system without in any way complicating the analysis, the proofs of many of these theorems are significantly simpler than they are in more conventional courses in either synthetic or analytic geometry. In particular, in the affine plane a coordinate system can always be chosen so that the vertices of a given triangle are the points $(0,0)$, $(1,0)$, $(0,1)$ or so that the vertices of a given parallelogram are the points $(0,0)$, $(1,0)$, $(0,1)$, $(1,1)$. Finally, it is possible at this point to make a substantial digression into the important topics of convex sets and linear programming.

The Euclidean Plane

Although affine geometry is sufficient for the investigation of many configurations in the Euclidean plane and of course is a legitimate subject for study by anyone interested in it for its own sake, it obviously falls short of providing a complete description of the external world as we think we observe it. In the affine plane, one can compare the lengths of segments on the same line but not on different lines. Hence it is meaningless to speak of isosceles triangles or the triangle inequality. Likewise, although there are parallel lines and intersecting lines, perpendicularity is not defined; hence it is impossible to discuss right triangles or rectangles. Finally, because in the affine plane there is no basis for measuring angles, questions of similarity cannot be explored.

Since the ultimate objective of the affine approach as here discussed is the Euclidean plane, it is now necessary to decide how to introduce the measurement of lengths and angles: *Are additional axioms required, or can this be done by means of suitable definitions?* It turns out that no new axioms are required. In other words, the Euclidean plane, in a sense, is latent in the affine plane and can be created from it by introducing suitable definitions of a perpendicularity relation and a distance function (as we illustrated in the second section of this chapter).

The notion of perpendicularity is introduced by defining a perpendicularity relation, $\perp$, among the lines of α to be any relation such that (6)—

1. no line is perpendicular to itself;
2. if a line l is perpendicular to a line m, then m is perpendicular to $l;$
3. for any point P and any line l in α, there is a unique line that contains P and is perpendicular to l.

From this it follows at once that lines perpendicular to the same line are parallel, that lines respectively parallel to two perpendicular lines are themselves perpendicular, and that if one of two parallel lines is perpendicular to a given line, so is the other.

The measurement of lengths in α is made possible by defining a distance function, d, to be any function with the following properties:

1. The domain of d is the set of ordered pairs of points of α, and the range of d is the set of nonnegative real numbers.
2. On each line of α, d is a linear distance function (as defined in the earlier discussion of the affine line).
3. Any parallel projection between parallel lines preserves the distance function, d.

4. If $\overrightarrow{PQ} \perp \overrightarrow{RQ}$, then $(PR)^2 = (PQ)^2 + (RQ)^2$.

It is interesting and important to note that from property 4 the Pythagorean relation is no longer a theorem but a property true by definition.

The Euclidean plane now emerges as any system consisting of an affine plane, α, a perpendicularity relation, $\perp$, in α, and a distance function, d, in α. In such a system the familiar distance formula of analytic geometry can now be established, and it can be shown (essentially as we did in the discussion of perpendicularity in the example in the second section of this chapter) that two lines having slopes are perpendicular if and only if their slopes, as previously defined in nonmetrical terms, are negative reciprocals. Finally, as the culmination of the development, the fundamental Euclidean structure theorem can be proved:

For each set of three noncollinear points, O, I, J, in an affine plane, α, there is a unique Euclidean plane (α, $\perp$, d) such that $\overleftrightarrow{OI} \perp \overleftrightarrow{OJ}$ and $d(\overline{OI}) = d(\overline{OJ}) = 1$.

Euclidean Geometry

In a sense, the affine approach to Euclidean geometry is now complete, and its exposition need be carried no further. However, the subsequent development of the properties of the Euclidean plane is strongly influenced by the methodology of the affine approach, and some further discussion is appropriate.

At the present stage it is now known (partly as a matter of definition, partly as a matter of proof) that a triangle ABC with sides a, b, c has a right angle at C if and only if

$$c^2 = a^2 + b^2.$$

This suggests that there may be a more general relation, valid for all triangles, that reduces to the Pythagorean relation when the triangle is a right triangle. If given an arbitrary triangle, $\triangle ABC$, a rectangular coordinate system is chosen so that the coordinates of the vertices are $C:(0,0)$, $B:(u,v)$, $A:(b,0)$, it is easily shown by the distance formula that

$$c^2 = a^2 + b^2 - 2ub$$

where a, b, c are, respectively, the lengths of the segments $\overline{BC}$, $\overline{CA}$, $\overline{AB}$. Although intuition suggests that the expression on the right should be symmetric in a and b, it does not appear to be. Hence it is natural to write

$$c^2 = a^2 + b^2 - \frac{2uab}{a}$$

and to consider the possibility of defining u/a to be cos $\angle ACB$. This is justified by showing that the ratio u/a is uniquely determined by the angle, independent of the choice of A and B on the sides of the angle and independent of which rectangular coordinate system is used to determine the coordinate, u, of the vertex that is not on the positive x-axis. Thus the cosine function is introduced and the law of cosines is obtained. It can now be shown that the range of the cosine function is the set of real numbers in the closed interval $[-1,1]$ and that given any number t such that $-1 \leq t \leq 1$ and a half-plane, H, determined by ray $\overrightarrow{OA}$, there is a unique ray $\overrightarrow{OB}$ in H such that cos $\angle AOB = t$. Further, it can be shown that if $\angle AOB$ is not a right angle, then the scale factor of the perpendicular projection from $\overleftrightarrow{OB}$ to $\overleftrightarrow{OA}$ is cos $\angle AOB$.

The notions of similarity and congruence are based on transformations known as similitudes, which are two-dimensional analogues of affinities: Let O, I, J and O', I', J' be the bases of two rectangular coordinate systems and let a be a positive constant. Then the transformation that pairs each point P, whose O, I, J coordinates are (x,y), with the point P', whose O', I', J' coordinates are $x' = ax$ and $y' = ay$, is called a *similitude*. If the constant a has the value 1, the similitude is called an *isometry*. From this definition it follows readily that if P and Q are any points and if P' and Q' are their images, then

$$P'Q' = a(PQ).$$

Two subsets, S and S', in the Euclidean plane are said to be similar if their points are in one-to-one correspondence and if there exists a positive number a such that for any two points, P, Q of S and their images, P', Q' in S', we have the distance relation

$$P'Q' = a(PQ).$$

Two figures that are similar with constant $a = 1$ are said to be congruent. From these definitions it follows that similarities and congruences are, respectively, restrictions of similitudes and isometries to particular subsets of the Euclidean plane. All the usual theorems involving similarity and congruence can now be proved, and, in particular, the significance of the cosine function is further underscored by the result that two angles are congruent if and only if they have the same cosine.

Translations and reflections in both lines and points can now be defined and studied and their relation to isometries determined. Finally, it is possible to prove the important result that every isometry is either a translation, the composition of two line reflections, or the composition of three line reflections.

The study of circles is expedited by the analytic techniques made available by the affine approach. Arc length is defined as the least upper bound of the lengths of appropriate portions of inscribed regular polygons of 2^{n+1} sides, and angle measures are defined in terms of the lengths of intercepted arcs. Although it is probably impossible at the high school level to give a completely satisfactory discussion of arc length on a circle, the affine approach permits a treatment at least as sound as that possible with other approaches.

As a final topic, the definition and computation of areas is undertaken by considering inner and outer sums of approximating squares. The ratio of the areas of similar figures is shown to be the square of the constant of the corresponding similarity, and formulas for the areas of the common geometric figures are obtained. The area of a circle is defined in terms of inner and outer sums of congruent approximating isosceles triangles. As in the case of arc length, not all the technical details can be completely explored. However, the level of clarity and precision that can be maintained should give the student an excellent background for the discussion of area in a rigorous course in calculus.

Conclusion

The preceding discussion has mainly been a description of the affine approach to Euclidean geometry as it appears in the text *Modern Coordinate Geometry* (4) prepared by the Wesleyan Coordinate Geometry Group. Although this approach differs in many respects from others, both conventional and unconventional, it also shares many features with them. Its main ideas, novel as they are, can be introduced through exploratory examples so that the student has an opportunity to observe the informal, inductive sides of mathematics and share in the sense of discovery that is part of the charm of mathematics. Formally, it proceeds deductively from a set of carefully stated axioms, and so the student acquires some familiarity with the axiomatic method. It makes early and continuing use of coordinates, both on a line and in a plane, yet many proofs are coordinate-free. Thus the student becomes familiar with both synthetic and analytic methods of proof. It makes such essential and effective use of transformations that those looking for a transformation approach to geometry might well consider it. And in its content, spirit, and natural level of precision and rigor, it is certainly close to the mainstream of contemporary mathematics.

The average teacher will probably find that there was little in his undergraduate training that prepared him specifically to teach the affine approach. However, the teachers in the experimental centers that tested the original

Wesleyan material experienced no great difficulty in adapting to it. Assistance, if needed, can be obtained from the text *Foundations of Geometry and Trigonometry* by Howard Levi (2), on which much of the Wesleyan development was based.

REFERENCES

1. Fishback, W. T. *Projective and Euclidean Geometry*. New York: John Wiley & Sons, 1962.
2. Levi, Howard. *Foundations of Geometry and Trigonometry*. Englewood Cliffs, N.J.: Prentice-Hall, 1960.
3. Snapper, Ernst. "Metric Geometry over Affine Spaces." Notes prepared by J. T. Buckley from the lectures given at Cornell University in the summer of 1964, sponsored by the Mathematical Association of America. Mimeographed.
4. Wesleyan Coordinate Geometry Group. *Modern Coordinate Geometry*. Boston: Houghton Mifflin Co., 1969.
5. Wylie, C. Ray. *Introduction to Projective Geometry*. New York: McGraw-Hill Book Co., 1970.
6. ———. "What Are Perpendicular Lines?" *Mathematics Teacher* 60 (January 1967): 24–30.

8

A Vector Approach
to Euclidean Geometry

STEVEN SZABO

IN ANY study of geometry, one is concerned with several sets of primitive objects and the relations among these objects. For example, in a "standard" treatment of Euclidean plane geometry, one is concerned with sets of primitive objects called *points* and *lines* and with relations among these objects, such as collinearity, incidence, and parallelism. For another example, in a "vector" treatment of Euclidean geometry (the subject under consideration in this chapter), one is concerned with sets of primitive objects called (sometimes) *points* and *vectors* and with relations among these objects, such as collinearity, dependence, and parallelism.

In an informal, or intuitive, study of geometry, one is usually concerned primarily with looking at models or pictures of geometric figures and trying to get some feeling about how such figures are "put together" in order to discover some of their properties. In such a study, one may discover many relations among various kinds of geometric objects and may even discover that certain relations may be derived from other relations. But the primary concern in an intuitive study of geometry is gathering information about, and gaining familiarity with, properties of the usual (and perhaps some not so usual) geometric configurations. Furthermore,

NOTE. The author is indebted to Herbert E. Vaughan and the late Max Beberman, without whose help and encouragement this chapter could not have been written.

such a study is important as a prelude to a more formal study of geometry at a later stage in one's mathematical education. Put another way, one should have a good deal of exposure to geometric ideas on an informal basis before he embarks on a formal study of geometry. (In support of this view, one can argue by analogy that a student should have many informal experiences with number relations prior to undertaking a formal study of the algebra of real numbers. It is not likely that one would begin to teach children a *formal* course in algebra unless those children had a great deal of information about arithmetic operations!)

By way of contrast, a formal study of geometry—especially a *first* study—involves making a selection of certain properties and relations as basic ones and using these as a foundation for the development of a mathematical system from which other properties and relations may be derived. Thus, although much of the subject matter may be the same and, indeed, although the processes of argumentation may be precisely the same as in an informal study, the premises from which conclusions are drawn are clearly specified—these are usually called *postulates* (or *axioms*), *definitions,* and *theorems*—and a logical organization of the subject is brought into focus.

The principal aim of this chapter is to discuss in detail one way in which a formal course in Euclidean geometry may be developed using vector methods. It is no doubt possible to do this in a variety of ways that are at once appealing and understandable to high school students and mathematically sound and fruitful. One plausible method of accomplishing this aim will be discussed in such a way that the reader can see how the formal aspects of the mathematical development unfold from the intuitive notions about the subject. (The mathematical content to be discussed here forms the basis for the geometry content of a two-year course developed by the UICSM. This course integrates algebra, geometry, linear algebra, trigonometry, and mathematical logic.) Then after the mathematical details are well fixed in mind, examples will be given of how problems in geometry may be treated by using the vector methods at hand. By tackling the problem in this way, we feel that we place the reader who wishes to use vector methods in teaching geometry in the firm position of making a choice either of preparing himself to give his students some particular full treatment of geometry based on vector methods or of choosing topics in geometry that he feels capable of treating with such methods.

It should be noted in passing that in the treatment of geometry to be described, we shall be dealing, for the most part, directly with the objects of our attention, for example, points, lines, and translations, rather than with canonical representations of these objects, such as ordered pairs or ordered triples of real numbers. In this sense, then, the treatment to be described is synthetic rather than analytic.

Following the discussion of the mathematical details of one viable treatment of Euclidean geometry with vector methods, it will be appropriate to summarize the pertinent ideas related to the philosophy and objectives of such a treatment of geometry. That it is more appropriate to do this after, rather than before, the mathematical details of such a treatment are discussed is clear considering both the help and perspective given to those who are unfamiliar with such a treatment and the lack of harm done to those who are familiar with such a treatment. Conversely, for those unfamiliar with how such a treatment might go, a discussion of philosophy and objectives runs the risk of "missing the boat" because of that unfamiliarity.

Before we go into the mathematical details of one possible treatment of Euclidean geometry with vector methods, it is appropriate to mention the kinds of students for whom such a treatment is appropriate and what might be reasonable to expect from them in the way of prerequisite mathematical knowledge and experiences.

Students for Whom the Treatment Is Appropriate

There is always the danger of feeling that just because a given subject or treatment is new, it is designed only for the very best of students. Of course, the best students will learn most of the subject matter to which they are exposed and will learn it in, or despite, the manner in which it is presented. Moreover, they will do it with relative ease and speed—that is what makes them "good students"! And in so doing, they tend to add support to the contention that the new material or new approach was designed just for the kind of students they are. What one tends to forget is that they would have also moved through the "tried and true" material in the same way. So, even though the best students are among those at whom a vector approach to Euclidean geometry may be directed, they are not the only ones.

In general, the students who show up in the college preparatory sections of high school geometry (if there are such sections) are those at whom a vector approach to geometry may, and perhaps should, be directed. Such students, typically, are capable of doing above-average work academically, have had at least one year of high school algebra, and will probably study mathematics in high school for at least three years. Also typically, some —perhaps many—will have a moderate amount of difficulty learning to cope with the subject matter; there still is no royal road to learning anything of substance in general, or to learning geometry in particular.

Many geometry classes in a cross section of schools in various regions of the United States have studied experimental text materials—some of these now available for general use are (5) and (6)—for a course developed along the lines of the one described in this chapter. The students enrolled were, more often than not, considered by their schools to be average rather than above-average or gifted students. The teachers were generally satisfied that the students involved learned at least as much geometry in the experimental course as they would have in the standard treatment of geometry offered in their schools. Further, the attrition rate was generally lower than in the regular sections, and the levels of achievement as measured on the usual nationally administered standardized tests (e.g., Preliminary Scholastic Aptitude Test, National Merit Scholarship Qualifying Test, College Entrance Examination Board, American College Testing) were reported by the user schools to be at the levels expected by them for the types of students involved in the program. (It might be argued, of course, that the teachers involved were exceptional ones, and, in truth, they probably would achieve above-average rankings on any rating instrument one could devise to identify "good" teachers. A common trait of these teachers was that of being able to get their students to become active participants in the on-going development of mathematical ideas. Put another way, they did not teach mathematics as a spectator sport.)

Remarks on Prerequisite Knowledge

Studying Euclidean geometry using vector methods amounts to studying the subject from the point of view of a special class of functions, called *translations,* acting on points. In order to derive full benefit from such a treatment, the student should have been exposed to a wide range of activities dealing with geometric objects in order to have accumulated a great deal of information on an intuitive level about how geometric figures work. The exposure to geometric notions that a student receives in most of the current elementary and junior high school mathematics programs should provide an adequate background of experiences on which to draw for any treatment of geometry similar to the one to be described here. The student also should have made a study of elementary algebra (of the real numbers) in which some attention was paid to the structural aspects of the real numbers—the field properties, order relations, and so on—and perhaps in which some elementary work was done with functions as mappings—introductory work with linear and quadratic functions, the operation of composition of functions,[1] and so on. Note that the particular

---

1. Composition of functions may be thought of as the operation of applying a first function and then applying a second function to that result. For example, let f be the function such that for each $x, f(x) = 3x - 2$, and let g be the function such that for each $x, g(x) = x^2$. Then g followed by f (or the composition of f on g, or $f \circ g$) is com-

algebra course studied by the student need not have brought out or stressed the *technical terms* such as "field" and "function composition," but rather it should have brought out the *ideas* for which these terms are referents. Any work done with using the basic field properties to derive other properties of the real numbers and in justifying the simplification of expressions and the solution of equations on the basis of the field properties will be very helpful to the student when he gets involved with the deductive aspects of the formal study of geometry. And owing to the fact that the vector methods used to obtain geometric results are algebraic in nature, the skills developed by the student in his study of algebra will be reinforced and enlarged in his study of geometry.

The prerequisite knowledge just described for a formal study of Euclidean geometry using vector methods is what a student should have in an *ideal* situation. Of course, not many students who come to the point in their high school program where they are ready for a first formal course in geometry are in the ideal position of having all the prerequisites mentioned here. Are such students to be barred from this treatment of geometry? Experience in a wide variety of experimental situations in public schools has indicated that they should not. If, for example, a student (or group of students) needs some introductory work with functions and function composition, no more than two weeks of work in this area seems adequate for the task of building up his skill and competence in working with the various necessary aspects of these concepts. Some of this introductory work may even be done in the context of geometric situations so that the student begins to make progress in the study of geometry while building up the skills needed to forge ahead. He will, of course, continue to make use of these concepts in his study of geometry and so will continue to strengthen his hold on them as he proceeds.

We now turn our attention to the mathematical details of a particular treatment of Euclidean geometry using vector methods. After giving the details of such a treatment, we shall make some remarks on a case for vector approaches to geometry. Matters concerning the training of teachers are dealt with in the final chapter of this book.

---

puted as follows:

$$[f \circ g](x) = f(g(x)) = f(x^2) = 3x^2 - 2$$

One may think of the function f as a "machine" that takes in an (argument) x, multiplies it by 3, subtracts 2, and gives out a result, called the *value of f at x*. Similarly, g is a machine that takes in an (argument) x and gives out x^2, the value.

Clearly, $f \circ g$ is a different function from $g \circ f$, for $[g \circ f](x) = (3x - 2)^2$. That is, squaring 2 less than 3 times a number is not always the same as finding 2 less than 3 times the square of the number. Composition of functions perhaps is not only the first nonarithmetic (mathematical) binary operation to be encountered by the student but also the first such operation that is not commutative.

A Plausible Treatment of Euclidean Geometry
Using Vector Methods

Preliminary remarks

For students who have been introduced to the notion of a function as a mapping, it is natural and most useful to exploit this notion in the study of geometry. Generated by intuitions about distance, parallel lines, translations, and perhaps reflections, the concept of distance-preserving mappings of Euclidean three-space onto itself can be developed. From the properties of these mappings, one can deduce all the standard (and much that is not now standard) high school geometry and do so in the context of the rest of high school mathematics.

In the treatment of Euclidean geometry to be discussed here, two of the principal objectives are to develop three-dimensional Euclidean geometry in the context of modern mathematics and to demonstrate for the student how one can make use of plausible intuitions to create a formal mathematical system. To these ends we shall adopt the strategy of studying a particular model for the given mathematical system and begin to abstract notions from the apparent behavior of this model. We shall then formulate the abstracted notions in mathematical language and work with these formulations to make inferences about the model and the mathematical system.

Intuitions about translations as slides

It is convenient, at the outset, to think of the "space of points" under consideration as the three-dimensional world in which we live. To study this space of points, it is convenient to look at translations of this space. A translation can be thought of as a sliding motion that moves points the same distance in the same direction. To get a feeling for what this means, consider the following experiment:

> Suppose that you have at your command all points in space and that you give the command to the points in your room to "move up three inches"—or "translate three inches up"—and also command *each* point to move the same way as the points in your room moved. Given this command, each point obeys.

It is easy, at least in your mind's eye, to see what happened when the command was given. All points moved, or translated, the same distance in the same sense. If you concentrated on certain of the points and paired them with the points to which they moved, that is, with their images under that translation, it is again easy, in your mind's eye, to see that (1) no point moved to *two* separate locations (and that therefore you are dealing with a function); (2) arrows drawn from points to their images are the same length and point the same way; and (3) any such arrow, in a sense, tells us all we need to know to find any point's image under the translation in question. For this reason it makes sense to talk about *the* translation from a first point to a second point as well as to visualize, or draw, an arrow from a first point to a second point and to say that this arrow "describes" a translation. (This is quite different from saying that a translation *is* an arrow.)

It is also easy to make use of tracings together with the notion of sliding in a prescribed way, as just described, to concentrate on certain collections of points in order to "see" what happens to these points under a given translation. The following sequence of pictures should be helpful in seeing how this might be done. To "see" what happens under the translation that takes P to Q, trace everything you see in figure 8.1 and mark all traced letters with primes, as in figure 8.2. Then slide the tracing along the line through P and Q until the dot for P' falls on the dot for Q (fig. 8.3). The points marked on the tracing sheet indicate the locations of the images of all the points in the figure under the translation that maps P on Q.

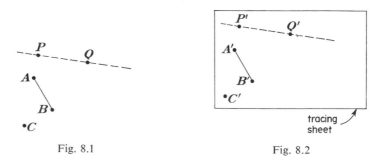

Fig. 8.1   Fig. 8.2

Many properties of translations can be made plausible by making use of this tracing-sheet model. For example, no point gets moved onto two different points under any given slide, from which it follows that translations (which are modeled by slides) are functions. It is also evident that no two points are slid onto the same point under a given slide, from which it follows that translations are one-to-one functions (that is, functions with inverses).

Other points about slides that are worth noting at the outset are (1) that there is a slide that takes a given first point to a given second point, (2) that each point is the image of some point under any given slide, and (3) that if C' is the image of C under the slide from P to Q, then Q is the image of P under the slide from C to C'. (See fig. 8.3.) From these three points

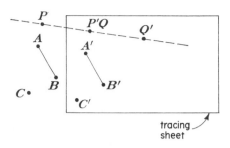

Fig. 8.3

it is evident that translations are functions, each of which has as its domain and range the set of points of space, and that a translation is determined completely by a given point and its image. This supports the convention of representing a translation by an arrow drawn from an argument (any point) to the image point of that argument. For example, in figure 8.3, an arrow drawn from P to Q (or from A to A', B to B', C to C') may be used to represent the translation from P to Q. (See fig. 8.4.) Drawing arrows to represent translations has the advantage of at once illustrating the intuitive notion of a mapping as a dynamic process as well as the intuitive notions of a translation preserving sense (arrows pointing the same way) and preserving distance (arguments such as A and C as far apart as their images A' and C').

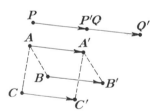

Fig. 8.4. Each arrow represents the *same* translation, namely the translation from P to Q. Note that the distances between pairs of points are the same as the corresponding distances between the images of those points ($AB = A'B'$, $AC = A'C'$, $PB = P'B'$, etc.).

The convention of drawing arrows to represent translations yields diagrams that are very suggestive of many of the important and basic algebraic properties of translations and the operation of function composition. Once the notion of sliding is fixed well enough in mind to provide a good mental picture of how translations operate on points, it is most helpful to enhance this mental imagery with the building of stick models that represent the actions of several translations acting successively on a point and its images. Such models also help to serve both as reminders that our basic model for

the space of points is the three-dimensional world in which we live and
as links between the algebraic operations on translations that will be formulated and the geometric concepts related to those operations.

Suppose that a first translation is represented by the arrow from P to Q
and that a second translation is represented by the arrow from R to S. (See
fig. 8.5.) Since each translation is a function (whose domain and range is
the set of all points), we know that the composition of any pair of translations is a function. Suppose that the translation from P to Q maps A_1,
A_2, and A_3 on B_1, B_2, and B_3, respectively, and that the translation from
R to S maps B_1, B_2, and B_3 on C_1, C_2, and C_3, respectively. Then the
resulting diagram suggests that the resultant of the first translation followed
by the second translation, that is, the composition of the second translation
on the first translation, *is also a translation,* namely, the one that maps A_1
on C_1 (or A_2 on C_2 or A_3 on C_3) and is represented by the arrows from
A_1. A_2, and A_3 to C_1, C_2, and C_3, respectively.

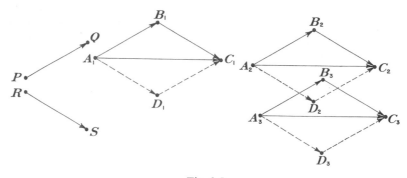

Fig. 8.5

The translation that maps R on S also maps A_1, A_2, and A_3 on some
points, say, D_1, D_2, and D_3, and the translation that maps P on Q also
maps D_1, D_2, and D_3 on some points. Appealing again to a diagram
(fig. 8.5), it is perfectly reasonable to expect, and accept, that the points
onto which D_1, D_2, and D_3 are mapped by the translation from P to Q are
C_1, C_2, and C_3, respectively. This suggests that the operation of composition of translations is commutative and makes the set of translations somewhat special and interesting, since composition of functions is not generally
commutative. Because composition of functions is associative, it follows
directly that composition of translations is associative. Even so, it is both
instructive and interesting to make diagrams or stick models that illustrate
this property. Using diagrams, the tracing-sheet model, and the previously
mentioned notion of a translation as a one-to-one mapping yields the
plausible notions of the identity translation and the inverses of translations.

These notions concerning translations and composition will be of special interest when we formulate properties of translations in algebraic terms.

Formalizing notions about translations

We summarize here some of the properties of translations that are easily plausible with the aid of the tracing-sheet model and most useful in the proposed development of geometry:

1. Given points P and Q, there is a translation that maps P on Q.

2. A translation is determined by any point and its image.

3. The composition of a pair of translations is a translation.

4. Composition of translations is both associative and commutative.

5. The identity mapping on the space of points is a translation.

6. Each translation has an inverse that is also a translation.

If we make use of these six properties of translations, it is not difficult to set up an algebra of points and translations. We have been discussing translations on our space of points as functions that map points on points and that have certain special observable and intuitively satisfying properties. To be better able to talk about points and translations and the relations among them, let us agree that ε is the set of all points in our space and that $\jmath$ is the set of all translations on ε. We shall use capital letters $(A, B, \ldots)$, with or without subscripts, as variables for points and bold-faced letters $(\mathbf{a}, \mathbf{b}, \ldots)$, with or without subscripts, as variables for translations. Later, as the occasion arises, we shall use the usual lowercase letters $(a, b, \ldots)$, with or without subscripts, as variables for real numbers. Adopting a convenient notation in which $B - A$ represents a translation that maps A on B and in which $A + \mathbf{a}$ is the point that results from applying translation $\mathbf{a}$ to point A, we formalize notions 1 and 2 in this postulate:

POSTULATE 0. (*a*) $B - A \in \jmath$
(*b*) $A + \mathbf{a} \in \varepsilon$

(Notice that "+" is used here to denote function application. That is, "$A + \mathbf{a}$" is being used to mean what "$\mathbf{a}(A)$" means. [See fig. 8.6.] Later, "+" will also be used to denote function composition, and so "$\mathbf{a} + \mathbf{b}$" will be used to mean what "$\mathbf{b} \circ \mathbf{a}$" means. It will always be clear from context just which meaning is intended, and it has been found to be readily understood and accepted by students.)

As an aside, we note that this postulate might be left in the background as an implicit agreement in much the same way that statements such as

$a + b \in \Re, \ a - b \in \Re, \ a \cdot b \in \Re$, and so on, are left in the background in ordinary developments of real number algebra. The sole purpose of such postulates, or "agreements," is the dual one of establishing what kinds of expressions make sense and what kind of thing (e.g., point, translation, or real number) is named by an expression that makes sense.

Given a translation **a**, it is the case that $A + $ **a** is a point (postulate $0(b)$) and, thus, that $(A + $ **a**$) - A$ is a translation (postulate $0(a)$). Looking at figure 8.6, we see that it is reasonable to accept that the translation **a** *is* the translation $(A + $ **a**$) - A$, that is, that **a** $= (A + $ **a**$) - A$.

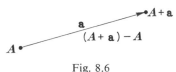

Fig. 8.6

Similarly, given points A and B, $B - A$ is a translation, and, therefore, $A + (B - A)$ is a point. Looking at figure 8.7, we see that it is reasonable to accept that the point $A + (B - A)$ is the point B, that is, that $A + (B - A) = B$. These notions are formalized in the following postulates, the first two really operational postulates of the system:

POSTULATE 1. $A + (B - A) = B$

POSTULATE 2.   **a** $= (A + $ **a**$) - A$

Even though our postulate system is not as yet a very rich one, it is reasonable to pause and see what we might learn about it. First, suppose that $A + $ **a** $= B$; that is, that the point $A + $ **a** is the point B. By postulate 2, we know that **a** $= (A + $ **a**$) - A$. So, since $A + $ **a** is B, it follows, just by substituting, that **a** $= B - A$. Hence, if $A + $ **a** $= B$, then **a** $= B - A$. Next, suppose that **a** $= B - A$. By postulate 1 we know that $A + (B - A) = B$. So it follows, again by substitution, that $A + $ **a** $= B$. Hence, if **a** $= B - A$, then $A + $ **a** $= B$. Clearly then, it follows that $A + $ **a** $= B$ *if and only if* **a** $= B - A$. This last result tells us that the point $A + $ **a** *is* B precisely when the translation **a** *is* $B - A$. Notice that this not only is intuitively obvious from the work with tracing sheets and diagrams such as figure 8.8, but it also follows from our postulates, and so it is a theorem.

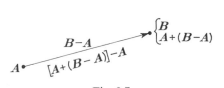

Fig. 8.7

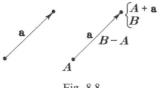

Fig. 8.8

As we saw earlier, it is intuitively clear that the set of translations is closed with respect to function composition, that is, that the compostion of any pair of translations is a translation. To arrive at a formulation of this notion to include among our postulates, we refer to what we have formalized about points and translations in conjunction with what we can learn from diagrams of translations acting on points. Consider three points, A, B, and C. (See fig. 8.9.) The translation that maps A on B is $B - A$, the one that maps B on C is $C - B$, and the one that maps A on C is $C - A$.

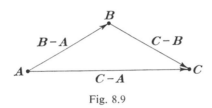

Fig. 8.9

Furthermore, it is clear that the composition of $C - B$ on $B - A$, that is, $(C - B) \circ (B - A)$, maps A on C. Therefore, we could formalize the notion that the composition of any pair of translations is a translation by accepting as a postulate the sentence "$(C - B) \circ (B - A) = C - A$." Instead of this, however, we adopt the convention of using "$+$" to denote function composition, agree that "$(B - A) + (C - B)$" means what "$(C - B) \circ (B - A)$" means, and state our next postulate as follows:

POSTULATE 3. $(B - A) + (C - B) = C - A$.

(This is sometimes referred to as the "bypass" postulate, for in a sense it tells us that we can use translations to get from a first point to a second point by going by way of any third point. We shall find much use for both the postulate and the underlying notion of "bypassing" in analyzing and studying properties of geometric figures.)

Now that we have some ideas on how translations and points behave, it is perhaps appropriate to comment in brief on the strategy that will be employed throughout the rest of this development to arrive at the notions of Euclidean geometry through the use of translations. We shall discuss in some detail those properties of translations that will enable us to classify the set of translations as a commutative (or Abelian) group. Then we shall discuss how we can arrive at the notion of real numbers operating on translations, concentrating on those properties of real numbers acting on translations that will enable us to classify further the set of translations as a vector space over the real numbers. (It will be the introduction of this concept that will identify the translations we have been working with as vectors, and it is the extension of this notion that will identify this approach to the study of geometry as a "vector" approach.) We shall then discuss the notions of linear dependence and linear independence of translations (vectors) and, in terms of these notions, arrive at those properties of geometric figures that are related to parallelism and ratio. In short, then, at this stage we shall have developed what is known as *affine geometry*. Furthermore, we shall demonstrate how we can make use of the notion of linear independence to fix the dimension of the space we wish to consider, and we shall fix the dimension of our space at three; therefore, in the formal system we are developing, we shall concentrate our efforts on three-dimensional geometry. Finally, we shall discuss the notion of the dot product (or inner product) of translations (vectors), and we shall demonstrate how to make use of the dot product to arrive at notions of perpendicularity and distance. Thus, in the end, we shall have accomplished the extension of our three-dimensional affine geometry to three-dimensional Euclidean geometry.

With this outline of events to come in mind, we are ready to proceed with the details of this particular vector approach to Euclidean geometry.

Identification of the group of translations

So far, we have formalized the first three properties of translations listed in the preceding section. In order to bring properties 4 through 6 into the formal system we are generating, we formulate the following postulates:

POSTULATE 4_0.　　(a)　　$\mathbf{a} + \mathbf{b} \in \mathfrak{I}$

　　　　　　　　　(b)　　$\mathbf{0} \in \mathfrak{I}$

　　　　　　　　　(c)　　$-\mathbf{a} \in \mathfrak{I}$

POSTULATE 4_1.　　$(\mathbf{a} + \mathbf{b}) + \mathbf{c} = \mathbf{a} + (\mathbf{b} + \mathbf{c})$

POSTULATE 4_2.　　$\mathbf{a} + \mathbf{0} = \mathbf{a}$

POSTULATE 4_3.　　$\mathbf{a} + -\mathbf{a} = \mathbf{0}$

POSTULATE 4_4.　　$\mathbf{a} + \mathbf{b} = \mathbf{b} + \mathbf{a}$

Some attention should be given to the various parts of the closure postulate, postulate 4_0. Part (a) of this postulate tells us, essentially, that $+$ is a binary operation on the set of translations. That is, given any translations **a** and **b**, their sum (or composition) is a translation. (It turns out that this result follows from postulates 0, 1, 2, and 3, but for reasons that will be evident later, we choose to list it among our postulates.) Part (b) of this postulate introduces a special translation, **0**, which by postulate 4_2 is the identity translation. Part (c) introduces a singulary (or unary) operation, $-$, on the set of translations, which by postulate 4_3 yields inverses of translations.

Postulate 4_1 tells us that $+$ is an associative operation on the set of translations and formalizes the first part of property 4. Of course, since translations are functions and since adding translations corresponds to composing them, this postulate is a perfectly natural result to include among our postulates. As previously mentioned, postulate 4_2 tells us that 0 is the identity translation, and postulate 4_3 tells us that $-$ is an operation that gives inverses, formalizing properties 5 and 6. Postulate 4_4 tells us that $+$ is a commutative operation on translations, formalizing the second part of property 4 and completing the job of including properties 1 through 6 in our system.

Perhaps a remark is in order on the acceptance of commutativity of composition (addition) of translations as a postulate. Its acceptance is completely analogous to, say, accepting that two points determine a line and is no less arbitrary or less appealing on intuitive grounds.

Notice that the postulates 4_0 through 4_4 deal only with translations and that the properties of translations formulated in these postulates correspond precisely with the basic properties of the real numbers under addition and "oppositing." (Oppositing is the singulary, or unary, operation, $-$, on the real numbers such that for each x, $-x$ is that number such that $x + -x = 0$). This observation may be used to stimulate a discussion about commutative (or Abelian) groups. Having introduced the notion of a commutative group and having become familiar with the basic properties of this sort of mathematical structure, we can greatly condense our list of postulates by replacing postulates 4_0 through 4_4 with the following single statement:

POSTULATE 4'''. $\mathfrak{I}$ *is a commutative group with respect to* $+$.

(It was our wish to easily identify translations with structures such as that of a group which prompted us to include the "closure" properties of 4_0 among our postulates.)

The principal advantage, from a practical point of view, of formally identifying the structure of a commutative group is that we are able to make use of our knowledge of theorems about the additive group of real numbers—or, for that matter, *any* commutative group—to suggest analogous theorems about translations. In addition to this practical reason, it also enables us to introduce the student to this very important mathematical structure in a relatively down-to-earth context related to notions with which he has become familiar.

Getting real numbers into the act

It is possible to give a natural interpretation to the product of a translation by a real number, and in doing so, it is easy to develop intuitions for several additional postulates for the formal system we are constructing to describe the space in which we live. For example, since it was our intention to study the (Euclidean) geometry of space, we can look at what our postulate system, in its current state of development, tells us about some familiar objects in geometry. One question we can raise is, Do we have enough in our system to enable us to describe lines as they probably appear to us? At this point, the best we can do is to look at the images of a given translation related to a given point. Consider, for instance, the non-$\mathbf{0}$ translation $\mathbf{a}$ and point A pictured in figure 8.10. The points that can reach A, together with those that can be reached from A, by successive applications of $\mathbf{a}$ clearly belong to the line through A and $A + \mathbf{a}$ and, just as clearly, do not constitute all the points of that line. We could reach more of the points of the line in question if instead of $\mathbf{a}$, we used the translation that moves points in the same direction that $\mathbf{a}$ does but only, say, half as far. We could use $\mathbf{a}\dfrac{1}{2}$ to designate such a translation, but this still will not do the job; we need more. What we really need is a way to bring into the act *all* such real multiples of $\mathbf{a}$. To do so is to accept the notion that $\mathbf{a}t$ is a translation for any real number t.

We can make use of this intuitive notion of multiplying translations by real numbers in conjunction with our convention for picturing translations to strengthen our intuitions on how this multiplication works. For example, given the translation $\mathbf{a}$ pictured in figure 8.11, we note that $\mathbf{a}2 + \mathbf{a}\dfrac{3}{2}$ is the same translation as $\mathbf{a}\dfrac{7}{2}$, that is, $\mathbf{a}\left(2 + \dfrac{3}{2}\right)$, for the arrows drawn to represent these translations are the same length and point in the same sense.

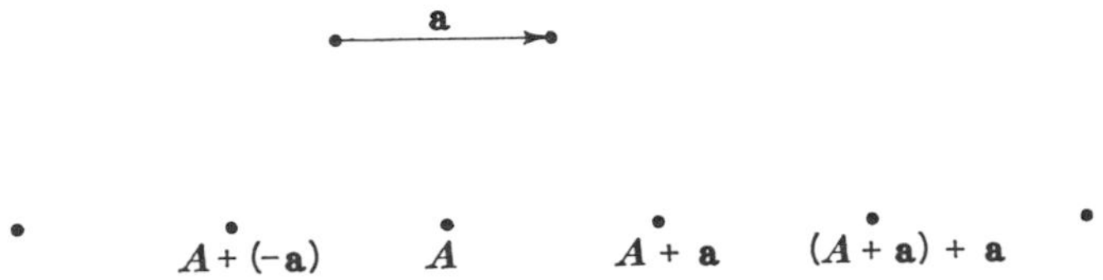

$$\bullet \!\!\xrightarrow{\hspace{1.2cm}\mathbf{a}\hspace{1.2cm}}\!\!\bullet$$

$$\bullet \qquad \overset{\bullet}{A + (-\mathbf{a})} \qquad \overset{\bullet}{A} \qquad \overset{\bullet}{A + \mathbf{a}} \qquad \overset{\bullet}{(A + \mathbf{a}) + \mathbf{a}} \qquad \bullet$$

Fig. 8.10

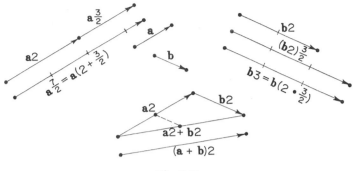

Fig. 8.11

For the same reason, it is evident that the $\frac{3}{2}$ multiple of **b2** is the same

translation as the $2 \cdot \frac{3}{2}$ multiple of **b**, that is, that $(\mathbf{b}2)\frac{3}{2} = \mathbf{b}\left(2 \cdot \frac{3}{2}\right)$.
And it is equally evident that $\mathbf{a}2 + \mathbf{b}2$ is the same translation as $(\mathbf{a} + \mathbf{b})2$.
Motivated by intuitions such as these, we add the following postulates to
our system:

POSTULATE 4_0.   (d) $\mathbf{a}t \in \mathfrak{z}$

POSTULATE 4_5. $\mathbf{a}1 = \mathbf{a}$

POSTULATE 4_6. $\mathbf{a}(s + t) = \mathbf{a}s + \mathbf{a}t$

POSTULATE 4_7. $(\mathbf{a} + \mathbf{b})t = \mathbf{a}t + \mathbf{b}t$

POSTULATE 4_8. $(\mathbf{a}t)s = \mathbf{a}(ts)$

Given these postulates, some would say that the group of translations
admits the real numbers as operators. (Formally, a group is said to admit a
field as operators when the group and field elements interact in such a way
as to satisfy the postulates $4_0(d)$ through 4_8. Sometimes the elements of
the field are called *scalars,* and the multiplication described in these postulates is called *scalar multiplication.* In the treatment of Euclidean geometry
being discussed here, we shall eventually make use of the *complete ordered
field* of real numbers. In a mathematically tight development, the field
structure of the real numbers would be developed prior to the introduction
of these operator postulates. Of course, students engaged in a study of
geometry from this point of view will bring to the course knowledge of the
algebra of real numbers and will therefore be familiar with the field properties, in principle if not in name.) Now, a commutative group that
admits the real numbers as operators is said to be *a vector space over the
real numbers,* and the elements of the commutative group that admits the
field of operators are called *vectors.* So, by adopting postulates $4_0(d)$

through 4_8, we have enriched the formal operational structure of translations to that of a vector space over the real numbers, and in this context the translations we have been discussing may properly be called *vectors*.[2] **This** suggests that after we become familiar with the basic properties of **this** enriched mathematical structure, it is again reasonable to condense our list of postulates about translations alone:

POSTULATE 4″. ℑ *is a vector space over the real numbers.*

(Note that we might have come to this point in the mathematical development of the concept of a vector space by simply concentrating on formulating properties of translations. Doing this would give us a system in which the point-translation relations described in postulates 0 through 3 are left hovering in the background. Whether or not we choose to formulate these properties in such a way as to become a formal part of the system, they will be used either implicitly or explicitly in generating geometric notions with vector methods. The choice to bring these properties out into the open in this discussion was dictated both by the objective to give a plausible treatment of geometry using vector methods—and thus not treading on the dangerous ground of leaving important notions unmentioned—and by successful experiences in using this sort of development with high school students and teachers. That others who wish to develop geometry with vector methods may choose not to bring out these relations is clear. The field is wide open to experimentation with alternate developments.)

Getting geometry formally into the act

As we saw earlier, it was not possible for us to describe a line—at least, not in a way that agreed satisfactorily with our intuitions about lines—in terms of a given translation without having real numbers as operators on translations. Before we formalize the notion of a line as well as other important and interesting subsets of our space of points, it will be both

---

2. The scope of what is said here may take some time to settle in place. Notice that we have *not* said that a vector is an ordered pair of real numbers or a directed line segment (or any other of the popular descriptions). We are simply saying that a vector is an element of a vector space; so the crux of understanding what vectors are is that of understanding what a vector space is. From this point of view it is more appropriate to say that an ordered pair is a vector *provided* the set of all those ordered pairs is furnished with a structure—for example, binary operation, field of operators, and so on— which satisfies postulates like those in the various parts of postulate 4″. It is interesting to note that when one says that a vector is a directed line segment and describes the appropriate addition and scalar multiplication operations to generate the properties of a vector space, it turns out that the elements of the vector space so generated are *not* the directed line segments one started with but rather are equivalence classes of so-called equipollent directed line segments. So, the things (i.e., directed line segments) that were originally called *vectors* are not even elements of the vector space being discussed!

informative and helpful to look into some intuitive notions about collinear points and coplanar points in order to see what information we can gather about translations from these notions. First, consider three points, A, B, and C. (See fig. 8.12.) Intuitively, these points are collinear when there is a line that contains them. Alternately, they are collinear when one of the translations determined by two of the points, $C - A$ for example, is a multiple of another such translation, say, $B - A$. Other ways of saying that $C - A$ is a multiple of $B - A$ are to say that

$$C - A \text{ is linearly dependent on } B - A$$

or that

$$C - A \text{ is a linear combination of } B - A$$

or that

$$C - A \text{ and } B - A \text{ are linearly dependent.}$$

$$\overset{\textstyle C-A \qquad B-A}{\underset{C \qquad\quad A \qquad\qquad B}{\longleftarrow\!\bullet\!\longrightarrow}}$$

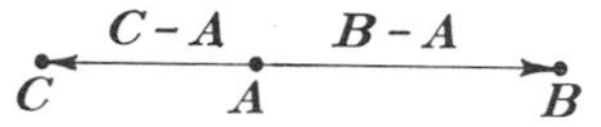

Fig. 8.12. C is collinear with A and B whenever $C - A$ is a multiple of $B - A$.

Next, consider three points, A, B, and C, that are noncollinear. Intuitively, it is evident that there is exactly one plane containing these three points, and furthermore, it is just as evident (see fig. 8.13) that one can reach any point D in this plane by a translation from A that is composed of some multiple of $B - A$ and some multiple of $C - A$. Other ways of saying that D is contained in the plane of A, B, and C are to say that

$$D - A \text{ is linearly dependent on } B - A \text{ and } C - A$$

or that

$$D - A \text{ is a linear combination of } B - A \text{ and } C - A$$

or that

$$D - A, B - A, \text{ and } C - A \text{ are linearly dependent.}$$

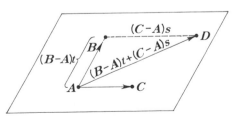

Fig. 8.13. D is coplanar with A, B, and C whenever $D - A$ is a linear combination of $B - A$ and $C - A$.

Motivated by considerations such as these, we find that it is not difficult to arrive at a formulation of the concept of linearly dependent translations. In general, we say that translations (vectors) $\mathbf{a}_1$, $\mathbf{a}_2$, . . . , $\mathbf{a}_n$ are linearly dependent if and only if there are numbers x_1, x_2, . . . , x_n, *not all zero*, such that $\mathbf{a}_1 x_1 + \mathbf{a}_2 x_2 + \ldots + \mathbf{a}_n x_n = 0$. To see that this is in agreement with what was said before, suppose that $\mathbf{c} = \mathbf{a}2 + \mathbf{b}3$, as shown in figure 8.14. Then, it is clear just from the vector algebra we have accepted in our system that $\mathbf{a}2 + \mathbf{b}3 + \mathbf{c} \cdot -1 = 0$. And since $-1 \neq 0$ (or since $2 \neq 0$ or $3 \neq 0$), it follows by definition that $\mathbf{a}$, $\mathbf{b}$, and $\mathbf{c}$ are linearly dependent.

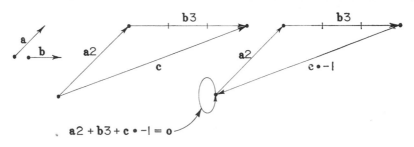

Fig. 8.14

Also, we say that translations (vectors) are *linearly independent* if and only if they are *not linearly dependent*. For an algebraic criterion for identifying linearly independent vectors, it should be clear that vectors $\mathbf{a}_1$, $\mathbf{a}_2$, . . . , $\mathbf{a}_n$ are linearly independent if and only if the *only* way to satisfy the equation

$$\mathbf{a}_1 x_1 + \mathbf{a}_2 x_2 + \ldots + \mathbf{a}_n x_n = 0$$

is to have each of x_1, x_2, . . . , x_n be zero. Otherwise, the vectors $\mathbf{a}_1$, $\mathbf{a}_2$, . . . , $\mathbf{a}_n$ are linearly dependent. This algebraic criterion for linear independence of vectors will be most useful to us in arriving at some of the more important relations in geometry.

We are now in a position to acquire an understanding of a significant amount of geometry within the framework of the formal system we have developed. Just as we used intuitions about space to arrive at the properties of translations we formulated in the postulates, we shall make use of such intuitions to develop some definitions of geometric notions. This strategy will give us strong and useful ties between our intuitive (and thus informal) notions about geometry and the formal mathematical system described by our postulates. These ties are important for a student both in rendering interpretations to statements that occur as conclusions of arguments and in gaining confidence in the use of the formal system to solve problems about the system.

Lines and planes

Given a point A and translation $B - A$, we consider the images of A under multiples of this translation. As we saw earlier, each multiple of $B - A$ gives rise to a point C that we would agree is collinear with A and B and to a translation $C - A$ that is linearly dependent on $B - A$. A definition of collinearity arises out of considerations such as this. We say that A, B, and C are collinear if and only if $B - A$ and $C - A$ are linearly dependent. In these terms, we say that a line is a set of points consisting of two distinct points together with all points that are collinear with those two points. See figure 8.15. With this definition, it is easy to establish that the line AB, through two points A and B, is the set of all points X for which there is a real number x such that

$$X = A + (B - A)x.$$

Similarly, a definition of coplanarity arises out of considering translations $D - A$ that are linearly dependent on $B - A$ and $C - A$ determined by the points A, B, and C. We say that A, B, C, and D are coplanar if and only if $B - A$, $C - A$, and $D - A$ are linearly dependent. In these terms, we say that a plane is a set of points consisting of three noncollinear points together with all points that are coplanar with those three points. See figure 8.16. As in the case with lines, it is easy to establish from this definition that the plane ABC, through three noncollinear points A, B, C, is the set of all points X for which there are real numbers x and y such that

$$X = A + (B - A)x + (C - A)y.$$

(Note the similarity of this equation to that of the line AB.)

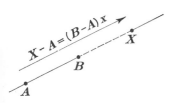

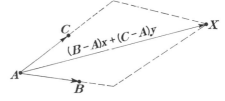

Fig. 8.15. X is in line AB if and only if $X = A + (B - A)x$.

Fig. 8.16. X is in plane ABC if and only if $X = A + (B - A)x + (C - A)y$.

Notice that it is now quite easy to show that if P and Q are two points of the plane ABC, then the line PQ is a subset of this plane. We do this as follows:

Given the plane ABC, we know that A, B, and C are noncollinear. For brevity, let $\mathbf{b} = B - A$ and $\mathbf{c} = C - A$. Then, if P and Q belong to the plane ABC, it follows that

$$P - A = \mathbf{b}p_1 + \mathbf{c}p_2 \quad \text{and} \quad Q - A = \mathbf{b}q_1 + \mathbf{c}q_2$$

for some real numbers p_1, p_2, q_1, and q_2. Therefore, since $Q - P = (Q - A) - (P - A)$ (by postulate 3 and the usual definition of subtraction of vectors),

$$Q - P = (\mathbf{b}q_1 + \mathbf{c}q_2) - (\mathbf{b}p_1 + \mathbf{c}p_2)$$
$$= \mathbf{b}(q_1 - p_1) + \mathbf{c}(q_2 - p_2).$$

Let S be any point of line PQ. Then $S = P + (Q - P)s$, for some real number s. Therefore, S belongs to line PQ if and only if, for some real number s,

$$S = P + [\mathbf{b}(q_1 - p_1) + \mathbf{c}(q_2 - p_2)]s$$

and, since $P = A + \mathbf{b}p_1 + \mathbf{c}p_2$, if and only if

$$S = A + \mathbf{b}[p_1 + (q_1 - p_1)s] + \mathbf{c}[p_2 + (q_2 - p_2)s].$$

By this last equation, we see that S belongs to the plane ABC. Therefore, if S belongs to the line PQ, then S belongs to the plane ABC. Hence, if P and Q belong to the plane ABC, then so does each point of the line PQ.

(One reason that this theorem was selected for discussion is that it is usually taken as a postulate in a formal development of three-dimensional Euclidean geometry. By the time we complete the formulation of the vector space "machinery" with which we are developing Euclidean geometry, we shall either have proved, or be in a position to prove, all the usual postulates for what is known as a Euclidean metric space. Thus we shall have the mathematical machinery to do *all* that could have been done in a standard geometry course, and we shall have done much more than become familiar with some formal aspects of geometry. We shall have become thoroughly familiar with the basic machinery of several fundamental mathematical structures (e.g., group, field, vector space, inner-product space), and through these structures, we shall have seen the close interrelationship between algebra and geometry and, if we follow through completely, trigonometry. It is these factors that make this type of approach to the teaching of geometry so appealing from almost any viewpoint, be it pedagogical, psychological, mathematical, or practical.)

Parallelism

We speak of the set of all linear combinations, that is, multiples, of a vector as the *direction* of that vector. Also, we may speak of the set of all linear combinations of a collection of vectors as the direction of that collection. In this context, we speak of the direction of a line as the direction of any translation determined by two points of that line (and, therefore, of any non-**0** translation that maps the line onto itself). Similarly, we speak of the direction of a plane as the direction of any two linearly independent translations determined by three noncollinear points of that plane.

Parallel lines are defined as lines that have the same direction; similarly, parallel planes are defined as planes that have the same direction. Clearly, then, a line is parallel to itself, and a plane is parallel to itself. A moment's thought about the tracing-sheet model should reveal that these definitions are in complete agreement with what one "sees" in the model and, in addition, provide one with an entry into the vector-space structure to deal with problems concerning parallelism. The proof that two parallel lines have an empty intersection illustrates quite nicely the power provided us by the established vector algebra:

Suppose that l_1 and l_2 are two lines with the same direction, say, that of **a**. Since $l_1 \neq l_2$—they are two lines—there is a point, say, L_1, on l_1 that is not on l_2. Let L_2 be any point of l_2. Then $L_1 \neq L_2$, since L_1 is not on l_2. Suppose that there is a point P in the intersection of l_1 and l_2. Then

$$P = L_1 + \mathbf{a}p \quad \text{and} \quad P = L_2 + \mathbf{a}q$$

for some p and q. But this means that

$$L_1 + \mathbf{a}p = L_2 + \mathbf{a}q$$

or, equivalently, that

$$L_1 = L_2 + \mathbf{a}(q - p).$$

By this last equation, L_1 is a point of l_2. This is a contradiction, for L_1 was chosen so that it was not on l_2. Hence, the intersection of l_1 and l_2 is empty. In other words, two parallel lines have an empty intersection.

The proof that two parallel planes have an empty intersection is completely similar. (It is interesting to note that at this stage in the formal development, we are not prepared to handle the converse of either of these theorems. The converse of the first is, of course, only a theorem in geometries of two or fewer dimensions, since skew lines exist in three or more dimensions. The converse of the second is only a theorem in geometries of three or fewer dimensions, since skew planes exist in higher dimensions. The formulation of the notion of dimension will be discussed later in this chapter.)

Segments and angles

Having defined lines and planes, we can see that the concepts of certain of their subsets, such as segment, ray, angle, triangle, and quadrilateral, arise in a natural way. For example, the segment with endpoints A and B—for short, segment AB—is the set of all points X for which there is a real number x, $0 \leq x \leq 1$, such that $X = A + (B - A)x$. In these terms, the triangle with vertices A, B, and C—for short, triangle ABC—is the union of the segments whose endpoints are the three noncollinear points A, B, and C. For another example, the ray from A through B—for short, ray AB—is the set of all points X for which there is a real number x, $0 \leq x$, such that $X = A + (B - A)x$. In these terms, angle BAC is the union of *noncollinear* rays AB and AC.

Given an angle BAC, with P between A and C and with Q between A and B, it is a somewhat difficult problem in a standard development of geometry to prove that the intervals BP and CQ intersect. And if one manages to accomplish the feat of proving that these intervals do intersect, he is still left with the lingering problem of just where they do intersect. It is, of course, intuitively obvious that the intervals in question really do intersect, and in the usual high school course it is taken for granted if it is considered at all. But with the vector methods we have at our disposal, we find that the solution of this problem is simply an exercise in elementary vector algebra that, when solved, will give us the additional information of precisely where the intervals intersect as a function of where P and Q are located on their respective rays. One solution to the problem, using the notation established in figure 8.17, follows:

> Note that R belongs to both intervals BP and CQ if and only if, for some real numbers $0 < b < 1$ and $0 < c < 1$,
> $$R = B + (P - B)b \quad \text{and} \quad R = C + (Q - C)c,$$
> that is,
> $$B + (P - B)b = C + (Q - C)c,$$
> or, equivalently,
> $$(B - C) + (P - B)b + (C - Q)c = 0.$$

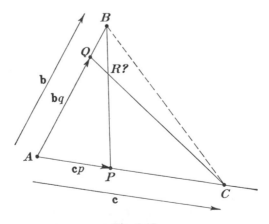

Fig. 8.17

(Look at fig. 8.17 and note that $(B - C) + (R - B) + (C - R) = 0$ to see why the last two equations are equivalent.) Our job will be finished if we can establish that such numbers b and c exist. Now in terms of **b** and **c**, the last equation tells us that

$$(\mathbf{b} - \mathbf{c}) + (\mathbf{c}p - \mathbf{b})b + (\mathbf{c} - \mathbf{b}q)c = 0,$$

so that

$$\mathbf{b}(1 - b - qc) + \mathbf{c}(-1 + pb + c) = 0.$$

Since **b** and **c** are linearly independent (otherwise, the rays AB and AC are collinear!), it follows that

$$1 - b - qc = 0 \quad \text{and} \quad -1 + pb + c = 0,$$

so that

$$b = \frac{1 - q}{1 - pq} \quad \text{and} \quad c = \frac{1 - p}{1 - pq}.$$

Note that the initial conditions on P and Q gave $0 < p < 1$ and $0 < q < 1$, so that both $0 < b < 1$ and $0 < c < 1$. Hence, intervals BP and CQ do intersect. Furthermore, given p and q, we know that the point of intersection is $B + (P - B)\left(\dfrac{1 - q}{1 - pq}\right)$, and also $C + (Q - C)\left(\dfrac{1 - p}{1 - pq}\right)$! So, not only have we established that the intervals do intersect, but we know just where they intersect.

Notice that we have a complete solution to the problem. For given values of p and q, it is now a simple matter of arithmetic to tell the precise location of the point R. For example, given that $p = \dfrac{1}{3}$ and $q = \dfrac{1}{4}$, it follows that $b = \dfrac{9}{11}\left(\text{and that } c = \dfrac{8}{11}\right)$, so that R is the point $B + (P - B)\dfrac{9}{11}$ —that is, R is $\dfrac{9}{11}$ of the way from B to P. (Later we shall have neater ways of saying this same thing.) Furthermore, if we remove the initial restrictions on the values of p and q, we see that the same work may be used to tell us under what conditions the lines BP and CQ will meet. And even the natural restrictions on the values of p and q that one obtains when solving for b and c (for example, $pq \neq 1$) may be studied for geometric significance. If $pq = 1$, then $q = 1/p$, from which it follows that lines BP and CQ are parallel. In the case where $p = 1$, we have $q = 1$, and lines BP and CQ coincide; so there is no *unique* point R of intersection. In the case where $p \neq 1$, we have $q \neq 1$, and lines BP and CQ are disjoint.

Another observation that can be made about this situation is that the line PQ is parallel to the line BC if and only if $p = q \neq 0$. This is easy to see, for the direction of the line BC is that of $\mathbf{c} - \mathbf{b}$, and the direction of line PQ is that of $\mathbf{c}q - \mathbf{b}p$. So, the lines in question are parallel if and only if $\mathbf{c}q - \mathbf{b}p$ is some non-0 multiple of $\mathbf{c} - \mathbf{b}$; that is, if and only if, for some $t \neq 0$,

$$\mathbf{c}q - \mathbf{b}p = (\mathbf{c} - \mathbf{b})t.$$

This is equivalent to

$$\mathbf{b}(t - p) + \mathbf{c}(q - t) = 0$$

and, since $\mathbf{b}$ and $\mathbf{c}$ are linearly independent, to

$$t = p \qquad \text{and} \qquad t = q,$$

so that $p = q \neq 0$.

Ratios and points of division

In order to simplify discussions of results obtained in working with segments, triangles, and the like, it is convenient to introduce the notions of ratio and point of division in terms of translations. We know that each point X of a line AB is such that $X = A + (B - A)x$, for some real number x. We say that the point X divides the segment from A to B in the ratio x to $1 - x$. This can be made plausible by thinking about the ratio of the *directed trip* from A to X to the *directed trip* from X to B as being the ratio in which X divides the segment from A to B. For example, if $X = A + (B - A)3$, as shown in figure 8.18, then

the ratio of the directed trip from A to X (3) to the directed trip from X to B (-2) is $3:-2$,

and we say that

X divides the segment from A to B in the ratio $3:(1 - 3)$, that is, $3:-2$.

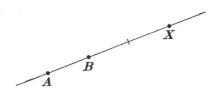

Fig. 8.18

For a second example, in the problem on angles related to figure 8.17, we found that when $p = \dfrac{1}{3}$ and $q = \dfrac{1}{4}$, the intervals BP and CQ intersect in the point R where $R = B + (P - B)\dfrac{9}{11}$ $\left(\text{and } R = C + (Q - C)\dfrac{8}{11} \right)$. In this case, we say that R divides the segment from B to P in the ratio $9:2$, that is, $\frac{9}{11} : \frac{2}{11}$, and that R divides the segment from C to Q in the ratio 8:3,

that is, $\frac{8}{11} : \frac{3}{11}$.

The consideration of ratios and of points of division gives rise to notions of the midpoint of a segment, the median of a triangle, the proportionality of segments intercepted by parallel lines, and the like. The midpoint M of a segment AB, for example, may be defined as the point that divides the segment from A to B in the ratio 1:1 so that $M - A = B - M$. This is equivalent to saying that $M = A + (B - A)\frac{1}{2}$, and both descriptions should agree with one's intuitive feeling about midpoints.

A median of a triangle is a segment whose endpoints are a vertex of the triangle and the midpoint of the side opposite that vertex. From what we learned about figure 8.17, we see that any two medians of a triangle intersect in a point that divides each median from vertex to side in the ratio 2:1. $\left(\text{Choose } p = q = \frac{1}{2} \text{ so that } b = c = \frac{2}{3}.\right)$ The result of this is that the medians of a triangle are concurrent in a point that divides each of them from vertex to side in the ratio 2:1.

Before moving on to the task of extending the formal system we are developing, we shall consider one more geometric situation to see what we can learn from it. Suppose that segments AB and CD are both parallel and noncollinear; that is, they are not contained in the same line. One question that can be raised is, Do the lines AD and BC have any points in common? One solution to this question is the following, in which the notation suggested in figure 8.19 is used:

We know that a point P belongs to both line AD and BC if and only if, for some real numbers p and q,

$$P = A + (D - A)p \quad \text{and} \quad P = B + (C - B)q,$$

that is,

$$P - A = (D - A)p \quad \text{and} \quad P - B = (C - B)q.$$

Since $(P - A) - (P - B) = B - A$, the latter is equivalent to saying that

$$B - A = (D - A)p - (C - B)q,$$

or, better, that

$$(B - A) + (C - B)q + (A - D)p = 0.$$

Now in terms of **b** and **c**, this means that

$$\mathbf{b} + (\mathbf{c} - \mathbf{b})q + (-\mathbf{b}d - \mathbf{c})p = 0,$$

so that

$$\mathbf{b}(1 - q - dp) + \mathbf{c}(q - p) = 0.$$

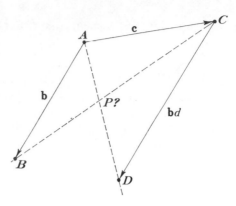

Fig. 8.19

Since **b** and **c** are linearly independent, it follows that

$$1 - q - dp = 0 \quad \text{and} \quad q - p = 0,$$

or, equivalently, that

$$p = \frac{1}{1+d} = q, \quad (d \neq -1).$$

Hence, we see that lines AD and BC have a point in common whenever $d \neq -1$. Furthermore, since $p = q$, this point divides the segments from A to D and from B to C in the same ratio, $1:d$.

We have, as we had with the geometric situation related to figure 8.17, a complete solution to the problem at hand. Not only do we know precisely when the lines in question meet, but we also know precisely where they meet in terms of d. For $d = 1$, we see that P is the midpoint of both segments AD and BC and that $ABDC$ is a parallelogram. In fact, it is easy to use this result, together with the definition of a parallelogram, to establish that a quadrilateral is a parallelogram if and only if its diagonals bisect each other. Further, we see that for any positive value of d, $\dfrac{1}{1+d}$ is between 0 and 1. Geometrically speaking, this tells us that a quadrilateral is a trapezoid if and only if its diagonals divide each other in the same ratio. (Since $1 > 0$, this suggests that a parallelogram is fair game to consider as a special kind of trapezoid, a not unreasonable suggestion supported by many mathematicians, among them E. Moise and H. Vaughan.) As a matter of fact, as illustrated in figure 8.20, geometric interpretations may be given for each of the various kinds of values for d, and even for $d = -1$, in which case it turns out that $ABCD$ is a parallelogram.

So far, we have discussed the use of vector methods to obtain results

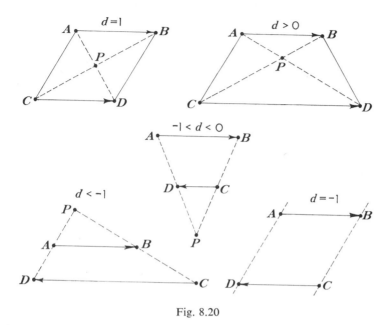

Fig. 8.20

in geometry that have to do solely with parallelism and ratio. Such matters are in the domain of what is called *affine geometry,* and our present postulate system is adequate for a thorough study of that subject. In order to extend our study to that of Euclidean geometry, we need to extend our formal structure to bring in matters concerned with perpendicularity and distance. To do this, we shall continue with the strategy of discussing intuitions about the matters to be formalized in order to build and maintain strong ties between the formal system being developed and the intuitively based model of the vector space of translations acting on points. But before we discuss a means to do this, we should do something about formalizing the notion of dimension for our space. And while we are doing this, we shall also bring into the act some work with coordinate systems, not because such work is important to this particular development of geometry, but rather because it will help to strengthen the already-established ties between algebra and geometry and will give us yet another means by which we may study properties of geometric figures.

The notion of dimension

In our intuitive discussion about ε and 3, we have kept alive the notions that ε is the set of points that we normally think of as the space in which we live and that 3 is the set of all translations acting on this space of points. Our original intention was to build a postulate system that would in a formal way describe $\mathfrak{I}$ and $\mathcal{E}$. Consequently, in our *formal* development of geometry, $\mathfrak{I}$ and $\mathcal{E}$ are simply sets that satisfy the postulates. It turns out that there are many ways of satisfying the present postulates. For example, if $\mathfrak{I}$ contains just the identity translation **0** and if $\mathcal{E}$ consists of a single point, then each of the present postulates is satisfied, and thus the present postulates describe this small, trivial, and generally uninteresting system. Further, if $\mathfrak{I}$ contains a non-**0** translation **a** and $\mathcal{E}$ contains a point A, then $\mathfrak{I}$ contains each translation that maps the line through A and $A + \mathbf{a}$ onto itself and $\mathcal{E}$ contains the points of that line, so that the present postulates describe the system where $\mathfrak{I}$ contains one, but not two, linearly independent vectors and where $\mathcal{E}$ consists of the points of a line in the direction of that vector. (A geometry such as this would be called a *one-dimensional* geometry.) Also, the present postulates describe the system where $\mathfrak{I}$ contains two, but not three, linearly independent vectors and where $\mathcal{E}$ consists of the points of a plane in the direction of those two vectors. (A geometry such as this is called a *two-dimensional,* or *plane,* geometry.) In general, the present postulates describe any system where $\mathfrak{I}$ contains n, but not $n + 1$, linearly independent vectors and where $\mathcal{E}$ consists of the points of a set that is in the direction of those n vectors, thus they can be used to describe a so-called *n-dimensional* geometry. Evidently, to describe the space we have in mind, namely, our three-dimensional space, we must add to our postulates. Let us add the following ones:

POSTULATE 4_9. $\mathfrak{I}$ *contains three linearly independent vectors.*

POSTULATE 4_{10}.   *Any four vectors in* $\mathfrak{I}$ *are linearly dependent.*

Postulate 4_9 tells us that $\mathfrak{I}$ is *at least* three-dimensional, and postulate 4_{10} tells us that $\mathfrak{I}$ is *at most* three-dimensional; together, they tell us that $\mathfrak{I}$ is three-dimensional. With this in mind, we may summarize our postulates about $\mathfrak{I}$ with the following:

POSTULATE 4′. $\mathfrak{I}$ *is a three-dimensional vector space over the real numbers.*

One theorem that is a consequence of this extended postulate system is that there are four noncoplanar points. Another is that if the intersection of two planes is empty, then the planes are parallel. The latter is the converse of a theorem established earlier, and with it, we now know that the space(s) described by our postulates cannot contain skew planes, that is, nonparallel planes that do not intersect. We do, however, have skew lines. Furthermore, the geometry now described by our postulates is called *three-dimensional,* or *three-space* geometry.

As mentioned earlier, it is not necessary to this development of geometry to go into the matter of coordinate systems. As a matter of fact, some would say that doing so takes us away from the beauty and rigor of pure geometry. (Of course, this whole development is doing this in the eyes of many!) We take the position that the more we are able to provide in the way of insight into the nature of the space we are studying and the more that we can provide the student in the way of concrete experience in associating one "structured" space with another as well as the more mathematical tools with which we can provide him to carry on the study, the better off we leave him in the long run. Therefore, we shall now take a side trip into an initial study of coordinate systems. For those who do not choose to delve into these matters, there will be no loss of continuity of the development of the postulate system by omitting the next section.

An introduction to coordinate systems

Given that **a**, **b**, and **c** are three linearly independent vectors in $\mathfrak{z}$, it is not difficult to show that each vector in $\mathfrak{z}$ may be uniquely expressed as a linear combination of **a**, **b**, and **c**. That is, given any **d**, there is a unique triple of real numbers (d_1, d_2, d_3) such that

$$(1) \qquad \mathbf{d} = \mathbf{a}d_1 + \mathbf{b}d_2 + \mathbf{c}d_3.$$

(That there is one such triple of numbers for a given **d** follows from postulates 4_9 and 4_{10}. That this triple of numbers is unique follows from the linear independence of **a**, **b**, and **c**.) Therefore, it is easy to see that the matching

$$\mathbf{d} \leftrightarrow (d_1, d_2, d_3)$$

determined by (1) is a one-to-one correspondence between $\mathfrak{z}$ and the set of all ordered triples of real numbers. (In the same way, a one-to-one correspondence may be established between the set of translations on an n-dimensional space and ordered n-tuples of real numbers. One simply must be guaranteed the existence of n, but not $n + 1$, linearly independent vectors—translations—in order to carry out the details of the argument.)

Also, given a point, say, O, there is a unique translation **d** that maps O onto any given point D. That is, for any point D, there is a unique translation **d** such that

$$(2) \qquad D = O + \mathbf{d}.$$

Therefore, it is easy to see that the matching

$$D \leftrightarrow \mathbf{d}$$

determined by (2) is a one-to-one correspondence between $\mathcal{E}$ and $\mathfrak{I}$. By composing the two one-to-one correspondences that we have established here, we obtain the one-to-one correspondence

$$D \leftrightarrow (d_1, d_2, d_3)$$

between the set $\mathcal{E}$ of points and the set $\mathfrak{R}_3$ of all ordered triples of real numbers. It should be clear that this one-to-one correspondence between $\mathcal{E}$ and $\mathfrak{R}_3$ is a function of both the basis $(\mathbf{a}, \mathbf{b}, \mathbf{c})$— that is, the ordered triple of linearly independent vectors $\mathbf{a}$, $\mathbf{b}$, and $\mathbf{c}$—and the point O. Any such one-to-one correspondence is called *a Cartesian coordinate system*. (Note that one has a rectangular Cartesian coordinate system if one chooses a basis whose terms are mutually orthogonal. This amounts to choosing a point as origin and three mutually perpendicular lines through that point as axes. We shall be in a position to do this when we discuss the notions of orthogonality of vectors and perpendicularity of lines later in this chapter.)

Choosing an origin and introducing a Cartesian coordinate system on the space of points $\mathcal{E}$ enables us to deal with planes and lines in terms of equations. For example, recall that the line AB is the set of all points X for which there is a real number t such that

$$X = A + (B - A)t.$$

Given that A and B have coordinates (a_1, a_2, a_3) and (b_1, b_2, b_3) respectively in the given coordinate system, it is easy to see that $A = O + \mathbf{a}a_1 + \mathbf{b}a_2 + \mathbf{c}a_3$ and that $B = O + \mathbf{a}b_1 + \mathbf{b}b_2 + \mathbf{c}b_3$, and so for any real number t, $(B - A)t = \mathbf{a}(b_1 - a_1)t + \mathbf{b}(b_2 - a_2)t + \mathbf{c}(a_3 - b_3)t$. Consequently, it is easy to see that if X has coordinates (x_1, x_2, x_3), then X belongs to the line AB if and only if, for some t,

$$(3) \qquad \left\{ \begin{array}{l} x_1 = a_1 + (b_1 - a_1)t \\ x_2 = a_2 + (b_2 - a_2)t \\ x_3 = a_3 + (b_3 - a_3)t. \end{array} \right.$$

Of course, (3) is easily recognized as parametric equations for the line AB.

It is interesting to note that the numbers $b_1 - a_1$, $b_2 - a_2$, and $b_3 - a_3$ characterize the translation $B - A$ in the sense that they are the components of a vector that is a basis for the direction of the line AB and, as such, determine the direction of that line. They are, in fact, sometimes called *direction numbers* of the line AB. Clearly, any triple of numbers (p_1, p_2, p_3) that is proportional to $(b_1 - a_1, b_2 - a_2, b_3 - a_3)$ forms the components of a vector that is a multiple (or linear combination) of $B - A$ and thus may be used as direction numbers for a line parallel to line AB. This notion gives us an easy procedure both for writing parametric equations of lines through given points in given directions and for determining whether or not two sets of parametric equations describe parallel lines. For example, the line l_1 through the point with coordinates $(3, -2, 1)$ and with direction numbers $(5, 7, -4)$ is given as follows:

$$l_1: \begin{cases} x_1 = 3 + 5t \\ x_2 = -2 + 7t \\ x_3 = 1 - 4t \end{cases} \qquad l_2: \begin{cases} x_1 = -6 + 10t \\ x_2 = 4 + 14t \\ x_3 = -2 - 2t \end{cases}$$

$$l_3: \begin{cases} x_1 = 5 + 10t \\ x_2 = 1 + 14t \\ x_3 = -3 - 8t \end{cases}$$

Furthermore, from the given parametric equations we see that l_1 and l_3 are parallel because the direction numbers $(5, 7, -4)$ of l_1 are proportional to the direction numbers $(10, 14, -8)$ of l_3, but neither l_1 nor l_3 is parallel to l_2 because $(5, 7, -4)$ is *not* proportional to $(10, 14, -2)$.

In a completely analogous way, we see that if point C with coordinates (c_1, c_2, c_3) is noncollinear with A and B, then X belongs to the plane ABC if and only if, for some numbers t and s,

$$(4) \qquad \begin{cases} x_1 = a_1 + (b_1 - a_1)t + (c_1 - a_1)s \\ x_2 = a_2 + (b_2 - a_2)t + (c_2 - a_2)s \\ x_3 = a_3 + (b_3 - a_3)t + (c_3 - a_3)s. \end{cases}$$

Equations (4) are easily recognized as parametric equations for plane ABC. As in the case of lines, the numbers $b_i - a_i$ and $c_i - a_i$ ($i = 1, 2, 3$), are the direction numbers of the two linearly independent basis vectors $B - A$ and $C - A$ for the plane. The question of the recognition of pairs of parametric equations that describe parallel planes is a bit more difficult than the corresponding problem for lines, but it is conceptually the same problem, namely, that of determining when two bases for planes determine the same (plane) direction. It turns out that if we solve the problem of making use of parametric equations of a plane to write a single equation that describes that plane, we shall find an ordered triple of numbers that, in a sense, characterizes the direction of the plane. Of course, the triple of numbers so obtained will be functions of the direction numbers of basis vectors for the plane but, interestingly, is virtually independent of the choice of basis vectors for the plane. To see that this is so, we solve the problem. As we proceed, it is worth noting that such work done with equations of lines and planes is reminiscent of solid analytical geometry.

Consider the parametric equations

$$\begin{cases} x_1 = a_1 + p_1 t + q_1 s \\ x_2 = a_2 + p_2 t + q_2 s \\ x_3 = a_3 + p_3 t + q_3 s \end{cases}$$

for the plane through the point A with coordinates (a_1, a_2, a_3) and in the direction of linearly independent vectors $\mathbf{p}$ and $\mathbf{q}$ with components (p_1, p_2, p_3) and (q_1, q_2, q_3), respectively. To say that $\mathbf{p}$ and $\mathbf{q}$ are independent is to say that (p_1, p_2, p_3) is not proportional to (q_1, q_2, q_3), and to say the latter is to say that *not all* the determinants $\begin{vmatrix} p_1 & p_2 \\ q_1 & q_2 \end{vmatrix}, \begin{vmatrix} p_2 & p_3 \\ q_2 & q_3 \end{vmatrix}, \begin{vmatrix} p_3 & p_1 \\ q_3 & q_1 \end{vmatrix}$ are zero. But this means that some two of the given parametric equations may be solved for t and s. Suppose that $\begin{vmatrix} p_1 & p_2 \\ q_1 & q_2 \end{vmatrix} \neq 0$. Then we may solve the first two of the parametric equations for t and s. Doing so, we find that

$$t = \frac{\begin{vmatrix} x_1 - a_1 & q_1 \\ x_2 - a_2 & q_2 \end{vmatrix}}{\begin{vmatrix} p_1 & q_1 \\ p_2 & q_2 \end{vmatrix}} \quad \text{and} \quad s = \frac{\begin{vmatrix} p_1 & x_1 - a_1 \\ p_2 & x_2 - a_2 \end{vmatrix}}{\begin{vmatrix} p_1 & q_1 \\ p_2 & q_2 \end{vmatrix}}.$$

Using these results together with the third of the parametric equations, we obtain, in turn:

$$x_3 - a_3 = p_3 \frac{\begin{vmatrix} x_1 - a_1 & q_1 \\ x_2 - a_2 & q_2 \end{vmatrix}}{\begin{vmatrix} p_1 & q_1 \\ p_2 & q_2 \end{vmatrix}} + q_3 \frac{\begin{vmatrix} p_1 & x_1 - a_1 \\ p_2 & x_2 - a_2 \end{vmatrix}}{\begin{vmatrix} p_1 & q_1 \\ p_2 & q_2 \end{vmatrix}}$$

$$(x_3 - a_3) \begin{vmatrix} p_1 & q_1 \\ p_2 & q_2 \end{vmatrix} - p_3 \begin{vmatrix} x_1 - a_1 & q_1 \\ x_2 - a_2 & q_2 \end{vmatrix} - q_3 \begin{vmatrix} p_1 & x_1 - a_1 \\ p_2 & x_2 - a_2 \end{vmatrix} = 0$$

$$(5) \quad (x_1 - a_1) \begin{vmatrix} p_2 & p_3 \\ q_2 & q_3 \end{vmatrix} + (x_2 - a_2) \begin{vmatrix} p_3 & p_1 \\ q_3 & q_1 \end{vmatrix} + (x_3 - a_3) \begin{vmatrix} p_1 & p_2 \\ q_1 & q_2 \end{vmatrix} = 0.$$

The last result gives us a single equation that describes the plane through A in the direction of the vectors $\mathbf{p}$ and $\mathbf{q}$. (Note that we obtain the *same* equation regardless of which two of the parametric equations are associated with a nonzero determinant.) Furthermore, any two independent vectors that are in the direction of $\mathbf{p}$ and $\mathbf{q}$—and thus are linear combinations of $\mathbf{p}$ and $\mathbf{q}$—have components whose corresponding determinants are proportional to the three determinants associated with $\mathbf{p}$ and $\mathbf{q}$. Therefore, two planes are parallel if and only if their single equations have proportional (determinant) coefficients.

We shall find later that the determinant coefficients in the single equation for a plane give us one more bit of information when the coordinate system being used is an orthonormal one, namely, that they are the direction numbers of any line that is perpendicular to the plane in question.

For those who like to express things neatly and compactly, the left side of the single equation (5) for the plane in question may be written in the form of a third-order determinant, in which case the equation looks like this:

$$\begin{vmatrix} x_1 - a_1 & x_2 - a_2 & x_3 - a_3 \\ p_1 & p_2 & p_3 \\ q_1 & q_2 & q_3 \end{vmatrix} = 0.$$

It is perhaps worth noting at this time that in an n-dimensional space, a Cartesian coordinate system associates each point with an ordered n-tuple of real numbers, and in such a space, the coordinates $(x_1, x_2, \ldots, x_n)$ of points on a line AB may be described by the parametric equations of the form

$$x_i = a_i + (b_i - a_i)t, \qquad (i = 1, 2, \ldots, n),$$

and the coordinates $(x_1, x_2, \ldots, x_n)$ of points on a plane ABC may be described by parametric equations of the form

$$x_i = a_i + (b_i - a_i)t + (c_i - a_i)s, \qquad (i = 1, 2, \ldots, n).$$

So even though we have restricted our attention to a study of geometry of three dimensions, all our methods may be applied to a study of the geometry of higher (or lower) dimensions. To formalize the results of any such study, we need only make the appropriate adjustments in postulates 4_9 and 4_{10}. The ease with which the adjustment of dimension may be made in this system, both formally and operationally, puts a development of geometry such as this one in sharp contrast to the usual synthetic developments of Euclidean geometry and has been found both to provoke lively discussions and to have great appeal among students.

It is easy to obtain the usual formulas for some of the common affine geometric notions in terms of the coordinates of the points involved, for as we have seen, the coordinates of points interact in the same way as the points themselves. For example, if A and B have coordinates (a_1, a_2, a_3) and (b_1, b_2, b_3) respectively, then the midpoint of the segment AB has coordinates $\left(a_1 + (b_1 - a_1)\dfrac{1}{2}, \; a_2 + (b_2 - a_2)\dfrac{1}{2}, \; a_3 + (b_3 - a_3)\dfrac{1}{2} \right)$, or, in simplest terms, $\left(\dfrac{a_1 + b_1}{2}, \; \dfrac{a_2 + b_2}{2}, \; \dfrac{a_3 + b_3}{2} \right)$. Using this result, we find it easy to show that the centroid of triangle ABC, that is, the point of intersection of the medians of triangle ABC, has coordinates

$$\left(\frac{a_1 + b_1 + c_1}{3}, \; \frac{a_2 + b_2 + c_2}{3}, \; \frac{a_3 + b_3 + c_3}{3} \right),$$

where C has coordinates (c_1, c_2, c_3).

In short, then, it is clear that we are in a position to do as much, or as little, coordinate geometry as desired. The methods used in such a study are precisely those that were developed in the basic, noncoordinate study of geometry. Having access to general vector methods, coordinate methods, and, indeed, synthetic methods—for we do have access to some of, and soon will have access to all, the usual theorems of a standard treatment of geometry—gives us a great deal of flexibility in working on problems.

Useful notions about perpendicularity

In order to extend our formal system to that of Euclidean geometry, we need to introduce the concepts of perpendicularity and distance. In any formal treatment of geometry with vector methods, introducing these concepts amounts to introducing a special kind of operation, called an *inner* (or *dot*) *product,* on the vector space of translations. As in the past, we can fall back on our intuitions about perpendicularity and distance to get some ideas on extending our formal system to handle these concepts. Intuitively, it is reasonable to accept that a line is perpendicular to a plane if, when the plane is horizontal, the line is vertical. Given a so-called vertical line and any point, it is evident that no two planes that contain the point can be horizontal, and, further, any horizontal plane will not be parallel to the given line and therefore will meet the line in exactly one point. Thus it is evident that

given a point P and a line l, there is exactly one plane π that contains P and is perpendicular to l.

The point of intersection of π and l is sometimes called *the foot of the perpendicular from P to l,* or *the projection of P on l* (see fig. 8.21).

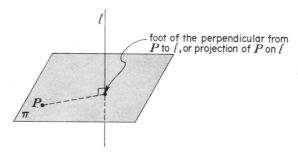

Fig. 8.21

It is equally evident that parallel planes are perpendicular to the same line (fig. 8.22) and that parallel lines are perpendicular to the same planes (fig. 8.23). These three notions concerning perpendicularity may be used

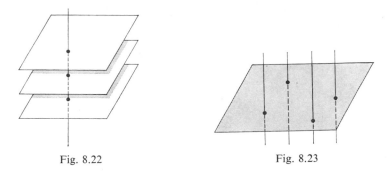

Fig. 8.22   Fig. 8.23

in a quite natural way to obtain, first, the projection of a vector on a given direction and, second, the component of one vector with respect to another.

To project a vector **a** on a given direction, say, that of a line l, choose any point P and project the points P and $P + $ **a** on l. (See fig. 8.24.) The translation (vector) from the projection on l of P to the projection on l of $P + $ **a** is called the *projection of* **a** *on the direction of* l. A moment's reflection on what the two pictures in figures 8.22 and 8.23 tell us will make it clear that the vector we are calling the "projection of **a** on the direction of l" is independent—that is, it is the same vector regardless—of the choice of point P as well as of whatever line we pick in the direction of (or parallel to) l. In figure 8.25, we illustrate that this projection is independent of whatever line we pick in the direction of l.

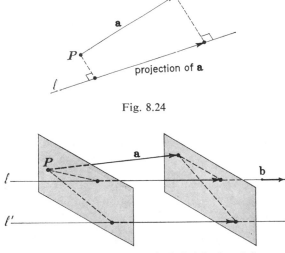

Fig. 8.24

Fig. 8.25. Each of the two heavy arrows included in l and l' represents the *same* vector, namely, the projection of **a** on the direction of l.

Now, given any non-**0** vector **b** in the direction of *l* (see fig. 8.25), it is clear that the projection of **a** in the direction of *l* is a vector that is a multiple of **b**, that is, a vector **b***c,* for some real number *c.* We call this number *c the component of* **a** *with respect to* **b**, or, for short:

$$\text{comp}_{\mathbf{b}}(\mathbf{a})$$

It should be clear, on intuitive grounds, that this "component" number is a function both of the length of **a** (that is, of how far **a** moves points) and of the length of **b**. (To see that this is true, think of what effect either "stretching" or "shrinking" **a** or **b** would have on the value of $\text{comp}_{\mathbf{b}}(\mathbf{a})$. See fig. 8.26 and consider some other cases for yourself.) Noting that this is true, we see that it is not too severe a restriction to limit our investigation to components of vectors with respect to given *unit* vectors, that is, to vectors of length 1.

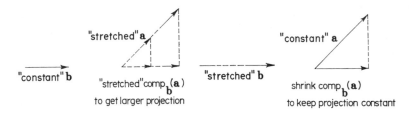

Fig. 8.26

Investigating the nature of "componenting" leads to a formulation of some of its properties. For example, figure 8.27 suggests the equivalence of the expressions

$$\text{comp}_{\mathbf{c}}(\mathbf{a}) + \text{comp}_{\mathbf{c}}(\mathbf{b})$$

and

$$\text{comp}_{\mathbf{c}}(\mathbf{a} + \mathbf{b}).$$

Thus it is acceptable on intuitive grounds that

(1) $\text{comp}_{\mathbf{c}}(\mathbf{a} + \mathbf{b}) = \text{comp}_{\mathbf{c}}(\mathbf{a}) + \text{comp}_{\mathbf{c}}(\mathbf{b}).$

As another example, it is evident from the ratio-preserving properties of parallelism that the component of a given multiple of **a** with respect to a unit vector **b** is precisely that multiple of the component of **a** with respect to **b**. (See fig. 8.28.) That is,

(2) $\text{comp}_{\mathbf{b}}(\mathbf{a}c) = (\text{comp}_{\mathbf{b}}\mathbf{a})c.$

(This is part of what is involved in the discussion centered on fig. 8.26.)

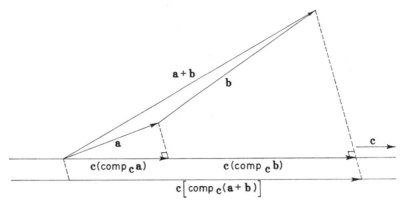

Fig. 8.27

On abbreviating expressions like

$$\text{comp}_{\mathbf{u}}(\mathbf{a})$$

to

$$\mathbf{a} \cdot \mathbf{u},$$

which we still read as "the component of **a** with respect to the unit vector **u**," we note that the results (1) and (2) take the forms

$$(\mathbf{a} + \mathbf{b}) \cdot \mathbf{c} = \mathbf{a} \cdot \mathbf{c} + \mathbf{b} \cdot \mathbf{c}$$

and

$$(\mathbf{a}c) \cdot \mathbf{b} = (\mathbf{a} \cdot \mathbf{b})c.$$

This suggests that "componenting" is a sort of multiplication, a suggestion that may be further supported by additional pictorially based investigations.

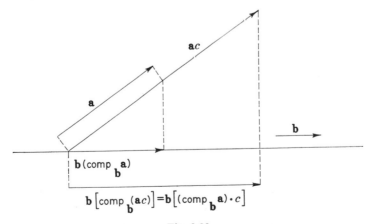

Fig. 8.28

As a matter of fact, from a thorough investigation of the basic properties of componenting, the postulates that describe the inner, or dot, product of vectors may be developed. In terms of the accepted dot abbreviation for componenting, these postulates may be formulated as follows:

POSTULATE 4_0.   (e) $\mathbf{a} \cdot \mathbf{b} \in \Re$

POSTULATE 4_{11}. $(\mathbf{a} + \mathbf{b}) \cdot \mathbf{c} = \mathbf{a} \cdot \mathbf{c} + \mathbf{b} \cdot \mathbf{c}$

POSTULATE 4_{12}. $(a\mathbf{c}) \cdot \mathbf{b} = (\mathbf{a} \cdot \mathbf{b})c$

POSTULATE 4_{13}. $\mathbf{a} \cdot \mathbf{b} = \mathbf{b} \cdot \mathbf{a}$

POSTULATE 4_{14}. $\mathbf{a} \cdot \mathbf{a} > 0$ if $\mathbf{a} \neq \mathbf{0}$

(Notice that the geometric intuitions that lead to the formulation and adoption of these "algebraic" types of postulates can serve to give us geometric interpretations to statements involving dot products and will be quite helpful in our continuing development of geometry with vector methods.)

As might be guessed, we are once again able to condense our list of postulates:

POSTULATE 4. $\Im$ *is a three-dimensional inner-product space.*

Having thus extended our formal system, we may begin to formulate geometric notions related to perpendicularity and distance in terms of this dot multiplication and in this way extend the formal structure of the geometry under consideration to that of Euclidean geometry of three dimensions. The major part of the remainder of this discussion will be concerned with this extension.

Theorems on perpendicularity

In order to arrive at the relation of perpendicularity in terms of dot products, it is convenient to fall back on our intuitive notion that a line l is perpendicular to a plane π if, when the plane π is considered to be horizontal, the line l is considered to be vertical. From this it is clear that the component of any translation $\mathbf{a}$ in the direction of l with respect to any non-$\mathbf{0}$ translation $\mathbf{b}$ in the direction of π is 0, and so $\mathbf{a} \cdot \mathbf{b} = 0$ (fig. 8.29). One way to arrive at perpendicularity in formal terms, then, is to base that notion on zero dot products. To do so will enable us to find properties of this relation by making full use of the algebra of dot products as an integral part of our algebra of points and translations.

We say that vectors $\mathbf{a}$ and $\mathbf{b}$ are orthogonal and write $\mathbf{a} \perp \mathbf{b}$ if and only if $\mathbf{a} \cdot \mathbf{b} = 0$. Further, we say that the direction of $\mathbf{a}$ is orthogonal to the direction of $\mathbf{b}$ if and only if each multiple of $\mathbf{a}$ is orthogonal to every

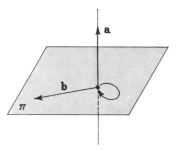

Fig. 8.29. The component of **a** with respect to **b** is 0, and so $\mathbf{a} \cdot \mathbf{b} = 0$.

multiple of **b**. Given lines l and m, we say that these lines are perpendicular and write $l \perp m$ if and only if the direction of l is orthogonal to the direction of m; that is, if and only if each vector in the direction of l is orthogonal to every vector in the direction of m (fig. 8.30). Similarly, a line l is said to be perpendicular to a plane π if and only if the direction of l is orthogonal to the direction of π. Two planes are said to be perpendicular if and only if there is a line in one of the planes that is perpendicular to the other plane.

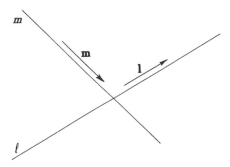

Fig. 8.30. $l \perp m$ if and only if $\mathbf{l} \cdot \mathbf{m} = 0$, where $\mathbf{l}$ is any vector in the direction of l and $\mathbf{m}$ is any vector in the direction of m.

Notice that nothing in the definition of perpendicular lines requires that the lines intersect. In fact, it should be clear that if a line l is perpendicular to a line m, then l is perpendicular to *each* line that is parallel to m, and in three-space many of these lines are skew to l (see fig. 8.31). Also, it is not difficult to establish that if a line is perpendicular to a plane, then it is perpendicular to *each* line in that plane. We do this as follows:

Suppose that l is a line and π is a plane such that $l \perp \pi$. Let A and B be two points of l and let P, Q, and R be three noncollinear points of π.

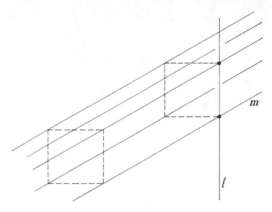

Fig. 8.31. Each of these lines is parallel to *m*, and so each has a direction orthogonal to *l* when *m* ⊥ *l*. Hence, each of these lines is perpendicular to *l*.

For convenience, let $\mathbf{b} = B - A$, $\mathbf{q} = Q - P$, and $\mathbf{r} = R - P$, as shown in figure 8.32. Then, since $l \perp \pi$, we have $\mathbf{b} \cdot \mathbf{q} = 0$ and $\mathbf{b} \cdot \mathbf{r} = 0$. Now, any line *m* that is a subset of π has the direction of $\mathbf{q}s + \mathbf{r}t$ for some real numbers s and t. It follows that

$$\mathbf{b} \cdot (\mathbf{q}s + \mathbf{r}t) = \mathbf{b} \cdot (\mathbf{q}s) + \mathbf{b} \cdot (\mathbf{r}t)$$
$$= (\mathbf{b} \cdot \mathbf{q})s + (\mathbf{b} \cdot \mathbf{r})t$$
$$= 0 \cdot s + 0 \cdot t$$
$$= 0.$$

Therefore, the direction of *l* is orthogonal to the direction of *m*, from which it follows that $l \perp m$, and the theorem is proved.

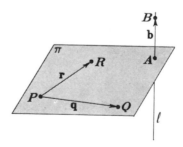

Fig. 8.32

Norms of vectors and distance

Given any non-**0** vector **u,** let $\mathbf{b} = \mathbf{u}b$, for some b. In a sense, the number b tells us about the distance that **b** moves points relative to the distance that **u** moves points. Noting that $\mathbf{b} \cdot \mathbf{b} = (\mathbf{u}b) \cdot (\mathbf{u}b) = (\mathbf{u} \cdot \mathbf{u})b^2$, we see that $\mathbf{b} \cdot \mathbf{b}$ is related to the *square* of this relative distance. We make use of this observation to generate the following definition of the norm of a translation:

The norm of **a**, *for short* $||\mathbf{a}||$, *is* $\sqrt{\mathbf{a} \cdot \mathbf{a}}$.

It should be clear from this definition, for example, that **a**5 and **a** · −5 each have norms that are 5 times that of **a**, and so both **a**5 and **a** · −5 have the same norm.

By definition, the norm of the translation $Q - P$, which maps P on Q, is $||Q - P||$, and this number tells us how far $Q - P$ moves P, that is, how far it is from P to Q. We use this notion to define the distance from P to Q, for short $d(P,Q)$, to be $||Q - P||$, the norm of the translation from P to Q.

With the definitions of norm and distance, we can make use of the algebra of dot products in conjunction with our algebra of points and translations to arrive at some fundamental theorems about distance. For example, it is easy to establish that $(\mathbf{a} + \mathbf{b}) \cdot (\mathbf{a} + \mathbf{b}) = \mathbf{a} \cdot \mathbf{a} + \mathbf{b} \cdot \mathbf{b} + 2(\mathbf{a} \cdot \mathbf{b})$ and that $(\mathbf{a} - \mathbf{b}) \cdot (\mathbf{a} - \mathbf{b}) = \mathbf{a} \cdot \mathbf{a} + \mathbf{b} \cdot \mathbf{b} - 2(\mathbf{a} \cdot \mathbf{b})$, from which it follows that

(1)
$\qquad$ **a** is the orthogonal to **b** if and only if
$\qquad$ $(\mathbf{a} + \mathbf{b}) \cdot (\mathbf{a} + \mathbf{b}) = \mathbf{a} \cdot \mathbf{a} + \mathbf{b} \cdot \mathbf{b} = (\mathbf{a} - \mathbf{b}) \cdot (\mathbf{a} - \mathbf{b})$.

In terms of norms of translations, this last result may be stated as follows:

$\qquad$ **a** is orthogonal to **b** if and only if
$\qquad$ $||\mathbf{a} + \mathbf{b}||^2 = ||\mathbf{a}||^2 + ||\mathbf{b}||^2 = ||\mathbf{a} - \mathbf{b}||^2$.

In terms of distance and perpendicularity, this last result has several important geometric interpretations, some of which are the following:

(a) In triangle ABC, let $\mathbf{a} = A - C$ and $\mathbf{b} = B - A$ (fig. 8.33). Then $B - C = \mathbf{a} + \mathbf{b}$, and from (1), we see that sides AC and AB are perpendicular if and only if

$$[d(A,C)]^2 + [d(A,B)]^2 = [d(C,B)]^2.$$

(This is easily recognized as the theorem of Pythagoras and its converse.)

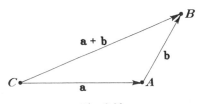

Fig. 8.33

(*b*) In parallelogram $ABCD$, let $\mathbf{a} = B - A$ and $\mathbf{b} = D - A$ (fig. 8.34). Then $B - D = \mathbf{a} - \mathbf{b}$ and $C - A = \mathbf{a} + \mathbf{b}$, and from (1), we see that sides AB and AD are perpendicular if and only if diagonals AC and BD have the same length. (This may be restated as follows: A parallelogram is a rectangle if and only if its diagonals are congruent.)

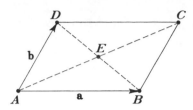

Fig. 8.34

(*c*) Since the diagonals of a parallelogram bisect each other, we know that AE is the median from A in $\triangle ABD$ in figure 8.34. So, another result we obtain from (1) is that sides AB and AD are perpendicular if and only if $d(A,E)$ is half of $d(B,D)$. (This may be restated as follows: A median from a given vertex of a triangle is half as long as the side to which it is drawn if and only if the triangle is a right triangle with right angle at the given vertex.)

(*d*) In triangle ABC, let D be the midpoint of side AC, and let $A - D = \mathbf{a}$ and $B - D = \mathbf{b}$ (fig. 8.35). Then $A - B = \mathbf{a} - \mathbf{b}$ and $B - C = \mathbf{a} + \mathbf{b}$. From (1) we see that sides AB and AC are congruent if and only if median BD is perpendicular to side AC. (This may be restated as follows: Triangle BAC is isosceles with base AC if and only if the median from B is perpendicular to the side AC.)

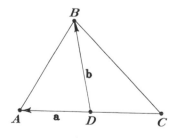

Fig. 8.35

As a final example of the methods we have at hand to handle problems on distance, consider triangle ABC with sides AB, BC, and CA having measures c, a, and b, respectively, as indicated in figure 8.36. For convenience, let $\mathbf{a} = C - B$, $\mathbf{c} = B - A$, and $\mathbf{b} = C - A$. Further, let D be any point on side BC, and let $D - A = \mathbf{d}$. A reasonable problem for investigation is the following:

Given that D divides the segment from B to C in the ratio $r:(1 - r)$ for $0 < r < 1$, compute the distance from A to D in terms of r, a, b, and c.

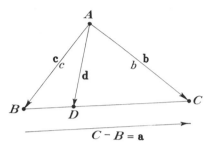

Fig. 8.36

Here is one solution:

Since D divides the segment from B to C in the ratio $r:(1 - r)$ where $0 < r < 1$, we know that $\mathbf{d} = \mathbf{b}(1 - r) + c\mathbf{r}$. So,

$$\mathbf{d} \cdot \mathbf{d} = (\mathbf{b} \cdot \mathbf{b})(1 - r)^2 + (\mathbf{c} \cdot \mathbf{c})r^2 + 2(\mathbf{b} \cdot \mathbf{c})[r(1 - r)],$$

which, in terms of norms, tells us that

$$||\mathbf{d}||^2 = b^2(1 - r)^2 + c^2r^2 + 2(\mathbf{b} \cdot \mathbf{c})[r(1 - r)].$$

Now, since $\mathbf{a} = \mathbf{b} - \mathbf{c}$ and $a^2 = b^2 + c^2 - 2(\mathbf{b} \cdot \mathbf{c})$, we know that $2(\mathbf{b} \cdot \mathbf{c}) = b^2 + c^2 - a^2$. So, we have

$$||d||^2 = b^2(1 - r)^2 + c^2r^2 + (b^2 + c^2 - a^2)[r(1 - r)].$$

Hence

$$[d(A,D)]^2 = a^2[r(r - 1)] + b^2(1 - r) + c^2r.$$

Thus, for example, if $a = 8$, $b = 9$, $c = 5$, and D is $\dfrac{1}{4}$ of the way from B to C so that D divides the interval from B to C in the ratio $\dfrac{1}{4}:\dfrac{3}{4}$, then

$$[d(A,D)]^2 = 64 \left(\frac{1}{4} \cdot \frac{3}{4}\right) + 81 \cdot \frac{3}{4} + 25 \cdot \frac{1}{4} = 55,$$

and so

$$d(A,D) = \sqrt{55}.$$

Given that D is the midpoint of side BC, we have the obvious corollary that the median from A is

$$\frac{\sqrt{2(b^2 + c^2) - a^2}}{2}.$$

An important inequality and its corollary

Consider the vectors **a** and **b** shown in figure 8.37. Let P be any point on the line QB. Then $P - Q = \mathbf{b}p$ for some p, and in terms of this, $A - P = \mathbf{a} - \mathbf{b}p$. So $(A - P) \cdot \mathbf{b} = \mathbf{a} \cdot \mathbf{b} - (\mathbf{b} \cdot \mathbf{b})p$. Now $A - P$ is orthogonal to **b** if and only if $(A - P) \cdot \mathbf{b} = 0$, and the latter is true if and only if $p = \dfrac{\mathbf{a} \cdot \mathbf{b}}{\mathbf{b} \cdot \mathbf{b}}$. One thing that we can say from this is that

P is the foot of the perpendicular from A to line QB if and only if

$$A - P = \mathbf{a} - \mathbf{b}\left[\frac{\mathbf{a} \cdot \mathbf{b}}{\mathbf{b} \cdot \mathbf{b}}\right].$$

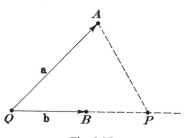

Fig. 8.37

If we concern ourselves for the moment with the task of computing the norm of $A - P$ in terms of **a** and **b**, we shall discover a most fundamental and useful relation between $\mathbf{a} \cdot \mathbf{b}$ and the norms of **a** and **b**. Notice that

$$\begin{aligned}
||A - P||^2 &= \left[\mathbf{a} - \mathbf{b}\left(\frac{\mathbf{a} \cdot \mathbf{b}}{\mathbf{b} \cdot \mathbf{b}}\right)\right] \cdot \left[\mathbf{a} - \mathbf{b}\left(\frac{\mathbf{a} \cdot \mathbf{b}}{\mathbf{b} \cdot \mathbf{b}}\right)\right] \\
&= \mathbf{a} \cdot \mathbf{a} + \frac{(\mathbf{a} \cdot \mathbf{b})^2}{\mathbf{b} \cdot \mathbf{b}} - 2\frac{(\mathbf{a} \cdot \mathbf{b})^2}{\mathbf{b} \cdot \mathbf{b}} \\
&= \frac{(\mathbf{a} \cdot \mathbf{a})(\mathbf{b} \cdot \mathbf{b}) - (\mathbf{a} \cdot \mathbf{b})^2}{\mathbf{b} \cdot \mathbf{b}} \\
&= \frac{||\mathbf{a}||^2\,||\mathbf{b}||^2 - (\mathbf{a} \cdot \mathbf{b})^2}{||\mathbf{b}||^2}.
\end{aligned}$$

So since $||A - P||^2 \geq 0$, it follows that

$$(||\mathbf{a}||\,||\mathbf{b}||)^2 - (\mathbf{a} \cdot \mathbf{b})^2 \geq 0;$$

that is, that

$$(||\mathbf{a}||\,||\mathbf{b}||)^2 \geq (\mathbf{a} \cdot \mathbf{b})^2.$$

Hence, since norms of translations are nonnegative, we have

(1) $$|\mathbf{a} \cdot \mathbf{b}| \leq ||\mathbf{a}||\,||\mathbf{b}||.$$

This tells us that the absolute value of the dot product of any two vectors is less than or equal to the product of the norms of those vectors. The inequality (1) is sometimes called the *Cauchy-Schwarz inequality*. (It turns out that equality holds in (1) if and only if **a** and **b** are linearly dependent.)

One consequence of the Cauchy-Schwarz inequality is the following:

Consider $\triangle ABC$, with $C - A = \mathbf{b}$, $B - C = \mathbf{a}$, **c** and $B - A = \mathbf{c}$ (fig. 8.38). Then $\mathbf{c} = \mathbf{a} + \mathbf{b}$, and so we have

$$||\mathbf{c}||^2 = ||\mathbf{a}||^2 + ||\mathbf{b}||^2 + 2(\mathbf{a} \cdot \mathbf{b}).$$

Since **a** and **b** are independent, $\mathbf{a} \cdot \mathbf{b} < ||\mathbf{a}|| \, ||\mathbf{b}||$. Therefore, it follows that

$$||\mathbf{c}||^2 < ||\mathbf{a}||^2 + ||\mathbf{b}||^2 + 2||\mathbf{a}|| \, ||\mathbf{b}|| = (||\mathbf{a}|| + ||\mathbf{b}||)^2.$$

Hence,

$$||\mathbf{c}|| < ||\mathbf{a}|| + ||\mathbf{b}||,$$

which tells us that the length of one side of a triangle is less than the sum of the lengths of the other two sides, a fact often called *the triangle inequality*.

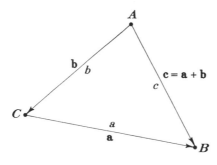

Fig. 8.38

From the discussion centered on figure 8.37, it is not difficult to see how one can make use of any basis for $\mathfrak{I}$, that is, any linearly independent triple of vectors, to generate a basis whose terms are mutually orthogonal. Doing so will establish, essentially, that rectangular coordinate systems exist as a formal part of our system. To this end, then, let $(\mathbf{a}, \mathbf{b}, \mathbf{c})$ be any

basis for $\mathfrak{I}$. As in the discussion on figure 8.37, let $\mathbf{b}' = \mathbf{b} - \mathbf{a} \left[\dfrac{\mathbf{a} \cdot \mathbf{b}}{\mathbf{a} \cdot \mathbf{a}} \right]$.

(See fig. 8.39.) Then **a** and $\mathbf{b}'$ are orthogonal and linearly independent. Now, let

$$\mathbf{c}' = \mathbf{c} - \mathbf{a} \left[\frac{\mathbf{a} \cdot \mathbf{c}}{\mathbf{a} \cdot \mathbf{a}} \right] - \mathbf{b}' \left[\frac{\mathbf{b}' \cdot \mathbf{c}}{\mathbf{b}' \cdot \mathbf{b}'} \right].$$

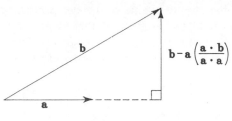

Fig. 8.39

(It is an easy exercise to show that $\mathbf{a} \cdot \mathbf{c}' = 0$ and $\mathbf{b}' \cdot \mathbf{c}' = 0$, from which it follows that $\mathbf{a}$, $\mathbf{b}'$, and $\mathbf{c}'$ are mutually orthogonal.) Then $\mathbf{a}$, $\mathbf{b}'$, and $\mathbf{c}'$ are both mutually orthogonal and linearly independent. Hence, if $(\mathbf{a},\mathbf{b},\mathbf{c})$ is a basis for $\Im$, then $(\mathbf{a},\mathbf{b}',\mathbf{c}')$ is an orthogonal basis for $\Im$. In short, orthogonal bases for $\Im$ exist if bases for $\Im$ exist.

Noting that each non-$\mathbf{0}$ vector gives rise to two unit vectors—for example, non-$\mathbf{0}$ $\mathbf{a}$ gives rise to unit vectors $\dfrac{\mathbf{a}}{\|\mathbf{a}\|}$ and $\dfrac{-\mathbf{a}}{\|\mathbf{a}\|}$—we see that it is also clear from what was just established that orthonormal bases (bases whose members are mutually orthogonal unit vectors) exist as a formal part of our system. Thus instead of working with coordinate systems defined in terms of arbitrary bases, we may choose to work with ordinary rectangular coordinates, which are defined in terms of orthonormal bases. In the next section, we shall make use of the latter notion to extend our knowledge of coordinate systems and geometry. Like our earlier discussion of coordinate systems, this notion may be omitted from the discussion without any loss in continuity.

Additional remarks on coordinate systems

Suppose that $(\mathbf{i}, \mathbf{j}, \mathbf{k})$ is an orthonormal basis for $\Im$. Then by definition, $\mathbf{i}$, $\mathbf{j}$, and $\mathbf{k}$ are unit vectors, and they are mutually orthogonal. Let $\mathbf{d}$ be any vector. Then for some d_1, d_2, and d_3,

$$\mathbf{d} = \mathbf{i}d_1 + \mathbf{j}d_2 + \mathbf{k}d_3.$$

From this and our definition of norms, we have

(1) $$\|\mathbf{d}\| = \sqrt{d_1{}^2 + d_2{}^2 + d_3{}^2}.$$

Furthermore, if $\mathbf{e}$ has components (e_1, e_2, e_3)—that is, if $\mathbf{e} = \mathbf{i}e_1 + \mathbf{j}e_2 + \mathbf{k}e_3$—then

(2) $$\mathbf{d} \cdot \mathbf{e} = d_1e_1 + d_2e_2 + d_3e_3.$$

(This is easily recognized as the usual definition of the dot product of two vectors in terms of their components, and our work lays bare the hidden assumption that underlies this definition, namely, that the components of vectors are given in terms of an orthonormal basis.)

Given a coordinate system with the basis $(\mathbf{i}, \mathbf{j}, \mathbf{k})$ and origin O, let A and B have coordinates (a_1,a_2,a_3) and (b_1,b_2,b_3), respectively. Then $B - A$ has components $(b_1 - a_1, b_2 - a_2, b_3 - a_3)$, so that, by (1) and our definition of distance,

$$(3) \qquad d(A,B) = \sqrt{(b_1 - a_1)^2 + (b_2 - a_2)^2 + (b_3 - a_3)^2}.$$

(This is easily recognized as the usual formula for the distance between two points in terms of their coordinates. Here again, this definition has the hidden assumption that the coordinate system is an orthonormal one.)

In our earlier discussion of coordinate systems, we found that a plane through a point A—with coordinates (a_1,a_2,a_3)—and in the direction of independent vectors $\mathbf{p}$ and $\mathbf{q}$—with components (p_1,p_2,p_3) and (q_1,q_2,q_3) respectively—may be described by an equation of the form

$$(4) \qquad (x_1 - a_1)m_1 + (x_2 - a_2)m_2 + (x_3 - a_3)m_3 = 0,$$

where m_1, m_2, and m_3 are the determinants $\begin{vmatrix} p_2 & p_3 \\ q_2 & q_3 \end{vmatrix}$, $\begin{vmatrix} p_3 & p_1 \\ q_3 & q_1 \end{vmatrix}$, and $\begin{vmatrix} p_1 & p_2 \\ q_1 & q_2 \end{vmatrix}$, respectively. Taking note of equation (2), we see that (4) may be interpreted as saying that

> (x_1,x_2,x_3) are the coordinates of a point that is in the plane through the point with coordinates (a_1,a_2,a_3) whenever the vector determined by these points is orthogonal to the vector whose components are the direction numbers of the plane.

In other words, when the coordinate system is an orthonormal one, the direction numbers of a plane are the components of a non-$\mathbf{0}$ vector in the direction of a line that is perpendicular to the plane. (See fig. 8.40.) Knowing how to compute dot products gives us an easy way to write an equation of a plane with given point and direction numbers.

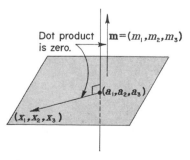

Fig. 8.40. The equation of the plane is $(x_1 - a_1)m_1 + (x_2 - a_2)m_2 + (x_3 - a_3)m_3 = 0$.

As an example of what we can now do with coordinates within the context of our formal system, let us find the equation of a plane π through points A, B, and C whose coordinates are $(-1,1,1)$, $(2,0,3)$, and $(1,1,2)$, respectively, and also determine the point of intersection of a line l that is perpendicular to π and contains the point D with coordinates $(1,2,3)$. First, let $\mathbf{p} = B - A$ and $\mathbf{q} = C - A$ so that $\mathbf{p}$ has components $(3,-1,2)$ and $\mathbf{q}$ has components $(2,0,1)$. Then π has direction numbers $\begin{vmatrix} -1 & 2 \\ 0 & 1 \end{vmatrix}$, $\begin{vmatrix} 2 & 3 \\ 1 & 2 \end{vmatrix}$, and $\begin{vmatrix} 3 & -1 \\ 2 & 0 \end{vmatrix}$, that is, -1, 1, and 2, and so an equation for π is

$$(x_1 - -1) \cdot -1 + (x_2 - 1) \cdot 1 + (x_3 - 1) \cdot 2 = 0.$$

Simplifying this, we have

(5) $$x_1 - x_2 - 2x_3 = -4.$$

Parametric equations for l are the following:

(6)
$$\begin{cases} x_1 = 1 + t \\ x_2 = 2 - t \\ x_3 = 3 - 2t \end{cases}$$

Using (5) and (6), we easily see that l intersects π in the point with coordinates $(1 + t, 2 - t, 3 - 2t)$ for some t, and $(1 + t) - (2 - t) - 2(3 - 2t) = -4$. From the latter, $t = \dfrac{1}{2}$; so, l and π intersect in the point with coordinates $\left(\dfrac{3}{2}, \dfrac{3}{2}, 2\right)$. (It should be clear that we are now also in a position to compute the distance from the point D to the plane π simply by using the result (3). Doing so, we find that $d(D,\pi) = \dfrac{\sqrt{6}}{2}$.)

More about angles and triangles

It was noted earlier that angle BAC is the union of noncollinear rays AB and AC (see fig. 8.41). Now, each of the rays AB and AC may be associated with a unit vector in its sense. For example, if $\mathbf{b} = B - A$ and $\mathbf{c} = C - A$, then $\mathbf{b}/||\mathbf{b}||$ and $\mathbf{c}/||\mathbf{c}||$ are the unit vectors in the senses of rays AB and AC, respectively. Making use of this fact, we associate with the angle BAC a real number, namely, the dot product of those unit vectors. This real number is called the *cosine* of the angle. Thus, if $\mathbf{u}_b$ and $\mathbf{u}_c$ are the unit vectors in the senses of the rays AB and AC of angle BAC, then the cosine of angle BAC (for short, $\cos BAC$) is defined as follows:

(1) $$\cos BAC = \mathbf{u}_b \cdot \mathbf{u}_c$$

Notice that cos is a function whose domain is the set of all angles. That

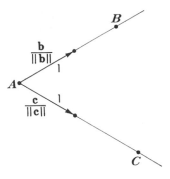

Fig. 8.41

its range is the set of real numbers between -1 and 1 is a consequence of the Cauchy-Schwarz inequality as it pertains to linearly independent vectors. (It is interesting, perhaps, to note that by definition (1), it follows that $\mathbf{b} \cdot \mathbf{c} = ||\mathbf{b}|| \, ||\mathbf{c}|| \cos BAC$. This equation is sometimes used as a defining property of the dot product. Of course, in order to learn to operate with the dot product under this definition, one must know how to operate with cosines of angles. Since our postulates enable us to operate formally with dot products, we are in a position to study and learn about relations that are defined in terms of dot products.)

From an intuitive point of view, an acute angle (fig. 8.42) is one that is, in some sense, "smaller" than a right angle (fig. 8.43), and an obtuse angle (fig. 8.44) is, in this same sense, "larger" than a right angle. Using what we can "see" about cosines in terms of projections of points on lines, we can reasonably define these terms as follows:

An angle is—

   1. *acute* if and only if its cosine is positive;

   2. *right* if and only if its cosine is zero;

   3. *obtuse* if and only if its cosine is negative.

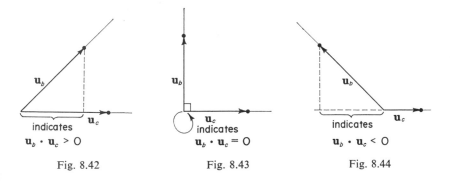

Fig. 8.42   Fig. 8.43   Fig. 8.44

(Note that the acceptance of this definition is completely analogous to, say, accepting that an angle is acute, right, or obtuse when its degree measure is less than, equal to, or greater than 90, respectively, and is no less arbitrary or less appealing on intuitive grounds than is the latter example.)

We can make use of what we know about dot products and cosines to learn more about triangles. Suppose that triangle ABC has sides of lengths a, b, and c, and that $B - A = \mathbf{c}$ and $C - A = \mathbf{b}$, as shown in figure 8.45. Then by definition $\|\mathbf{c} - \mathbf{b}\| = a$, and so

$$a^2 = (\mathbf{c} - \mathbf{b}) \cdot (\mathbf{c} - \mathbf{b})$$
$$= \mathbf{c} \cdot \mathbf{c} + \mathbf{b} \cdot \mathbf{b} - 2\mathbf{c} \cdot \mathbf{b}$$
$$= c^2 + b^2 - 2\mathbf{c} \cdot \mathbf{b}.$$

And since $\mathbf{c} \cdot \mathbf{b} = cb \cos BAC$, it follows that

(2) $$a^2 = c^2 + b^2 - 2cb \cos BAC.$$

This result is often called the *law of cosines* and is a completely general result about triangles.

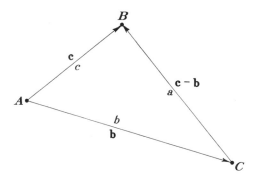

Fig. 8.45

Notice that by equation (2) we know that $a^2 = c^2 + b^2$ if and only if $\cos BAC = 0$; that is, if and only if angle BAC is a right angle. So, one consequence of the law of cosines is the Pythagorean theorem and its converse, which we had established earlier. (See (1) in the section on norms of vectors and distance.) Also, by (2) we see that in right triangle ABC with right angle at A,

$$\cos ABC = \frac{c^2 + a^2 - b^2}{2ac} = \frac{2c^2}{2ac} = \frac{c}{a}.$$

Similarly, $\cos ACB = b/a$. From these results we not only have a "neat" method for computing the cosines of the other angles but can easily see that each of these other angles is acute (for both $c/a > 0$ and $b/a > 0$). Hence, we have the following easy corollary:

> No triangle can have more than one right angle, and if one angle of a triangle is right, the remaining angles are acute.

As an example of the usefulness of the law of cosines in doing computations related to questions concerning a given triangle, suppose that sides AB, BC, and AC of triangle ABC have measures 13, 14, and 15, respectively (fig. 8.46). Then

$$\cos ABC = \frac{14^2 + 13^2 - 15^2}{2 \cdot 14 \cdot 13}$$
$$= \frac{140}{2 \cdot 13 \cdot 14}$$
$$= \frac{5}{13}.$$

Therefore, angle ABC is acute. Furthermore, if D is the foot of the perpendicular from A to line BC, we know that D is on the ray BC, for angle ABC is acute. So, angle ABD *is* angle ABC, and in right triangle ABD, $\cos ABD = 5/13$. But $\cos ABD = BD/AB$, and since $AB = 13$, it follows that $BD = 5$. (That we are also in a position to compute the length of the altitude from A as well as to make computations concerning perpendiculars from B and C should be clear.)

Using the notation established in figure 8.45, we find that another quite useful relation among cosines of angles in a triangle may be obtained. Note that $\cos BAC = \dfrac{\mathbf{b} \cdot \mathbf{c}}{bc}$ and that $\cos ABC = \dfrac{(\mathbf{b} - \mathbf{c}) \cdot -\mathbf{c}}{ac}$. So we obtain, in turn, the following:

$$(bc) \cos BAC + (ac) \cos ABC = (\mathbf{b} \cdot \mathbf{c}) + (\mathbf{b} - \mathbf{c}) \cdot -\mathbf{c}$$

(3)
$$c(b \cos BAC + a \cos ABC) = \mathbf{c} \cdot \mathbf{c} = c^2$$
$$b \cos BAC + a \cos ABC = c$$

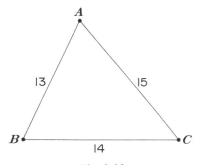

Fig. 8.46

Equation (3) is sometimes called the *projection theorem,* and it should be clear from this theorem that no two angles of a triangle can have negative cosines, for if this happens, one of the sides would have a negative measure, which is impossible. Thus a triangle can have at most one obtuse angle. Later, when we establish the law of sines for triangles, we shall have occasion to make use of (3) to derive one of the two interesting and useful relations among the three angles of a triangle.

In the discussion that centered on figure 8.37 and led to the Cauchy-Schwarz inequality, we saw that for a given angle $AQB,$

(4) $$||A - P||^2 = \frac{||\mathbf{a}||^2||\mathbf{b}||^2 - (\mathbf{a} \cdot \mathbf{b})^2}{||\mathbf{b}||^2}$$

where P is the foot of the perpendicular from A to line QB and $\mathbf{a}$ and $\mathbf{b}$ are as shown in figure 8.47. Looking at the numerator of the fraction in (4), we see that

$$||\mathbf{a}||^2||\mathbf{b}||^2 - (\mathbf{a} \cdot \mathbf{b})^2 = ||\mathbf{a}||^2||\mathbf{b}||^2 \left[1 - \frac{(\mathbf{a} \cdot \mathbf{b})^2}{||\mathbf{a}||^2||\mathbf{b}||^2} \right]$$

$$= ||\mathbf{a}||^2||\mathbf{b}||^2 \left[1 - \left(\frac{\mathbf{a}}{||\mathbf{a}||} \cdot \frac{\mathbf{b}}{||\mathbf{b}||} \right)^2 \right]$$

$$= ||\mathbf{a}||^2||\mathbf{b}||^2 [1 - (\cos AQB)^2].$$

From this, it is easy to see that

$$AP = ||\mathbf{a}|| \sqrt{1 - (\cos AQB)^2}.$$

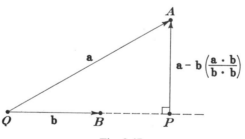

Fig. 8.47

The number $\sqrt{1 - (\cos AQB)^2}$ is always positive and is clearly a function of angle AQB. It is convenient to call this number "sin AQB" (read as "the sine of angle AQB"); that is, for any angle AQB,

(5) $$\sin AQB = \sqrt{1 - (\cos AQB)^2}.$$

(In the example related to fig. 8.46, we have

$$\sin ABC = \sqrt{1 - \left(\frac{5}{13} \right)^2} = \frac{12}{13}.$$

Note that this computation is related directly to that needed to compute the length of the altitude from A.)

It follows immediately from (5) that for any angle, say BAC, $(\cos BAC)^2 + (\sin BAC)^2 = 1$. Also, given triangle ABC with sides labeled in the usual way (for example, as in fig. 8.45), it follows from the law of cosines that

$$1 + \cos BAC = \frac{(c+b)^2 - a^2}{2cb}$$
$$= \frac{(c+b+a)(c+b-a)}{2cb}$$

and that

$$1 - \cos BAC = \frac{a^2 - (c-b)^2}{2cb}$$
$$= \frac{(a+c-b)(a-c+b)}{2cb}.$$

Since $(1 + \cos BAC)(1 - \cos BAC) = 1 - (\cos BAC)^2 = (\sin BAC)^2$, it follows that

$$(6) \quad \sin BAC = \frac{\sqrt{(a+b+c)(a+b-c)(a-b+c)(-a+b+c)}}{2bc}.$$

Letting

$$R = \sqrt{(a+b+c)(a+b-c)(a-b+c)(-a+b+c)},$$

we see from (6) that

$$R = (2bc)\sin BAC.$$

Similarly,

$$R = (2ac)\sin ABC$$

and

$$R = (2ab)\sin BCA.$$

Hence,
$$(7) \qquad bc \sin BAC = ac \sin ABC = ab \sin BCA.$$

From (7), it follows immediately that in any triangle ABC,

$$(8) \qquad \frac{\sin A}{a} = \frac{\sin B}{b} = \frac{\sin C}{c},$$

where A refers to angle BAC, B refers to angle ABC, and C refers to angle BCA. (This last result is sometimes called the *law of sines* and is a completely general result about triangles.)

As an example of one use to which we can put the law of sines, recall the computations done in relation to the triangle in figure 8.46. We found that $\cos B = 5/13$ and that $\sin B = 12/13$. So by the law of sines, $\dfrac{\sin C}{13} = \dfrac{\sin B}{15}$, and so $\sin C = \dfrac{13}{15} \cdot \dfrac{12}{13} = \dfrac{4}{5}$. Similarly, $\sin A = \dfrac{56}{65}$.

The law of sines together with sentence (3), the projection theorem, can be used to find an interesting relation among the three angles of a triangle. Using our familiar notation, we know by (3) that in any triangle ABC

$$c = b \cdot \cos A + a \cdot \cos B.$$

So

$$c \cdot \sin C = (b \cdot \sin C) \cdot \cos A + (a \cdot \sin C) \cdot \cos B.$$

Now, by the law of sines, $b \cdot \sin C = c \cdot \sin B$, and $a \cdot \sin C = c \cdot \sin A$. Thus, by substitution, we have

$$c \cdot \sin C = c \cdot \sin B \cdot \cos A + c \cdot \sin A \cdot \cos B;$$

that is,

(9) $$\sin C = \sin B \cos A + \sin A \cos B.$$

By this last result, we see that if we know the cosines of two angles of a triangle, then we are in a position to compute the sine of the third angle of the triangle.

Finding the corresponding result for cosine, that is, computing the cosine of the third angle of a triangle when given the cosines of two of the angles of the triangle, is a bit more complicated. But it is just this result that will enable us to conclude that if two angles of a first triangle are congruent to two angles of a second triangle, then the remaining angles are congruent. To this end, then, we shall first solve a problem about angles and then show how the result can be used to give us what we need in a triangle.

Consider angle BAC where, for simplicity, B and C are the unit points on the rays AB and AC, respectively. (See fig. 8.48.) Let D be any unit point in the interior of angle BAC. Let **b**, **c**, and **d** be the *unit vectors* from A to B, C, and D, respectively. We shall refer to angle BAC as A, to angle BAD as A_1 and to angle CAD as A_2. Clearly, if E and F are the feet of the perpendiculars from B and C, respectively, to line AD, then—

1. $B - E = \mathbf{b} - \mathbf{d}(\cos A_1)$ and $C - F = \mathbf{c} - \mathbf{d}(\cos A_2)$;
2. $BE = \sin A_1$ and $CF = \sin A_2$;
3. $B - E$ and $C - F$ are oppositely sensed.

The first two of these results are immediate consequences of our work related to figure 8.47, but the last of these results comes basically from what it means for a point to be interior to an angle. Now, look carefully in figure 8.49 at the angles BAB' and BAC', where line $B'C'$ is parallel to line BE and line CF. Since they share the ray AB and have oppositely

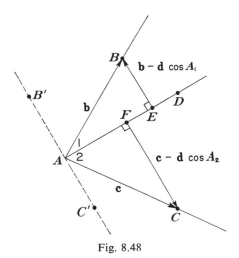

Fig. 8.48

sensed nonshared sides, it should be clear that their cosines are *opposites!*
(Put another way, angles BAB' and BAC' are supplements.)

Now,

$$\cos BAB' = \mathbf{b} \cdot [\mathbf{b} - \mathbf{d}(\cos A_1)]\left(\frac{1}{\sin A_1}\right)$$

$$= \frac{(\mathbf{b} \cdot \mathbf{b}) - (\mathbf{b} \cdot \mathbf{d})\cos A_1}{\sin A_1}$$

$$= \frac{1 - (\cos A_1)^2}{\sin A_1}$$

$$= \sin A_1$$

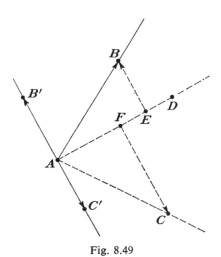

Fig. 8.49

and

$$\cos BAC' = \mathbf{b} \cdot [\mathbf{c} - \mathbf{d}(\cos A_2)]\left(\frac{1}{\sin A_2}\right)$$

$$= \frac{(\mathbf{b} \cdot \mathbf{c}) - (\mathbf{b} \cdot \mathbf{d})\cos A_2}{\sin A_2}$$

$$= \frac{\cos A - \cos A_1 \cdot \cos A_2}{\sin A_2}.$$

Since $(\cos BAB') + (\cos BAC') = 0$ (they are opposites), then

$$\sin A_1 + \frac{\cos A - \cos A_1 \cdot \cos A_2}{\sin A_2} = 0.$$

Simplifying this last result and solving for $\cos A$, we see that

(10) $\cos A = \cos A_1 \cos A_2 - \sin A_1 \sin A_2.$

To see how this result can be related to a triangle, consider triangle ABC in figure 8.50, with exterior angle EBC at B. Let D be the point in the interior of angle EBC such that ray BD is in the same sense as ray AC. Then by (10) we have $\cos EBC = \cos B_2 \cos B_1 - \sin B_2 \sin B_1$. Angles A and B_2 are congruent (look at the unit vectors in the senses of the rays of these angles), as are angles C and B_1. So, $\cos EBC = \cos A \cos C - \sin A \sin C$. But $\cos EBC$ and $\cos ABC$ (or $\cos B$) are opposites, for the angles EBC and ABC are supplementary. Hence

(11) $\cos B = \sin A \sin C - \cos A \cos C.$

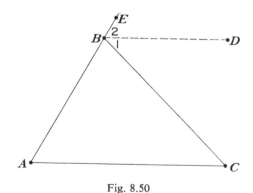

Fig. 8.50

Thus we are able to compute the cosine of the third angle of a triangle as soon as we know the cosines (and, therefore, the sines) of the first two angles of the triangle; that is, given two angles of a triangle, the third is fixed.

Three standard triangle congruence theorems

Congruent angles are angles whose cosines are equal, and congruent triangles are triangles whose vertices can be matched in such a way that corresponding angles and corresponding sides are congruent. Using these notions, we can put the law of cosines to good use in order to derive two of the three "standard" triangle congruence theorems. The first of these— the so-called side-side-side congruence theorem—may be derived as follows:

Suppose that $ABC \leftrightarrow DEF$ is a matching of vertices of triangle ABC with those of triangle DEF such that the corresponding sides are congruent. Letting a, b, and c be the measures of sides BC, CA, and AB, respectively, of triangle ABC and d, e, and f the corresponding measures of triangle DEF, we have $a = d$, $b = e$, and $c = f$. So, by the law of cosines,

$$\cos BAC = \frac{c^2 + b^2 - a^2}{2cb}$$
$$= \frac{f^2 + e^2 - d^2}{2fe}$$
$$= \cos EDF.$$

Thus angle BAC is congruent to angle EDF. Similarly, angle ABC is congruent to angle DEF, and angle ACB is congruent to angle DFE; so corresponding angles are congruent. Hence triangles ABC and DEF are congruent.

Since the converse of this result is clearly a theorem, what we have established here is the following standard theorem:

THEOREM 1. *A first triangle is congruent to a second triangle if and only if there is a matching of vertices of the first triangle with those of the second such that corresponding sides are congruent.*

The second of the standard triangle congruence theorems—the so-called side-angle-side congruence theorem—may be derived as follows:

Suppose that $ABC \leftrightarrow DEF$ is a matching of vertices of triangle ABC with those of triangle DEF such that two sides and the included angle— sides BA and AC and angle BAC, for example—of triangle ABC are congruent to the corresponding two sides and included angle of triangle DEF. Using the notation of the previous proof, we have $c = f$, $b = e$, and $\cos BAC = \cos EDF$. By the law of cosines,

$$a^2 = c^2 + b^2 - 2cb \cos BAC$$
$$= f^2 + e^2 - 2fe \cos EDF$$
$$= d^2.$$

Since side measures are nonnegative numbers, it follows that $a = d$. So, by theorem 1, triangles ABC and DEF are congruent.

Since the converse of this result is clearly a theorem, what we have established here is the following standard theorem:

THEOREM 2. *A first triangle is congruent to a second triangle if and only if there is a matching of vertices of the first triangle with those of the second such that two sides and the included angle of the first are congruent to the corresponding two sides and included angle of the second.*

The third standard triangle congruence theorem—the so-called angle-side-angle congruence theorem—may be established by making use of the law of sines. Here is one such proof:

Suppose that $ABC \leftrightarrow DEF$ is a matching of vertices of triangle ABC with those of triangle DEF such that two angles and the included side— angles A and B and side AB, for example—are congruent to the corresponding two angles and included side of triangle DEF. From this, it follows that $\cos A = \cos D$, and so $\sin A = \sin D$. Similarly, $\cos B = \cos E$, and $\sin B = \sin E$. So, by (11) of the preceding section, we have

$$\cos C = \sin A \sin B - \cos A \cos B$$
$$= \sin D \sin E - \cos D \cos E$$
$$= \cos F.$$

Thus angles C and F are congruent. Furthermore, by the law of sines,

$$a = \frac{c \cdot \sin A}{\sin C} = \frac{f \cdot \sin D}{\sin F} = d$$

and

$$b = \frac{c \cdot \sin B}{\sin C} = \frac{f \cdot \sin E}{\sin F} = e.$$

Hence, all pairs of corresponding sides and corresponding angles are congruent, and the triangles are congruent.

As before, the converse of this result is also a theorem. Hence, we have established our third theorem:

THEOREM 3. *A first triangle and a second triangle are congruent if and only if there is a matching of vertices of the first with those of the second such that two angles and the included side of the first are congruent to the corresponding two angles and included side of the second.*

General remarks on congruence and isometries

Congruent figures are figures that have "the same size and shape." Congruence, in this sense, is a relation among geometric figures. One way to think about establishing that two figures are congruent is somehow to "move" one of the figures into the spot occupied by the other in such a way as not to "destroy" its shape. This could be accomplished by keeping the figure "rigid," that is, by keeping all distances between points of the figure "fixed" while the move is effected. From a mathematical point of view, this sort of fixed-distance move can be looked on as a mapping that preserves distances, and any such mapping is called an *isometry* (the prefix *iso* means *equal* and *metry* means *measure*).

It is quite a simple task to show that any translation is an isometry. Let **a** be any translation and let A and B be any two points, as seen in figure 8.51. Then, since $B - A = (B + \mathbf{a}) - (A + \mathbf{a})$, we have

$$\begin{aligned} d(A,B) &= ||B - A|| \\ &= ||(B + \mathbf{a}) - (A + \mathbf{a})|| \\ &= d(A + \mathbf{a}, B + \mathbf{a}). \end{aligned}$$

Thus, **a** preserves distances, and so **a** is an isometry.

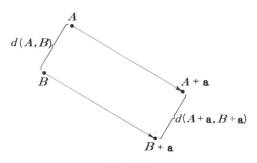

Fig. 8.51

Another mapping of points to points that is convenient to investigate is that of a plane reflection. In terms of translations and perpendiculars, we can define this sort of mapping as follows:

DEFINITION 1. f *is a reflection in the plane* π *if and only if* $f(P) = P + (M - P)2$, *where M is the foot of the perpendicular from P to π.*

Imagining that we are viewing the plane π "end on," we may picture what the reflection f in π does to points, as shown in figure 8.52. After a fashion, the reflection f may be thought of as a mapping that pulls points from one side of the plane π to the other and leaves the points of π fixed. That any such reflection is an isometry follows immediately from what we already know about translations, midpoints, and distance. One argument to establish this fact follows.

Using the notation in figure 8.52, let $\mathbf{m} = M - P$ and $\mathbf{n} = N - Q$. Then $M - f(P) = -\mathbf{m}$ and $N - f(Q) = -\mathbf{n}$, and so, by the bypass postulate,

$$Q - P = \mathbf{m} + (N - M) + -\mathbf{n} = (\mathbf{m} - \mathbf{n}) + (N - M)$$

and

$$f(Q) - f(P) = (\mathbf{n} - \mathbf{m}) + (N - M).$$

Noting that $\mathbf{m} - \mathbf{n}$ and $\mathbf{n} - \mathbf{m}$ are both orthogonal to $N - M$, it follows that

$$d(P,Q) = \sqrt{||\mathbf{m} - \mathbf{n}||^2 + ||N - M||^2}$$

and

$$d(f(P), f(Q)) = \sqrt{||\mathbf{n} - \mathbf{m}||^2 + ||N - M||^2}.$$

Thus, since $||\mathbf{m} - \mathbf{n}|| = ||\mathbf{n} - \mathbf{m}||$, it follows that

$$d(P,Q) = d(f(P), f(Q)),$$

that is, the reflection f is an isometry.

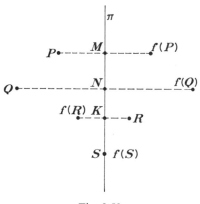

Fig. 8.52

Some obvious results about isometries, which we shall simply list, follow:

1. Composition of any two isometries is an isometry.
2. Each isometry has an inverse (which is itself an isometry).
3. The identity mapping, $\mathbf{0}$, is an isometry.

(These may be summarized by simply noting that the set of isometries is a group with respect to composition.)

One way to organize one's knowledge about congruence is to formulate some basic notions about isometries and then derive some useful theorems about some basic geometric figures, such as segments, angles, and triangles. To this end, we may define congruent figures as follows:

DEFINITION 2.   *A first set is congruent to a second set if and only if there is an isometry that maps the first set on the second.*

Thus, to show that two figures are congruent, all we need do is establish that some isometry maps one of the figures on the other. (That congruence as a relation is symmetric, reflexive, and transitive is convenient to note and is a consequence of the fact noted above—that isometries form a group under composition.) By concentrating our attention on certain classes of point sets, such as segments, rays, angles, and triangles, we may begin to collect some useful congruence theorems about these figures.

First, we establish the following congruence theorem about segments:

THEOREM.   *A first segment is congruent to a second if and only if the measures of the segments are equal.*

To establish this, suppose first that segments AB and CD in figure 8.53 have the same measure; that is, that $d(A,B) = d(C,D)$. Let $\mathbf{a}$ be the translation that maps A on C. Since $\mathbf{a}$ is an isometry, $d(C,B + \mathbf{a}) = d(A,B)$. Of course, where $B + \mathbf{a} = D$, $\mathbf{a}$ maps segment AB on segment CD, and we are finished, for in this event segments AB and CD are congruent. Where $B + \mathbf{a} \neq D$, let f be the reflection in the plane π through C and perpendicular to the line containing $B + \mathbf{a}$ and D. It follows that $f(B + \mathbf{a}) = D$, and so the isometry $f \circ \mathbf{a}$, that is, f composed with $\mathbf{a}$, maps segment AB on segment CD. Hence, in any case, segments AB and CD are congruent.

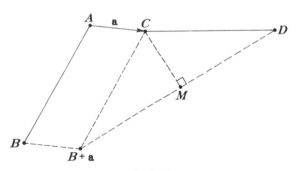

Fig. 8.53

In precisely this same way, it is easy to show that any two lines or two rays are congruent. With just a bit more work, it can be established that two angles are congruent if and only if they have the same cosine measure; out of this will fall the side-side-side and side-angle-side triangle congruence theorems.

Many properties of isometries are of considerable interest and importance from a mathematical as well as an organizational point of view. It is particularly interesting that *any* isometry of three-dimensional space can be described as the resultant of composing a sequence of plane reflections. For example, any translation can be described in terms of reflections in two parallel planes; any rotation is the resultant of reflections in two intersecting planes; and, although it is a nontrivial matter, it turns out that *any* isometry of three-dimensional space is the resultant of at most four plane reflections. So, in a sense, plane reflections are the most basic of all isometries of three-dimensional space. But the amount of material one has to understand or at least have a feeling for weighs against making use of plane reflections as a starting point in a formal development of isometries, at least in a *first* attempt at formalization.

Much is to be said for basing the notion of congruence on isometries, and a formal development of this notion (and that of similarity) in a course in geometry has much merit in it. This will no doubt be discussed in some detail in other sections of this book. The purpose of this brief excursion into the realm of isometries was simply to demonstrate that one could make use of this very general notion within the framework of a vector approach to Euclidean geometry.

Spheres and circles

From an intuitive point of view, a sphere is the set of points each of which is a given distance from a given point. A circle is a set of points *in a plane* each of which is a given distance from a given point in that plane. In each case, the given distance is called the *radius* and the given point is called the *center*. Both of these notions are easy to define in terms of norms of translations:

DEFINITION. *The sphere with center C and radius $r > 0$ is the set of points P such that $r = \|P - C\|$.* (See fig. 8.54.)

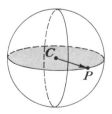

Fig. 8.54. Sphere with center C and radius $\|P - C\|$

DEFINITION. *The circle of π with center C in π and radius $r > 0$ is the set of points P in π such that $r = \|P - C\|$.* (See fig. 8.55.)

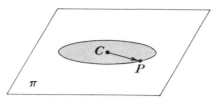

Fig. 8.55.   Circle with center C and radius $\|P - C\|$

It is not difficult to establish that two spheres are congruent if and only if they have the same radius:

> Let K_1 be a sphere with center A and radius a, and let K_2 be a sphere with center B and radius b. Now, any isometry that maps K_1 on K_2 will map A on B. If $a \neq b$, then no isometry can map K_1 on K_2 because points of K_2 will not be the same distance from B as points of K_1 are from A. So, K_1 is not congruent to K_2 if $a \neq b$. If $a = b$, the translation (isometry) $B - A$ maps K_1 on K_2. So, K_1 is congruent to K_2 if $a = b$. Hence, K_1 is congruent to K_2 if and only if $a = b$.

Other important properties of spheres and circles, such as

> (a) congruent chords of congruent circles subtend congruent minor arcs

and

> (b) semicircular arcs of congruent circles are congruent,

may be established through the use of isometries. The usual incidence relations between a line and a circle, a line and a sphere, and a plane and a sphere may be developed in the usual ways because we already have, in addition to the vector algebra, all the usual mathematical machinery related to perpendicularity and distance.

Concerning angles inscribed in a circle, one can be fairly well convinced that two angles inscribed in the same circular arc are congruent by performing the following experiment:

> Make a tracing out of an inscribed angle ACB and cut out a wedge-shaped piece of stiff paper from the tracing, as suggested by figure 8.56.

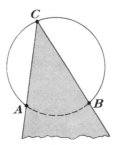

Fig. 8.56

Mark a point P on one edge of the wedge. Place the wedge on the figure so that the edge marked P just covers the point A and so that the vertex of the wedge is on the circle. Notice that the other edge just passes over B. Do this in at least two ways. Next, place the wedge so that both edges just cover the points A and B and notice that the vertex of the wedge is now on the circle. Do this at least two times.

The proof that this is true (using vector algebra methods) is somewhat long, and so for the purposes of this chapter, it is probably best to omit it.[3] Needless to say, the algebraic solution will lay bare the complete solution of the problem, which may be stated as follows:

An angle inscribed in a major arc of a circle is half as large as its corresponding central angle.

An angle inscribed in a minor arc of a circle is the supplement of an angle half as large as its corresponding central angle.

A concluding remark

It is hoped that by now the reader has a fairly good idea of how one can make use of vector methods in developing Euclidean geometry within the context of contemporary mathematics. The attempt has been made to give the reader not only a feeling for the formal vector methods that may be applied both in the solutions of geometry problems and in the discovery of properties of geometric figures but in addition a feeling for the under-lying intuitions about points and translations that are so important to rendering geometric interpretations to the results of any formal work with vector methods.

The manner in which this particular kind of course evolves is, of course, typical of the evolution of all theoretic knowledge. The scheme employed involves beginning with a general (but probably vague) principle, finding a significant instance where the notion may be given a very concrete and precise meaning, and using that instance as a stepping-stone to advance gradually toward generality. In this advance, we are guided by mathematical construction and abstraction. If we are lucky, as we seem to be in this course, we end up with an idea no less universal than the one from which we started. Gone, at least from the surface, may be much of the emotional appeal, but present is the even greater unifying power of thought —and it is exact rather than vague (7, p. 6).

---

3. Some persons may say at this point, "Aha! The really difficult problems are swept under the rug." This is not so, for the difficulties are due, for the most part, to the length of the argument. The proof of this theorem in *any* formal system has some underlying difficulties. The usual proof involving degree measures of arcs has, from a formal mathematical point of view, the difficulties that are inherent in the establishment and assignment of such measures. And the latter is much more difficult to handle formally than is the vector algebra needed to solve this same problem in the formal system being discussed here.

A Case for a Vector Approach
to Euclidean Geometry

Vector spaces are important mathematical structures in their own right and as such deserve some attention in the high school curriculum. But the fact that vector spaces serve as underpinnings for many applications of mathematics in a wide variety of fields—for example, the physical and social sciences, business, and engineering—is perhaps an even more compelling reason for devoting some attention to them in the high school program. It seems to be a fact of life that most high school students study mathematics for two years—taking a year of algebra and a year of geometry—in order to meet the requirements for a high school diploma. So if we are to include some work with vectors and vector spaces in the high school program in order to introduce these notions to the mass of students who take only two years of the program, we must begin this work during the first two years of the program. Because of the strong intuitive links one can build between a vector space and geometric notions, it is natural to look to the high school geometry course as a place in which to develop the formal notion of a vector space and, so, of a vector. Thus let us accept for the moment the premises that it is reasonable to introduce vectorial notions into the high school curriculum and that it is reasonable to do this within the framework of the geometry course.

Keep in mind that one of the principal objectives of the high school geometry program is that of developing a formal structure, that is, postulates, definitions, theorems, and so on, with which to make a more careful study of Euclidean geometry than that which is taught in the elementary and junior high school programs. That this has come to mean that geometry is the subject where one "learns to write two-column proofs," at least in the eyes of the students who have studied geometry, is the focal point of much of the criticism that has been leveled at high school geometry. It is not so much that the study of geometry within the framework of some formal structure is under attack; rather, it is the fact that the axiomatic structure is given to the student in an ad hoc way, and most of the emphasis in the course of study appears to be on constructing detailed and apparently logically tight proofs. Clearly, the emphasis must be shifted toward developing insights into how geometric intuitions may be linked to a formal structure in order to give one the power of rendering geometric interpretations to problematic situations.

In support of this view, one need only turn to the oft-quoted—or misquoted—Jean Dieudonné in his now famous "Euclid must go!" speech at the Royaumont conference in 1959. It is appropriate to try to set the record straight here regarding Dieudonné's remark, which has been interpreted as meaning that the teaching of Euclidean geometry should be abolished. Nothing could be farther from the truth. The context in which he made the remark was that of pointing out the nature of the patchwork revisions of the high school curriculum and appealing for a much deeper reform to be undertaken. The following excerpts from his speech should make clear his true feelings on the subject and should also serve as a reminder of the grave danger of misinterpretation when a phrase is quoted out of context (2, p. 35):

> To prevent any misunderstanding, I also want to emphasize that, although I may have seemed sharply critical of geometry, I have no desire to minimize its importance. Never before have the language and ideas taken from geometry played a greater part in higher mathematics, and it is so obvious that applied mathematics has its background in geometry that it is hardly necessary to mention it. . . .
>
> . . . My quarrel is therefore not with the purpose, but with the *methods* of teaching geometry, and my chief claim is that it would be much better to base that teaching not on artificial notions and results which have no significance in most applications, but on the basic notions which will command and illuminate every question in which geometry intervenes. . . .
>
> . . . I therefore consider that it is one of the main tasks of secondary schools to train and develop the intuition of space in students, and at the same time to place it in the logical frame which will enable them to use it later. Nothing should be spared to reach that goal as soon and as completely as possible.

There seems to be no way that these remarks can be interpreted to mean that Dieudonné is advocating that Euclidean geometry be cast out of the high school curriculum. Rather, he is making a strong appeal that Euclidean geometry be taught *so as to be useful within the framework of modern mathematics.* That this can be done by means of a vector approach has been demonstrated in the main body of this chapter. Furthermore, from this discussion it should be clear that rather than wishing to abolish Euclidean geometry from the high school curriculum, those who advocate enriching the high school curriculum with matters dealing with vector spaces are supportive of, and intent on, promoting the subject quite strongly.

It should be granted, then, that from the point of view of mathematical progress, not only is it reasonable to introduce vectorial notions into the high school curriculum within the framework of the geometry course, but basing the formal study of geometry at this level on the underlying vector space of translations will make the study of geometry an integral part of the study of modern mathematics. But what about the student? Will not a shift in the foundations on which his course is based affect the end product and how he views it? Will not such a shift result in previously emphasized, and thus important, standard topics receiving considerably less attention? Will not such a shift result in subject matter too sophisticated for his level being introduced into the course, thereby making the learning of the subject too difficult for the student? Will not such a shift diminish the historical significance of the course? We shall discuss each of these questions in turn.

What about the student? Certainly, in any course, we as teachers must concern ourselves with helping the student cope with the ideas that are basic to the course. The extent to which these ideas can be related to the student's background of experiences is related directly to the success we shall have in helping to make the subject matter an integral part of the student's store of applicable knowledge. The effect of the standard treatments of geometry has been—although it may not need to have been—to formalize the subject in isolation from the rest of modern mathematics, and so, in consequence, the subject matter of geometry has *not* become an integral part of the student's store of applicable mathematical knowledge. It is the intent of this chapter to demonstrate that a vector approach to geometry can correct this situation at the same time that it introduces the important subject of vectors and vector spaces into the high school curriculum. Therefore, for the student a vector approach to geometry not only introduces him to a new and important mathematical structure in a very concrete way, but it also provides him with a formal introduction to geometry that will be useful to him.

Will a shift in the foundations on which the geometry course is based affect the end product? It is certainly hoped so! Any shift that does not affect the end product will simply result in another way in which to develop geometry in isolation from the rest of modern mathematics. The whole purpose of seeking alternate foundations from which to develop Euclidean geometry is to enable us to present the subject in its true light as an integral part of modern mathematics. Whether one does this in a way that maintains identifiable one-year segments of "algebra" and "geometry" (and thereby does not "buck the system") or seeks other ways to do this is a matter of choice. That it behooves us to do so in *some* way is clear.

Will such a shift result in less emphasis being placed on some topics currently popular in standard developments? This will no doubt happen for very obvious reasons. In the first place, one cannot shift the basis of development, bring in new subject matter, and still expect to place the same degree of emphasis on all the old "pet" topics. Secondly, a shift to develop geometry as an integral part of the rest of mathematics will, or should, cause a shift away from devoting a lot of effort, say, to writing two-column proofs about the congruence of triangles that happen to share a common side or a common angle, to rendering geometric interpretations to problematic situations, and to learning to give arguments within the framework of the formal system about conclusions related to those situations. In this sense, the process of argumentation (or giving proofs) is put in its proper secondary position to the process of geometric formulation of problematic situations. This is not to say that learning to give mathematical arguments is unimportant. It is to say that giving mathematical arguments is not the sole concern in the study of geometry but rather pervades all mathematics and its applications.

Will such a shift be too sophisticated for the student to handle? This could, but need not, happen. However, if one keeps in mind that what is being proposed is that the formal structure on which geometry is to be developed should grow in a natural way out of the student's experiences and intuitions about physical space, there should be little danger of this happening. Certainly, the standard developments of geometry are at least as abstract in their foundations as a vector approach to Euclidean geometry is. And the formal system underlying the standard developments has components that are at least as arbitrary and much less obviously tied in an applicable way to the geometrization of problematic situations than a vector approach is.

Finally, will such a shift diminish the historical significance of the subject? There is little doubt that exactly the contrary should occur. Remember that Euclidean geometry is at the heart of all applied mathematics— Albert Blank has said, "Sometimes we forget that Euclidean geometry is applied mathematics" (1, p. 15)—and that both the formal structure and the present emphasis in standard developments of the subject leave it standing apart from the rest of modern mathematics, withering on the vine, so to speak. So by shifting the basis for the development of the subject to a vector approach, we not only bring the subject into the fold as an integral part of modern mathematics but thereby enhance its historical significance as well. For not only was it the subject matter first organized as a deductive system some twenty-three hundred years ago, but it pervades all of modern-day applications of mathematics. In what better way can the importance and historical significance of a subject be measured?

In some general ways, we have been alluding to the advantages of a vector approach to Euclidean geometry from the point of view of mathematical structure and the accessibility of applications within the framework of modern mathematics. But what are some advantages of such an approach from the point of view of the student? As we see it, there are several such advantages, some of which follow. (No attempt has been made to list them in order of importance.) A vector approach to Euclidean geometry—

1. acquaints the student with important kinds of algebraic structure by working with relatively concrete geometrical examples of them;

2. places some emphasis on the use of transformations from one space of objects to another as a useful tool for analyzing relations, which serves as an excellent foundation for applying mathematics to the analysis of physical situations as well as for any subsequent formal study of linear algebra and its applications;

3. provides some introductory work with function spaces;

4. integrates plane and solid geometry in a natural way, which, in addition, makes it easy to point out the possibility of studying higher-dimensional Euclidean geometry and gives the student a feeling for the way in which these higher-dimensional geometries are, in a sense, richer than the lower-dimensional ones;

5. integrates, in a real sense, the subjects of algebra and geometry and enables the student to develop some insight and understanding for the unity of mathematics;

6. provides a sound mathematical basis for the application of vector methods in the sciences and other fields;

7. affords the student a continuum of both the development and practice of algebraic skills and the rendering of geometric interpretations to problematic situations;

8. helps to develop the student's awareness that mathematics can be used to model physical phenomena;

9. affords the student considerable versatility of attack on problems because of the relatively early stage at which both the so-called synthetic and analytic methods can be brought to bear on a situation whenever they can compete successfully with the vector methods at hand.

Of course, no matter what advantages one development of a subject has over others, it is still true that the teacher must help to create a classroom atmosphere in which students can actively participate in the on-going development of the subject. As Henry Pollack has so aptly put it, you cannot "teach mathematics to anybody as a spectator sport. To get any feeling for it at all you have to participate" (3, p. 74).

The case for reform in the teaching of geometry has been made. To teach students a vector approach to Euclidean geometry, we clearly need to get them actively involved in learning about vector methods in geometry.

And to do this, we need to have teachers who have acquired a good understanding of how vector methods may be used to develop geometry. Further, we need school systems that are not so wrapped up in preserving the traditional mathematics courses that they prevent this much-needed reform from taking place. These are matters involved with the training of teachers, the subject of the concluding chapter of this book. Of course, school systems could, and should, support in-service programs for improving their curricular offerings, and it would help if many field-tested and proved model texbooks with detailed teacher's commentaries were published so that such in-service groups would have a basis for the study and implementation of new ideas. With all that might be, we can only hope that enough people will see the need to act and will work to achieve the worthwhile goal of bringing Euclidean geometry into its rightfully prominent place in the modern mathematics program.

BIBLIOGRAPHY

1. Blank, Albert. "The Use and Abuse of the Axiomatic Method in High School Teaching." In *The Role of Axiomatics and Problem Solving in Mathematics*, pp. 13–19. Boston: Ginn & Co., 1966.
2. Dieudonné, Jean. "New Thinking in School Mathematics." In *New Thinking in School Mathematics*, edited by Howard Fehr, pp. 31–46. Organization for European Economic Cooperation: Paris, 1961.
3. Pollack, Henry. "Panel Discussion." *Educational Studies in Mathematics* 1 (May 1968): 61–79.
4. Szabo, Steven. "An Approach to Euclidean Geometry through Vectors." *Mathematics Teacher* 59 (March 1966): 218–35.
5. Vaughan, Herbert E., and Steven Szabo. *A Vector Approach to Euclidean Geometry: Vector Spaces and Affine Geometry*. Vol. 1. New York: Macmillan Co., 1971.
6. ———. *A Vector Approach to Euclidean Geometry: Inner Products, Euclidean Geometry, and Trigonometry*. Vol. 2. New York: Macmillan Co., 1973.
7. Weyl, Hermann. *Symmetry*. Princeton, N.J.: Princeton University Press, 1952.

9

Geometry in an Integrated Program

HARRY SITOMER

THE PRECEDING chapters describe a variety of approaches to school geometry. Some chapters, however, do not limit themselves to a single approach. For instance, chapter 7, "An Affine Approach to Euclidean Geometry," employs Euclid's synthetic methods (described in chapter 4), coordinates (described in chapter 5), and transformations (described in chapter 6). Each additional approach introduces the methods and content of other branches of mathematics, thus revealing close relationships between geometry and these other branches. As more and more new approaches are used, the organization of geometry and its "texture " change. When no restrictions (other than pedagogic) are placed on the number of approaches, geometry takes its place in an integrated, or unified, program of teaching mathematics.

In this chapter I shall—

1. trace the history of integrated approaches;
2. briefly describe some early attempts to teach geometry in an integrated program;
3. describe in some detail the latest, and probably the most thoroughgoing, attempt at an integrated approach, namely, the six-year course written by the Secondary School Mathematics Curriculum Improvement Study (SSMCIS);
4. discuss some major pedagogic issues that emerge out of an integrated approach, in particular, the place of axiomatics in secondary school mathematics.

303

Early History

Before the invention of coordinate geometry, mathematicians used almost exclusively the synthetic methods of Euclid. Some exceptions, however, did occur—in the researches of Appollonius and Archimedes, in the algebraic applications to geometry of Vieta, Ghetaldi, and Oughtred, and in the works of some Arabian mathematicians. But it was not until 1637 in his *Géométrie* that René Descartes (1596–1650) first made systematic use of numbers in the study of geometry. Credit must also be given to Pierre de Fermat (1601–1665), whose *Varia Opera Mathematica,* published in 1679, used methods similar to those of Descartes. The major contribution of Descartes and Fermat was the principle that an equation of two variables corresponds to a geometric locus in a plane (1, p. 175).

Analytic geometry as we know it today was developed by others with the formal introduction of coordinate axes and coordinates of points. The immediate result of this new approach to geometry was the discovery of many new theorems and the simplification of the proofs of some old theorems.

With mathematical inventions such as transformations, vectors, complex numbers, matrices, groups, and vector spaces, geometry flourished beyond Euclid's wildest dreams. Mathematicians in general felt free to use any available methods in solving their problems. Furthermore, methods and theorems of geometry were used to solve problems in algebra, and the boundary between algebra and geometry blurred. Sometimes it was difficult to tell whether one was working in algebra or geometry.

Despite these newly discovered methods, which were exploited to such good advantage by mathematicians, teachers of geometry in secondary schools were either oblivious to these developments or unwilling to introduce them into the classroom, and so until the end of the nineteenth century, school geometry, with some minor changes that altered neither the spirit nor the methods, was still Euclid's.

Felix Klein

Felix Klein, the celebrated mathematician at Göttingen, Germany, furnished an impetus to the teaching of geometry in an integrated program. At the University of Göttingen in 1908, he gave a series of lectures designed for teachers and prospective teachers in German secondary schools. These lectures were eventually printed in three volumes, each called *Elementary Mathematics from an Advanced Standpoint.* Two of them, subtitled *Arithmetic, Algebra, Analysis* and *Geometry,* were translated into English in 1932 and 1939, respectively. In the preface to *Elementary Mathematics from an Advanced Standpoint: Geometry,* Klein clarifies his position on viewing mathematics as a whole:

> Allow me to make a last general remark, in order to avoid a misunderstanding which might arise from the nominal separation of this "geometric" part of my lectures from the first arithmetic part. In spite of this separation, I advocate here, as always in such general lectures, a tendency which I like best to designate by the phrase *"fusion of arithmetic and geometry"*— meaning by arithmetic, as is usual in the schools, the field which includes not merely the theory of integers, but also the whole of algebra and analysis. [3, p. 2]

Indeed, Klein followed this program of "fusion" in his lectures.

Again, in the course of one of his lectures, he reaffirms his belief in the integration of mathematics:

> From time to time, it has been proposed that geometry, as an independent subject of instruction, be separated from mathematics, and that, generally speaking, mathematics, for purposes of instruction, be resolved into its separate disciplines. In fact, there have been created, especially in foreign universities, special professorships for geometry, algebra, differential calculus, etc. From the preceding discussion, I should like to draw the inference that the creation of such narrow limits is not advisable. On the contrary, the greatest possible living interaction of the different branches of the science which have a common interest should be permitted. Each single branch should feel itself, in principle, as representing mathematics as a whole. Following the same idea, I favor the most active relations between mathematicians and the representatives of all the different sciences. [3, p. 56]

In his lectures to teachers, Klein presented mathematics, not as isolated disciplines, but as an integrated, living organism. Most people feel intuitively that the objects of geometry (points and lines) and the objects of algebra (numbers) are distinct and should be studied separately. But when points and numbers were organically related, as in analytic geometry, an explosion in mathematical discovery resulted both in geometry and in algebra. Klein argued that if algebra and geometry can illuminate each other for mathematicians engaged in research, it would seem reasonable to expect a similar advantage for students if they were to learn mathematics as an integrated whole.

John A. Swenson

As far as the writer knows, the first systematic attempt in the United States to teach integrated mathematics was made by John A. Swenson (1880–1944). He wrote five books, four of which were published at his own expense. They were designed for grades 9 through 12 and satisfied the requirements of the New York State Syllabi and Examinations. They contain a mine of creative ideas for teaching mathematics in an integrated manner. The novelty of his approach is illustrated by the following two examples, in which proofs have been modified to conform to modern notation.

1. To prepare for the theorem that an angle inscribed in a circle has half as many degrees as its intercepted arc, he proved the following theorem (11, p. 400):

THEOREM. *A secant forming an angle of $x°$ with a radius at its outer extremity cuts off a minor arc containing $180 - 2x$ degrees.*

Proof.

$$m \angle A = x = m \angle B \text{ (see fig. 9.1).}$$
$$\therefore m \angle O = 180 - 2x = y.$$

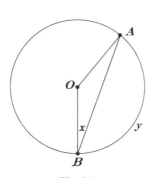
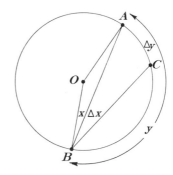

Fig. 9.1   Fig. 9.2

To prove $m\angle ABC = \frac{1}{2} m\widehat{AC}$, the conclusion of the theorem, let $x = m\angle ABC$ (see fig. 9.2). Then $y = m\angle AOB = 180 - 2x$.

$$\therefore \frac{\Delta y}{\Delta x} = -2 \text{ (!!), or } |\Delta x| = \left|\frac{1}{2}\Delta y\right|.$$

Since $|\Delta y| = m\widehat{AC}$ and $|\Delta x| = m\angle ABC$, $m\angle ABC = \frac{1}{2}m\widehat{AC}$.

Note the use of increments. Swenson introduced this concept in his book 1 (grade 9), where it is *not* related to derivatives.

2. The triangle congruence theorems are introduced by an experiment (12, p. 85) on $\triangle ABC$ where $\angle A$ and $\overline{AB}$ are fixed and $\angle B$ is allowed to vary (see fig. 9.3). The study seeks to relate the magnitudes of $\overline{AC_i}$ and $\angle ABC_i$, $(i = 1,2, \ldots)$. The results are shown in the graph in figure 9.4.

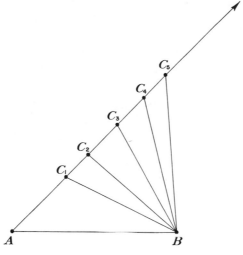

Fig. 9.3

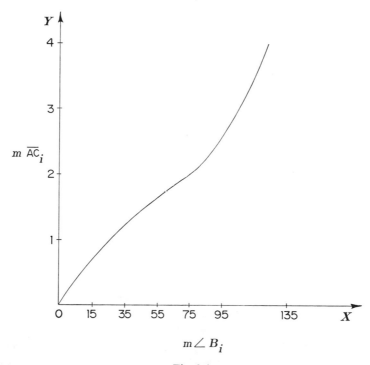

Fig. 9.4

Note that the function involved is monotonic increasing. (Swenson describes this by saying that $\angle B$ and $\overline{AC}$ vary in the same sense.) There is a one-to-one correspondence between the set of angles $\{B_i\}$ and the set of opposite sides $\{AC_i\}$. Thus each angle B_i in the set determines an opposite side AC_i, and conversely, each side AC_i determines an angle B_i. He concludes with the following experimentally verified statement:

> If in a triangle a side and one adjacent angle are constant, then the other adjacent angle and its opposite side are so related that a given value of one determines the other uniquely. [11, p. 86]

This leads easily to the conclusion that if the parts, ASA, of a triangle are constant, then the other parts of the triangle are determined. For if in $\triangle ABC$ (see fig. 9.5), $\angle A$, $\overline{AB}$, and $\angle B$ are constant, then the side opposite $\angle B$, namely, AC, is determined, as well as the side opposite $\angle A$, namely, $\overline{BC}$. Since $\angle A$ and $\overline{AB}$ are constant, $\angle C$, the angle opposite AB, is determined.

Thus, if in two triangles ASA = ASA, then the remaining corresponding parts are congruent, and hence the triangles are congruent. It is a short step to show that if $\overline{AB}$, $\angle B$, and $\overline{BC}$ of $\triangle ABC$ are constant, then the angles opposite $\overline{AB}$ and $\overline{BC}$ are determined, as well as the side opposite $\angle B$. Thus one proves the SAS theorem from the ASA assumption.

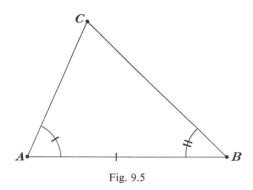

Fig. 9.5

Two observations concerning Swenson's treatment are in order. First, it exploited the function concept, which was defined as a "dependence" of one variable on another. Second, his books included all the items in the New York State Syllabus, and since this syllabus included geometry separately, he wrote a separate geometry text.

From my talks with him, I gathered that this did not represent Swenson's opinion of what *should* be taught. He made the point that although a new course of study had been introduced in the 1930s, the teaching of geometry (and other branches of mathematics) could be improved by allowing the use of methods and concepts from other branches of mathematics.

The word *integrated* is sometimes used to describe a sequence of courses in mathematics in which each course is constructed in the traditional way. Notable here is the series of four textbooks by Price, Peak, and Jones called *Mathematics: An Integrated Series,* in which book 2 has the subtitle *Lines/Planes/Space* (4). There are many excellent features in this geometry. However, there is no attempt to use more than one approach to geometry (except for the use of coordinates near the end of the book, where the meaning of the word *integrated* is restricted to that of multiple approaches).

Secondary School Mathematics Curriculum Improvement Study

The most thoroughgoing attempt known to the writer to use an integrated approach is the six-year course for grades 7 through 12 written by the Secondary School Mathematics Curriculum Improvement Study (SSMCIS). This course will be described in detail.

First, a bit of the history of this study might be of interest. *Course 1* for the seventh grade was planned and written in the summer of 1966. Since then, an additional course has been written each successive summer following a June planning conference by mathematicians and teacher-writers. The teachers did the actual writing of the chapters; the mathematicians reviewed them. Also during each summer, a corps of teachers was given in-service training to acquaint them with the course content, to help them gain in-depth understanding, and to discuss teaching procedures.

Teachers who received in-service training then taught the courses to above-average students (estimated as the upper 15 percent) and reported their experiences. These reports served as a basis for the revision of each course. Thus each course had an experimental and a revised edition. The experimental edition of *Course 6* was written in the summer of 1971. The student texts are called *Unified Modern Mathematics.* Each is accompanied by a teacher's commentary.

The study was sponsored by Teachers College, Columbia University, and was directed by Howard F. Fehr. It was funded first by the Office of Education (HEW) and later by the National Science Foundation.

Course 1 (grade 7)

Transformations play a most important role in SSMCIS. The first reference to a transformation occurs in chapter 1, "Finite Number Systems," where the rotations of a regular hexagon through multiples of 60° are used intuitively to interpret operations in a finite operational system with six numbers. These rotations are restricted to the vertices of regular polygons. With operations and mathematical systems as background, chapter 3, "Mathematical Mappings," discusses mappings on the set of integers, with much attention to arrow diagrams on natural number lines. For instance, the mapping $n \rightarrow n + 2$ is shown in figure 9.6. In chapter 4,

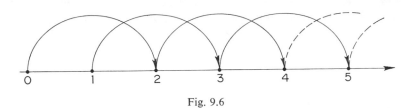

Fig. 9.6

"The Integers and Addition," the resemblance to a translation prompts the introduction of translations on the integer line in order to verify some properties of addition of integers. Thus, for instance, the translations $n \rightarrow n + 2$ and $n \rightarrow n + -2$ are inverses, since their composition is $n \rightarrow n + 0$, the identity translation. Following an informal discussion, these translations verify those properties of addition with natural numbers that are to be retained.

In chapter 6, "Multiplication of Integers," a similar discussion leads to dilations on the integer line. For instance, the dilation $n \rightarrow 2n$ is shown in figure 9.7. Properties of multiplication are verified by examining these dilations. For instance, the dilation $n \rightarrow -2n$, (considered as the composition of the dilation $n \rightarrow 2n$ and the reflection, or half-turn, in the origin) verifies that the product of two negative integers is a positive integer.

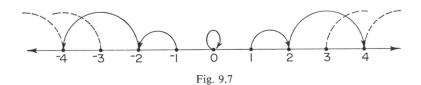

Fig. 9.7

A first step toward coordinates in a plane is taken in chapter 7, "Lattice Points in the Plane." This chapter extends the study of integers to the set of ordered pairs of integers and to conditions on these pairs. After some study of coordinates of lattice points, compound conditions, and graphs, the student is ready to see how mappings from Z to Z may be represented by a set of lattice points. A brief introduction to lattice points in space follows, as well as an introduction to some new terms, such as *parallele-*

piped. The chapter ends with both a look at translations from $Z \times Z$ to $Z \times Z$ and how to represent them by coordinate rules designated by $(x,y) \rightarrow (x + a, y + b)$ and a brief introduction to dilations from $Z \times Z$ to $Z \times Z$ with the rule $(x,y) \rightarrow (ax,ay)$, where a is a nonzero integer. Throughout the chapter, students are asked to explore informally and to seek properties of translations, dilations, and compositions.

These experiences with lattice points are then extended in chapter 9, "Transformations of the Plane," to all points of the plane through paper-folding activities, making diagrams, observing, and learning the vocabulary that seems to express clearly and vividly any observed geometric relations. The transformations treated here are reflections in a line, reflections in a point (or half-turns), rotations, and translations. Several geometric terms are also introduced (or reviewed), such as *line, ray, segment, perpendicular lines, parallel lines, angle,* and *triangle.* The notations are those of the School Mathematics Study Group (SMSG). Throughout this introduction no effort is made to prove any of the assertions. Most of the work is accomplished by a sequence of *activities,* such as the following:

> *Activity 6*: On a sheet of unlined paper draw a line *m* and joint the non-collinear points *A, B, C,* as shown in figure 9.8. The figure formed is called triangle *ABC.* Find the reflection of triangle *ABC* in *m.* Compare the angles at *A, B,* and *C* with those at *A', B', C'.* Then compare the lengths of $\overline{AB}$, $\overline{AC}$, and $\overline{BC}$ with the lengths of $\overline{A'B'}$, $\overline{A'C'}$, and $\overline{B'C'}$.

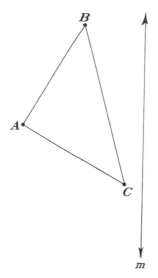

Fig. 9.8

Cut out triangle ABC. See if you can make it fit on $A'B'C'$. Did you have to turn ABC over before making it fit? Will it always be necessary to turn it over? If not, when will it be unnecessary? Try this experiment again with a different triangle. Try special kinds of triangles. [5, pt. 2, p. 88]

These activities suggest a search for invariants which eventually becomes the prime motivation for showing how transformations agree or differ and why they may be classified as isometries.

The following exercise at the end of this chapter in the review exercises suggests the variety of concepts considered (5, pt. 2, p. 112):

Fill in the table with "Yes" if the mapping always has the property and "No" if it does not.

Mapping preserves	Reflection in a line	Symmetry in a point	Translation	Rotation
distances (isometry)				
collinearity				
betweenness				
midpoint				
angle measure				
parallelism				

This informal introduction to isometries is then made more precise in chapter 10, "Segments, Angles, and Isometries." The subsets of a line are introduced by the line separation principle; namely, any point P on a line l separates the rest of l into two disjoint sets such that—

1. if A and B are two distinct points in one of these sets, then all points between A and B are in this set;

2. if A is in one set and C is in the other, then P is between A and C.

(No attempt is made here to make precise the meaning of betweenness for points.) This principle leads directly to the ideas of half-lines, rays, and, by analogy, to the plane separation principle—thence to half-planes, quadrants of a plane, coordinate systems, and angles.

The measure of a segment, when related to translation and half-turns (changing the direction) of the ruler, is seen to be an invariant under both these transformations and gives the student a better understanding of (1) how to use a ruler and (2) the formula for distance, $|x_1 - x_2|$. Line coordinate systems are given a more formal status; midpoints and other points of division are handled with the aid of coordinates.

Plane coordinate systems and coordinates result from the composition of two translations—one parallel to the *x*-axis, the other to the *y*-axis. This, together with the midpoint formula, permits the student to see a parallelogram as a quadrilateral whose opposite sides are parallel or whose diagonals bisect each other.

Perpendicular lines are reviewed, this time in terms of line reflections. Two lines *l* and *m* are perpendicular if $l \neq m$ and *l* is invariant under a reflection in *m*. This paves the way for rectangular coordinate systems (as contrasted with affine) and coordinate rules for reflections in the *x*- or *y*-axis. This enables the student to review some properties of line reflections with the aid of coordinate rules $(x,y) \rightarrow (x,-y)$ or $(x,y) \rightarrow (-x,y)$.

Angles are defined as the intersection of two half-planes and are measured with the aid of a semicircular protractor. Boxing the compass relates direction to angles. Through drawings and measurements, the angle is seen to be invariant under the reflection in the line containing its midray. This leads to isosceles triangle and "kite" theorems. Properties of an angle are studied by noting its image under various isometries and assuming that an angle has the same measure as its image. For instance, since translations map lines onto parallel lines—and because of the parallel postulate—it is easy to see why a transversal across parallel lines forms a pair of corresponding angles whose measures are equal. The proof of the theorem that states that *the sum of the measures of the angles of a triangle is 180* gives the flavor of informal proof used (5, pt. 2, p. 178):

First consider the translation that maps *A* onto *C*. (See fig. 9.9.) This translation maps *C* onto *R* and *B* onto *S*. What are the images of $\overrightarrow{AB}$ and $\overrightarrow{AC}$ under this translation? Do you see that this translation maps $\angle CAB$ onto $\angle RCS$?

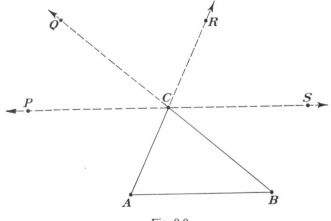

Fig. 9.9

Examine the translation that maps B onto C. Under this translation what is the image of $\overrightarrow{BA}$? Of $\angle ABC$?

The third mapping is a point reflection in C. Under this mapping what is the image of $\angle ACB$?

The remainder of the proof is quite obvious. Note the close resemblance between this proof using translations and the experimental proof involving the cutting of angles and the rearranging of them along a line. It is also suggested that students would find it interesting to find their own combination of isometries to prove this theorem. For instance, this might include the reflections in the midpoint of $\overline{CB}$.

The chapter ends with a study of measures of angles in special kinds of triangles and in convex polygons. Note that several times during the year, objects of geometry—points, lines, rays, segments—have been presented on different levels of sophistication, and they have been closely related to transformations of the plane. These transformations are used to write informal proofs based on isometries.

As *Course 1* indicates, the emphasis on congruence of triangles as the basic tool is replaced by isometries. This development becomes more prominent in subsequent courses. The following example is an exercise from the review exercises of *Course 1* (5, pt. 2, p. 189):

Exercise. In figure 9.10, $AB = AC$ and $DB = DC$. Using a line reflection, prove $m\angle DAB = m\angle DAC$.

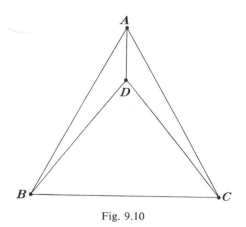

Fig. 9.10

The student is expected to use the reflection in the bisector of $\angle BAC$, which, being perpendicular to $\overline{BC}$, contains D. Hence, the image of $\angle DAB$, under this reflection, is $\angle DAC$. Hence the desired conclusion.

Course 2 (grade 8)

This course starts with a discussion of "Mathematical Language and Proof," including basic notions of logic and rules of inference. Illustrations involve number systems and geometric objects that appeared in *Course 1*.

Chapter 3, "An Introduction to Axiomatic Affine Geometry," introduces the notions of undefined terms, which are assigned properties by three initial axioms:

AXIOM 1. (*a*) *Plane* π *is a set of points, and it contains at least two lines.*

(*b*) *Each line in* π *is a set of points containing at least two points.*

AXIOM 2. *For every two points in plane* π *there is one and only one line in* π *containing them.*

AXIOM 3. *For every line m and point E in plane* π, *there is one and only one line in* π *containing E and parallel to m.*

After giving him the proofs of two theorems (mainly about the existence of points and lines) and asking him to cite reasons for two additional proofs, the student is then asked to write proofs for ten theorems, of which only two are really challenging. To emphasize the abstract nature of points, lines, and the plane in this system, a number of nongeometric models are displayed. These include some finite geometries. Parallelism between lines is then proved to be an equivalence relation, and a special study of parallel projections lays one of the foundations needed for coordinate systems. The chapter ends with an introduction to vectors as an equivalence class of directed segments and relates them to the group of translations of a plane. (Groups are discussed in chapter 2 of *Course 2*.) The development is frankly intuitive, suggesting that the student will not be confined by the rigorous requirements of axiomatic constructions. It is also suggested that there will be alternate treatments—some formal, others informal.

The following exercises suggest what might be expected of students (6, pt. 1, p. 190):

Consider the following "six-point geometry":

Plane P: The set of six vertices, $\{A, B, C, D, E, F\}$.

Line: Any pair of distinct points.

Point: Any one of the six vertices.

(*a*) Indicate which of the axioms 1, 2, 3 are true in this model and which are not true.

(*b*) Consider the line $\{A,B\}$ and the point F not in this line. How many lines are there which contain F and are parallel to $\{A,B\}$? Name them.

(*c*) Name all lines that contain A and are parallel to the line $\{F,B\}$.

The next major advance in geometry occurs in chapter 6, "Coordinate Geometry." It adds three more axioms to the three stated in chapter 3.

AXIOM 4. *For each pair of distinct points A and B of a line, there is exactly one coordinate system on that line in which A corresponds to 0 and B corresponds to 1.*

AXIOM 5. *If (A,B) and (A',B') are bases for coordinate systems on a line, then there is a relation $x' = ax + b$ with $a \neq 0$, which, for each point X of the line, relates its A,B-coordinate x to its A',B'-coordinate x'.*

The relation $x' = ax + b$ is seen to be a composition of a dilation and a translation on a line and is known as an affine mapping on a plane.

AXIOM 6. *Let f be a parallel projection from line l to line l'. Let A and B be distinct points of l and let A' and B' be their images under f. Then for every point X of l, the A'B'-coordinate of its image, X', is the same as the A,B-coordinate of X.*

Axioms 5 and 6 have the immediate effect of assigning equations to lines in a plane. Let l and l' be intersection lines (coordinate axes) with x- and y-coordinate systems, respectively, such that their point of intersection has coordinate 0 in both systems. Now let m be any line intersecting l and l'. The projection from l to m parallel to l' gives each point of m the coordinate x of the corresponding point of l. Similarly the projection from l' to m, parallel to l, gives each point of m the coordinate y of the corresponding point of l'. Hence by axiom 5 the pairs of coordinates of points on l are related by $y = ax + b$. (For lines parallel to a coordinate axis, one of the coordinates is constant.) This makes possible a study of intersections of lines and, by the slope a in $ax + b$, parallelism of lines. If we add the formula for midpoints, we have enough tools to prove some theorems with coordinates, as illustrated by the following theorem (6, pt. 2, p. 312):

The diagonals of a parallelogram bisect each other.

Proof. The A,B,D-coordinates of A, B, and D are (0,0), (1,0), and (0,1), respectively. (See fig. 9.11.) Since C is on the parallel to $\overleftrightarrow{AD}$ that

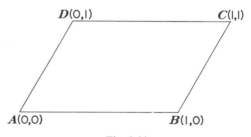

Fig. 9.11

passes through B, its x-coordinate is 1. Since C is on the parallel to $\overleftrightarrow{AB}$ that passes through D, its y-coordinate is 1. An equation for $\overleftrightarrow{AC}$ is

(1) $$y = x.$$

An equation for $\overleftrightarrow{BD}$ is

(2) $$x + y = 1.$$

Solving (1) with (2) gives the solution set $\{(\frac{1}{2}, \frac{1}{2})\}$. We conclude that the diagonals bisect each other. Explain why.

Note the convenience of affine coordinate systems, wherein oblique lines may be chosen as axes and any point on each axis, other than the origin, as unit point. It is this convenience that simplifies the first steps in studying the analytic geometry of the plane.

The transition from an affine to the Euclidean plane takes place by making plausible the Pythagorean property of right triangles, assuming that the area of a square region is the square of the length of a side. Four exact copies of a right triangle are arranged in a square region in two different ways, as shown in figure 9.12.

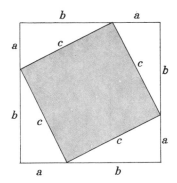 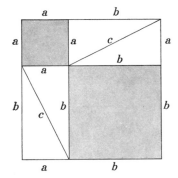

Fig. 9.12

The uncovered parts of the large squares (the shaded regions) must have the same area. Hence $c^2 = a^2 + b^2$.

With the introduction of rectangular coordinate systems with base (O,I,J) —where $\overleftrightarrow{OI} \perp \overleftrightarrow{OJ}$ and $OI = OJ = 1$—and the Pythagorean property, the distance formula can be deduced, and the student is in a position to prove some theorems that compare the lengths of segments on different lines.

Chapter 9, "Transformations in the Plane: Isometries," continues the distance-preserving transformations on a more sophisticated level. They are reviewed and extended to a more systematic treatment of invariants, including fixed points and fixed lines. Much use is made of coordinates to study inverse isometries and compositions of isometries (including glide reflections). The set of isometries of the plane is shown to be a group having as subgroups the set of direct isometries (rotations and translations), of which the set of translations is a subgroup. All isometries can be shown to be the composition of not more than three line reflections—a fact that is of special interest. Important too is the relationship among the ideas of isometry, congruence, and symmetry. The definitions of congruence and symmetry are quoted as follows:

> Two figures are called *congruent* if there is an isometry that maps one of the figures onto the other. [6, pt. 2, p. 173]

A figure is defined as any set of points. Thus it makes sense to talk about congruent lines, rays, segments, circles, or sets of disconnected points.

> A figure is *symmetric* if there is an isometry, other than the identity, that transforms the figure onto itself. [6, pt. 2, p. 175]

The following symmetric figures are of special interest in geometry: angles, isosceles triangles, parallelograms, isosceles trapezoids, and circles. Many of their properties follow from their symmetries.

The chapter ends with a section on dilations and similarities in order to avoid the impression that all transformations are isometries. A similarity is the composition of a dilation and an isometry.

The following sampling of exercises from chapter 9 may give some idea of the scope of this chapter. (In order to abbreviate them, liberties have been taken with longer problems that were "programmed" into parts.)

1. Given collinear points A, B, C. Show that their images under a line reflection are also collinear.

2. Show that the composition of translations is commutative.

3. Let P and Q be distinct points. Show that the composition of half-turns in P and Q is a translation from P toward Q, whose distance is $2PQ$.

4. Show that $(x,y) \rightarrow (y,x)$ is the coordinate rule for the reflection in the line with equation $y = x$.

5. Let F be the midpoint of $\overline{AB}$. Prove $H_A \circ H_F \circ H_B \circ H_F = I$, the identity transformation. (H_X is the half-turn in X.)

6. Does the set of opposite isometries form a group? (An isometry is opposite if it reverses the orientation of three noncollinear points.)

7. Prove that if f and g are isometries, then $(f \circ g)^{-1} = g^{-1} \circ f^{-1}$.

8. Show that the congruence relation is an equivalence relation.

9. Draw a picture of a dilation with scale factor $\frac{1}{2}$, followed by a translation, as it affects three noncollinear points.

Chapter 10, "Length, Area, Volume," starts with a discussion of four basic measurement principles: unit, congruence, additive, and real non-negative number properties. These principles are first applied to segments, plane regions, and solid regions in situations where counting suffices to measure; that is, where measures are integral or rational. For a rectangular region one of whose dimensions is an irrational number, an informal discussion concerning least upper bounds of a sequence of rectangular regions that are subsets of the given region is presented. A similar discussion for measuring segments—and in which least upper bounds are defined—occurs in chapter 5, "The Real Number System" (6, pt. 1, pp. 236–38). Formula areas are then derived for regions bounded by triangles, parallelograms, and trapezoids. To measure a more general region, the region is overlaid with a square grid, the "inner" squares are counted, and then both the "inner" and the "border" squares are counted. These two numbers furnish a lower and upper approximation of the area sought. By reducing the size of the squares in the grid, a second pair of lower and upper approximations is found. This process produces two sequences of numbers. If both sequences approach the same number, that number is the area of the region. The method is applied to derive the circumference of a circle and the area of a circular region.

Course 2 has an appendix called "Mass Points." A mass point is defined as an ordered pair (a,A) designated aA, where a is a positive number called the *weight of the mass point* and A is a point in a plane (or space). As for all ordered pairs, two mass points are equal iff their corresponding components are equal.

The concept of mass points is relatively easy. It is described in detail by Hausner in a book intended for college use (2). It was originally a chapter in *Course 2* of *Unified Modern Mathematics,* where it was well received by both teachers and students. But subsequently, because the topic was deemed not to be in the mainstream of mathematics, it was relegated to the appendix. No doubt it will receive less attention there than other chapters. It is described here for those who may not have easy access to *Unified Modern Mathematics, Course 2*.

The operation of addition of two mass points is given by the definition

$$aA - bB = (a + b)C$$

where C is a point between A and B such that $a(AC) = b(BC)$, with AC

and BC the lengths of $\overline{AC}$ and $\overline{BC}$. C is called the *center of masses* (or *centroid*). This definition is inspired by the law of levers, where C is the fulcrum (balancing point) between A and B. Since the definition assigns a unique sum to each pair of mass points, the set of mass points and addition constitute an operational system.

Addition is shown to be commutative and assumed (after experimentation) to be associative whether the points in three mass points are collinear or not. For the special case where $A = B$, $aA + bA = (a + b)A$.

It is easily proved that $aA + aB = 2aC$ iff C is the midpoint of $\overline{AB}$.

The following theorems are proved to illustrate how mass points are used.

THEOREM 1. *The medians of* $\triangle ABC$—$\overline{AD}$, $\overline{BE}$, $\overline{CF}$—*meet in a point, G, such that* $\dfrac{AG}{GD} = \dfrac{BG}{GE} = \dfrac{CG}{GF} = \dfrac{2}{1}$.

Proof. Assign weights of 1 to each vertex A, B, C, as shown in figure 9.13. Then

$$1A + 1B = 2F,$$
$$1B + 1C = 2D,$$
$$1C + 1A = 2E.$$

By the associative and commutative properties, then, $1A + 1B + 1C$ can be computed as

$$(1B + 1C) + 1A = (1C + 1A) + 1B = (1A + 1B) + 1C$$

or

$$2D \quad + 1A = \quad 2E \quad + 1B = \quad 2F \quad + 1C.$$

Mass points are equal iff their corresponding components are equal. Hence FC, DA, EB meet in a point, G, such that $CG = 2FG$, $AG = 2DC$, and $BG = 2EG$, or $\dfrac{AG}{GD} = \dfrac{BG}{GE} = \dfrac{CG}{GF} = \dfrac{2}{1}$.

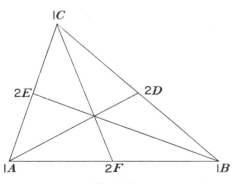

Fig. 9.13

The next theorem illustrates how mass points can be used to prove a theorem about quadrilaterals.

THEOREM 2. *The midpoints of the sides of a quadrilateral are vertices of a parallelogram.*

Proof. Assign a weight of 1 to each vertex A, B, C, D. Let E, F, G, and H be midpoints as shown in figure 9.14. Then

$$S = 1A + 1B + 1C + 1D = (1A + 1B) + (1C + 1D) = 2E + 2G = 4K$$
$$S = (1A + 1D) + (1B + 1C) = 2H + 2F = 4K.$$

Therefore $\overline{HF}$ and $\overline{GE}$ bisect each other, whence $EFGH$ is a parallelogram.

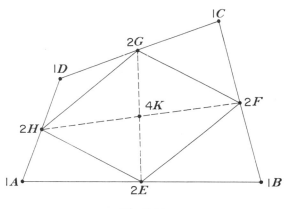

Fig. 9.14

The third theorem illustrates the method and value of "splitting" a mass point.

THEOREM 3. *The line joining the midpoints of two sides of a triangle bisects every segment between their common vertex and a point in the third side.*

Proof. Assign weights b to B, c to C, and $b + c$ to A, as shown in figure 9.15. Then $(b + c)A = bA + cA$. We split $(b + c)A$ into its addends, bA, which is added to bB, and cA, which is added to cC. The respective sums are $2bD$ and $2cE$, where D and E are respective midpoints of $\overline{AB}$ and $\overline{AC}$. Consider $(b + c)A + bB + cC$. It is equal to $(bA + bB) + (cA + cC) = 2bD + 2cE = (2b + 2c)G$, where $DG:GE = 2c:2b$. It is also equal to $(b + c)A + (bB + cC)$. If $bB + cC = (b + c)F$, then F is appropriately chosen in $\overline{BC}$ and varies with varying values of b and c. For all b and c,

$$(b + c)A + (bB + cC) = (b + c)A + (b + c)F = (2b + 2c)G.$$

Therefore G is the midpoint of $\overline{AF}$, for all F in $\overline{BC}$.

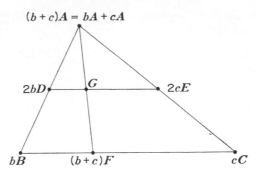

Fig. 9.15

A fourth theorem demonstrates the power of mass points for theorems in three-space.

THEOREM 4. *Let $A_1A_2A_3A_4$ be a tetrahedron with midpoint M_{12} in $\overline{A_1A_2}$, midpoint M_{23} in $\overline{A_2A_3}$, and so on (midpoint M_{ij} in $\overline{A_iA_j}$). Then segments $M_{12}M_{34}$, $M_{13}M_{24}$, and $M_{14}M_{23}$ bisect each other.*

Proof. Assign a weight of 1 to each of the vertices A_1, A_2, A_3, and A_4 as shown in figure 9.16. Then

$$1A_1 + 1A_2 + 1A_3 + 1A_4 = (1A_1 + 1A_2) + (1A_3 + 1A_4)$$
$$= (1A_1 + 1A_3) + (1A_2 + 1A_4)$$
$$= (1A_1 + 1A_4) + (1A_2 + 1A_3)$$

or

$$2M_{12} + 2M_{34} = 2M_{13} + 2M_{24} = 2M_{14} + 2M_{23} = 4H,$$

whre H is the midpoint of each of the segments, $M_{12}M_{34}$, $M_{13}M_{24}$, $M_{14}M_{23}$.

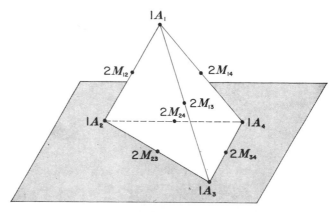

Fig. 9.16

Two remarks are in order concerning these mass-point proofs: (1) they use some theorems that are proved by other approaches (particularly parallelograms) and (2) the mass-point approach is restricted to affine geometry.

Course 3 (grade 9)

Chapter 1, "Introduction to Matrices," illustrates how matrices can be useful in several situations, one of which is transformations. For instance, the rule for a reflection in the x-axis

$$x' = 1x + 0y$$
$$y' = 0x - 1y$$

can be written as the matrix $\begin{bmatrix} 1 & 0 \\ 0 & -1 \end{bmatrix}$. The image of any point (a,b) under this reflection can be found from the product $\begin{bmatrix} 1 & 0 \\ 0 & -1 \end{bmatrix} \begin{bmatrix} a \\ b \end{bmatrix}$. This helps to develop the definition of matrix multiplication. The rule of the composition of a half-turn in the origin of a coordinate system with rule $\begin{bmatrix} -1 & 0 \\ 0 & -1 \end{bmatrix}$ and the dilation about 0 with rule $\begin{bmatrix} 3 & 0 \\ 0 & 3 \end{bmatrix}$ is the product $\begin{bmatrix} 3 & 0 \\ 0 & 3 \end{bmatrix} \begin{bmatrix} -1 & 0 \\ 0 & -1 \end{bmatrix} = \begin{bmatrix} -3 & 0 \\ 0 & -3 \end{bmatrix}$. With such examples and counterexamples, it can be seen why matrix multiplication is not commutative, nor is composition of transformations. Multiplication of a matrix by a scalar, such as $3 \begin{bmatrix} 1 & 0 \\ 0 & -1 \end{bmatrix} = \begin{bmatrix} 3 & 0 \\ 0 & -3 \end{bmatrix}$, can also be interpreted in terms of transformations.

Geometric topics per se are not considered until chapter 8, "Circular Functions." The chapter ends with the law of cosines and the law of sines, which are applied to some geometric problems.

The major advance in this grade occurs in chapter 9, "Informal Space Geometry." It does not pretend to axiomatize the subject but starts with a set of "observations" that include what are sometimes taken as axioms and other basic properties. Some of these follow:

1. A plane is a set of points with the property that whenever two points are in the set, the line containing them is in the set.
2. Given three noncollinear points, there is one and only one plane that contains them.
3. Not all points lie in the same plane.
4. If two planes have a point in common, they have a line in common.

5. If P is a point and π_1 is a plane, there is exactly one plane π_2 containing P and parallel to π_1.

Each observation is formulated only after an activity in which points, lines, and planes are represented physically and the relations among them are easily observed. In exercises, students are asked to tell whether such statements as "If $l_1||\pi$ and $l_2||\pi$, then $l_1||l_2$" are true or false. They are also asked to find physical models that illustrate such statements as "Skew lines do not intersect" and to make diagrams that indicate such properties. A short section shows how, if the observations are taken as axioms, other properties can be deduced. This suggests the possibility that space geometry can be axiomatized.

Coordinates in space are then introduced, relative to base points (O,I,J,K) in which axes need not be perpendicular and OI, OJ, and OK need not be equal. This leads to the coordinate description of planes parallel to the coordinate planes, for instance, $\{P(x,y,z) : z = 3\}$ and half-spaces such as $\{P(x,y,z) : z > 3\}$. Using two conditions permits the coordinate description of lines parallel to axes, such as $\{P(x,y,z) : x = 2, y = 3\}$. It also enables students to represent vertices of a tetrahedron by $(0,0,0)$, $(1,0,0)$, $(0,1,0)$, and $(0,0,1)$. Adding $(0,1,1)$, $(1,0,1)$, $(1,1,0)$, and $(1,1,1)$ produces a parallelepiped.

After an informal discussion of the perpendicularity of lines and planes based on activities, two more observations follow (7, pt. 2, pp. 239–41):

6. If a line m is perpendicular to each of two intersecting lines in plane π at point P, then m is perpendicular to plane π at P.

7. If a line l is perpendicular to a plane π, then any plane containing l is perpendicular to the plane π.

Rectangular coordinate systems in space are then introduced. The midpoint and distance formulas, presented as generalizations of corresponding formulas in a coordinatized line and plane, are verified and then used in exercises. The chapter ends with the sphere as the set of points $\{(x,y,z) : x^2 + y^2 + z^2 = r^2\}$, the right circular cylinder $\{(x,y,z) : x^2 + y^2 = r^2\}$, and the right circular cone $\{(x,y,z) : x^2 + y^2 = z^2\}$. Students are expected to sketch these figures in three-space.

Course 4 (grade 10)

In chapter 2, "Quadratic Equations and Complex Numbers," a short section relates transformations of the plane to complex numbers. Addition of complex numbers is related to translations by the isomorphism $(a + bi) \to (a,b)$, where $(a_1,b_1) + (a_2,b_2)$ is defined to be $(a_1 + a_2, b_1 + b_2)$. After defining the conjugate of $z = x + iy$ to be $z = x - iy$, the student is asked to name the familiar transformations now denoted by $z \to \bar{z}$, $z \to -z$, $z \to -\bar{z}$, $z \to iz$, and $z \to -iz$. The transformation $z \to az$, $a \in R$, $a \neq O$, is seen to be a dilation with scale factor a, whereas $z \to aiz$, $a \in R$, $a \neq O$, is the composition of a dilation and a quarter-turn. This is generalized to the spiral similarity $z \to tz$, where t is a complex number, and which is the composition of a dilation and a rotation.

Chapter 3, "Circular Functions," uses complex numbers, matrices, and circular functions to advance the student's understanding of transformations, particularly rotations and spiral similarities. The matrix $\begin{bmatrix} \cos\theta & -\sin\theta \\ \sin\theta & \cos\theta \end{bmatrix}$ is shown to be associated with a rotation about the origin through angle θ. The composition of two rotations—the first through angle α and the second through angle β—both about the origin is given by

$$\begin{bmatrix} \cos\beta & -\sin\beta \\ \sin\beta & \cos\beta \end{bmatrix} \cdot \begin{bmatrix} \cos\alpha & -\sin\alpha \\ \sin\alpha & \cos\alpha \end{bmatrix}$$

$$= \begin{bmatrix} \cos\beta\cos\alpha - \sin\beta\sin\alpha & -\cos\beta\sin\alpha - \cos\beta\cos\alpha \\ \sin\beta\cos\alpha + \cos\beta\sin\alpha & -\sin\beta\sin\alpha + \sin\beta\cos\alpha \end{bmatrix}.$$

Since $\begin{bmatrix} \cos\alpha & -\sin\alpha \\ \sin\alpha & \cos\alpha \end{bmatrix}\begin{bmatrix} 1 \\ 0 \end{bmatrix} = \begin{bmatrix} \cos\alpha \\ \sin\alpha \end{bmatrix}$,

$\begin{bmatrix} \cos\beta & -\sin\beta \\ \sin\beta & \cos\beta \end{bmatrix}\begin{bmatrix} \cos\alpha \\ \sin\alpha \end{bmatrix}$ represents $\cos(\alpha + \beta)$, $\sin(\alpha + \beta)$.

The formulas for $\cos(\alpha + \beta)$ and $\sin(\alpha + \beta)$ fall out quite easily.

Spiral similarities now follow readily. The last in a sequence of examples is quite instructive (8, pt. 1, p. 156):

Example 4. Describe the spiral similarity associated with the matrix

$$\begin{bmatrix} \frac{3}{2}\sqrt{3} & -\frac{3}{2} \\ \frac{3}{2} & \frac{3}{2}\sqrt{3} \end{bmatrix}.$$

The given matrix is equal to the product

$$\begin{bmatrix} 3 & 0 \\ 0 & 3 \end{bmatrix}\begin{bmatrix} \frac{1}{2}\sqrt{3} & -\frac{1}{2} \\ \frac{1}{2} & \frac{1}{2}\sqrt{3} \end{bmatrix}.$$

The "3" in the first matrix was obtained as

$$\sqrt{\left(\frac{3\sqrt{3}}{2}\right)^2 + \left(\frac{3}{2}\right)^2}.$$

Therefore the spirial similarity is the composition of both a rotation that is a one-twelfth–turn and a dilation that has a scale factor of 3.

Chapter 5, "The Algebra of Vectors," and chapter 9, "Subspaces of Affine and Euclidean Spaces," are a radical departure from traditional geometry. They have two major objectives: the more important one is to introduce the notion—and some basic theorems—about vector spaces; the less important one is to teach the elementary analytics of three-dimensional geometry. These objectives are consistent and advantageously linked because affine and Euclidean geometry are both derived from vector spaces.

The development starts with the set of n-tuples of reals (mainly triples, but also couples and quadruples), together with rules for adding them and also multiplying by a real (scalar). The student is told to regard these n-tuples as possibly capable of representing a variety of states; for instance, the triple (volume, pressure, temperature) might represent the state of a gas, or the quadruple (temperature, air pressure, humidity, pollution factor) might represent a state of weather. But these n-tuples are not vectors if the two operations vector addition and scalar multiplication are not used. Couples may be conceived as coordinates of points in a plane or triples as coordinates of points in three-space if these operations are interpreted in the context of these points. Thus the operations on vectors are organically related to corresponding operations on points. This defines the program of these chapters.

Starting with a nonzero vector, say (a_1, a_2, a_3), the set of scalar multiples $\{X : X = r(a_3, a_2, a_1), r \in R\}$ is called a *vector line*. It can be interpreted geometrically as a line in three-space passing through the origin of a coordinate system, since all such vector lines necessarily contain the zero vector $(0,0,0)$, denoted $\mathbf{0}$. Addition of two vectors can be related to the diagonal of a parallelogram. But SSMCIS prefers to give more weight to the interpretation of a translation by which the sum of vectors $\mathbf{A} + \mathbf{B}$ is the vector $\mathbf{C}$, such that $\mathbf{C}$ is the image of $\mathbf{B}$ under the translation that maps $\mathbf{0}$ onto $\mathbf{A}$. With this interpretation, the sum of a fixed vector $\mathbf{A}$ with the set of vectors in a vector line $\{X : X = r\mathbf{B}\}$ is represented by $\{X : X = A + r\mathbf{B}\}$, an affine line. When each vector in one of two distinct vector lines $\{X : X = r\mathbf{A}\}$ is added to each vector in the other $\{X : X = s\mathbf{B}\}$ (linear combinations), the sums form a vector plane $\{X : X = r\mathbf{A} + s\mathbf{B}, r, s \in R\}$. If each vector in a vector plane is increased by a fixed vector $\mathbf{C}$, the sums constitute an affine plane $\{X : X = \mathbf{C} + r\mathbf{A} + s\mathbf{B}\}$.

Another relation emerging from addition is parallelism. Two affine lines are parallel to each other if they are "translates" of the same vector line. Similarly, two affine planes are parallel to each other if they are "translates" of the same vector plane. And an affine line is parallel to an affine plane iff it is parallel to an affine line in the affine plane.

To illustrate these relations, let $\mathbf{A} = (1,2,3)$, $\mathbf{B} = (0,1,2)$ and $\mathbf{C} = (1,0,3)$.

1. The vector line l containing $\mathbf{A}$ is given by $X = r(1,2,3)$, $r \in R$. This can also be written as three equations:

$$x_1 = r, \quad x_2 = 2r, \quad x_3 = 3r.$$

2. The affine line parallel to l and containing $(0,1,2)$ is given by $X = (0,1,2) + r(1,2,3)$, or

$$x_1 = r, \quad x_2 = 1 + 2r, \quad x_3 = 2 + 3r.$$

3. The vector plane π_1 containing $\mathbf{A}$ and $\mathbf{C}$ (and necessarily $\mathbf{0}$) is given by $X = r(1,2,3) + s(1,0,3)$, or

$$x_1 = r + s, \quad x_2 = 2r, \quad x_3 = 3r + 3s.$$

4. The affine plane parallel to π_2 and containing $\mathbf{B}$ is given by $X = (0,1,2) + r(1,2,3) + s(1,0,3)$, or

$$x_1 = r + s, \quad x_2 = 1 + 2r, \quad x_3 = 2 + 3r + 3s.$$

When generalized, the structures of vector lines and vector planes lead to the abstract vector space. After displaying a variety of models of this abstract vector space and proving that affine spaces containing the set of n-tuples over the reals are indeed models of vector spaces, the student sees that a number of basic theorems concerning properties of the abstract vector space are proved. When these theorems are applied to R^3 and subspaces of R^3, the student then sees how many of the theorems he learned about generating vector lines and vector planes are special cases of the abstract vector-space theorems. The theorem that *the solution set of a system of homogeneous linear equations is a subspace of R^n (where n is the number of variables in the system)* and the theorem that *the sum of two subspaces of a vector space is also a subspace of that vector space* are of particular interest.

Chapter 9, "Subspaces of Affine and Euclidean Spaces," deals with additional—that is, other than parametric—ways of representing subspaces. These techniques are used to facilitate the representation of intersections of subspaces. It is a natural step at this stage to introduce such concepts as linear dependence, linear independence, bases, and natural bases. By defining inner products, the student recognizes that affine spaces become Euclidean spaces in which distances between points and angles are measured.

Orthogonal vectors and normal lines lead to perpendicularity relations and, in turn, to orthogonal complements. Although this description does not make it quite clear, it should be noted that the emphasis in this chapter is on geometry as a model of a vector space over the set of real numbers. However, this does not reduce the content of an elementary introduction to the analytic geometry of space.

Course 5 (grade 11)

The study of vector spaces leads naturally to an examination of linear mappings. Chapter 6, "Linear Mappings," opens with the following problem (9, pt. 2, p. 1):

A Food Mix Problem

1 pound of bread contains 300 units of calories and 20 units of proteins.
1 pound of meat contains 500 units of calories and 24 units of proteins.
1 pound of potatoes contains 60 units of calories and 4 units of proteins.
How many pounds of bread, meat, and potatoes contain C units of calories and P units of proteins where C and P are given numbers?

This problem is seen related to a mapping from $R^3 = \{(b,m,p)\}$, where b, m, and p represent pounds of bread, meat, and potatoes, to $R^2 = \{(C,P)\}$. A number of questions are asked about this mapping. Is it "onto"? Is it one-to-one? If not, what is the range? What is the kernel? Answers to these questions are given in such terms as vector lines, vector planes, and so on. Thus the language of geometry continues to be used even when the substance is no longer geometry. Geometry has served so well that it has become part of something larger than itself!

Course 6 (grade 12)

Chapter 2 of *Course 6*, "The Conics and Polar Coordinates," starts with a discussion of parabolas. It highlights the property that, for each parabola, there exist a point (focus) and a line (directrix) such that each point of the parabola is equidistant from them. This suggests the general definition of a conic as the locus of points the ratio of whose distances to a focus and a directrix is a nonnegative constant. Translations are used to simplify equations of conics, and in exercises students are asked to eliminate by rotation terms containing the product xy. Separate sections treat each of the three other conics (ellipse, circle, and hyperbola); differentiation is used to study tangents of conics. In the section on circles, the student is asked to prove some of the standard properties of chords and tangents and to show that angles inscribed in semicircles are right angles. The chapter ends with polar coordinates and polar-coordinate equations for conics.

The second half of *Course 6* has the novel organization under which students are asked to undertake their own study of one or more of three chapters. One of these chapters is "Matrices, Determinants, and Eigenvalues." The properties of determinants are developed axiomatically from three basic properties that are easily interpreted (1) for areas, when the vectors in a certain function are couples, and (2) for volumes, when the vectors are triples. In addition to other uses, eigenvalues are used in geometry to discover fixed points and fixed lines under linear mappings.

Trends in SSMCIS

The following trends seem to characterize integrated mathematics instruction in *Unified Modern Mathematics:*

1. Geometry is not taught intensively in one year; its topics are spread cyclically over many years.
2. Though the meaning of axiomatics is taught, some topics in geometry are developed informally, others formally; some topics not in geometry, such as probability and algebraic structures, receive some formal, or axiomatic, treatment also.
3. Many topics of the traditional geometry courses are not included.
4. Many approaches are used, including coordinates (both affine and rectangular), transformations, complex numbers (only somewhat), and vectors—first intuitively as displacements, and later as n-tuples of reals of a vector space, leading to linear mappings.

An evaluation of these trends is needed in order to see how they measure up in achieving some commonly held objectives for teaching geometry, namely—

1. to equip students with a knowledge of the facts of geometry that are useful in a thinking and working life;
2. to teach the nature of proof and to train students in the writing of proofs;
3. to teach the nature of, and appreciation for, axiomatic systems.

Though important, hoped-for intangibles are not listed. These are mainly intellectual attitudes and aesthetic reactions. They are extremely difficult to evaluate in any program of teaching and are primarily imparted through the scholarship, sincerity, and enthusiasm of the teacher.

The first of the listed objectives can reasonably be achieved by an intensive one-year course, integrated or not, or by a course spanning several years. A priori, it would seem that exposure over a longer period of time should be more effective in teaching the facts of geometry, but this judgment needs to be verified empirically.

Proof and Axiomatics

We are concerned here with the second and third objectives, namely, proofs and axiomatics. The mathematician knows that these two objectives are not separable, since a proof cannot be devised that is not ultimately based on axioms. But in *Unified Modern Mathematics,* axioms not explicitly stated are basic to some of the proofs. Many geometry textbooks have such proofs. Therefore the question is, Can we teach students to recognize validity in proofs and to write proofs that are based on tacit axioms? Let us for convenience call such proofs *informal.*

An example of an informal proof occurs for the theorem that states that the diagonals of a parallelogram bisect each other. The mathematician will first try to show that the diagonals intersect. This is not done in many high school geometry texts. To the student it is so obvious that he sees no need to prove it; but that diagonals bisect each other is not obvious to him, since he knows measurements entail errors and doubts. To remove these doubts there is need for deductive proof.

SSMCIS meets this problem by introducing informal proofs as early as in *Course 1.* For instance, the basic properties of isosceles triangles and parallelograms are proved *by* isometries while assuming, without fuss, such statements as this: Under any isometry the measure of an angle is the same as the measure of its image angle. Such proofs are not confined to geometry. In chapter 11, "Elementary Number Theory," of *Course 1,* there is a proof of the statement about natural numbers, "If a divides bc and a and b are relatively prime, then a divides c."

Chapter 1, "Mathematical Language and Proof," in *Course 2* introduces such concepts as logical connectives, conditional and biconditional statements, quantified statements, rules of inference, and direct and indirect mathematical proofs. These notions find immediate applications in chapter 2, "Groups," where after some examples of models, the formal definition of an abstract group sets the stage for proofs of theorems. These theorems are then applied to models of groups, not all of them consisting of numbers.

As already mentioned, the notion of an axiomatic system is introduced in *Course 3,* where, naturally, theorems are proved deductively.

It should be noted that the definition of a group supplies the axioms. Moreover, the multiplicity of models of a group makes it easier for the student to understand the abstract nature of what is being defined. The terms in this definition are undefined.

Geometry, unfortunately, does not have a multiplicity of models. Unlike a group, each proof in geometry has a diagram. The set of diagrams is the only model, and the student finds it easy to confuse the model with the abstract system in which some terms are undefined. As an example, consider again the proofs offered by a mathematician and by a student for the theorem that the diagonals of a parallelogram bisect each other. Why does the mathematician prove that the diagonals intersect, but the student does not? The mathematician is talking about an abstract or mathematical parallelogram. The student is talking about the physical diagram of a parallelogram.

Such examples in the traditional geometry course abound. Many involve properties of half-planes and the order of points on a line. They are to be found, for instance, in the topics of alternate interior or corresponding parallel angles of lines and sums of measures of angles of a polygon. Even with the improvements in the two geometries of the School Mathematics Study Group (SMSG), tacit assumptions in the traditional course still occur. A geometry text written otherwise would be far beyond the capabilities and interest of most high school students, even those who are college bound. *Modern Coordinate Geometry,* a carefully written geometry that has a complete set of axioms, appeals *only* to a small number of students interested in mathematics.

Seen in this light, a good argument can be made for SSMCIS: it teaches the nature of proof without being rigorous in developing geometry axiomatically. However, SSMCIS does not neglect the notion of axiomatics. Indeed, after some informal preparatory treatment, it axiomatizes the notion of groups (as already mentioned), fields, vector spaces, probability, finite geometries, and the laws of exponents; but when a proof is too difficult for most students, they are asked to accept the theorem if it is made plausible by other methods.

As we need to recognize, little of the spirit or the method of axiomatics is present in the traditional tenth-grade geometry course. It cannot therefore be justly charged that a program of teaching geometry as part of an integrated course over a number of years weakens the objective of teaching axiomatics. Moreover, the spirit of axiomatics is better taught in small systems, not in long, drawn-out ones where the overwhelming number of trees (theorems) interferes with the keen awareness of the forest (system). If axiomatics is to be appreciated, a system having a variety of models should be used to clarify its abstract nature.

Seen in this light, proof may be accepted as a method either to convince (to prove true) or to interrelate a set of statements in order to display a logical structure (to prove valid), and consequently it has a part to play in both informal and formal mathematics. In either case, proof has a real place in a course in integrated mathematics.

Summary

The teaching of geometry as part of mathematics conceived as a unified, or integrated, whole is made possible by the multiplicity of approaches: the synthetic methods of Euclid, coordinates, transformations, mass points, matrices, complex numbers, and vector spaces. It is possible—perhaps advisable—to begin with affine geometry and work into Euclidean geometry through inner products, perpendicularity, and a distance function on pairs of points.

The trend toward the integration of geometry with other branches of mathematics was late in getting started. It was given a significant impetus by Felix Klein and tried in the United States by John A. Swenson. This integration has encouraged the inclusion of coordinate geometry in the tenth grade.

A thoroughgoing attempt at teaching geometry as part of a unified mathematics is made by SSMCIS in their six courses in *Unified Modern Mathematics* for grades 7 through 12. Characteristic features of this text series are—

1. the spread of geometry topics over six years;
2. the central position of transformations;
3. the early introduction of matrices;
4. the omission of some of the traditional topics in geometry;
5. the early introduction of coordinates;
6. the early introduction of three-space geometry, which is presented informally;
7. the early introduction of the meaning of axiomatics but no commitment to use the method exclusively (in fact there is frequent use of informal methods);
8. the use of geometry as a model of an abstract vector space.

REFERENCES

1. Cajori, Florian. *A History of Mathematics.* 2d ed., rev. and enl. New York: Macmillan Co., 1929.
2. Hausner, Melvin. *A Vector Space Approach to Geometry.* New York: Prentice-Hall, 1965.
3. Klein, Felix. *Elementary Mathematics from an Advanced Standpoint: Geometry.* Translated from the 3d German edition by E. R. Hedrick and C. A. Noble. New York: Macmillan Co., 1939. Reprint. New York: Dover Publications, n.d.
4. Price, H. Vernon, Philip Peak, and Phillip S. Jones. *Mathematics: An Integrated Series, Lines/Planes/Space*, bk. 2. New York: Harcourt, Brace & World, 1965.
5. Secondary School Mathematics Curriculum Improvement Study (SSMCIS). *Unified Modern Mathematics, Course 1*, pts. 1 and 2. 2d rev. ed. New York: Teachers College, Columbia University, 1968.
6. ————. *Unified Modern Mathematics, Course 2*, pts. 1 and 2. 2d rev. ed. New York: Teachers College, Columbia University, 1969.

7. ———. *Unified Modern Mathematics, Course 3*, pts. 1 and 2. 2d rev. ed. New York: Teachers College, Columbia University, 1970.

8. ———. *Unified Modern Mathematics, Course 4*, pts. 1 and 2. 2d rev. ed. New York: Teachers College, Columbia University, 1971.

9. ———. *Unified Modern Mathematics, Course 5*, pts. 1 and 2. Rev. ed. New York: Teachers College, Columbia University, 1971.

10. ———. *Unified Modern Mathematics, Course 6*, pts. 1 and 2. Experimental ed. New York: Teachers College, Columbia University, 1971.

11. Swenson, John A. *Integrated Mathematics*, bk. 2. Ann Arbor, Mich.: Edwards Bros., 1934.

12. ———. "Graphic Methods of Teaching Congruence in Geometry." In *The Teaching of Geometry*. Fifth Yearbook of the National Council of Teachers of Mathematics. New York: Bureau of Publications, Teachers College, Columbia University, 1930.

10

An Eclectic Program in Geometry

JACK E. FORBES

THE SYSTEMATIC solution of any curriculum problem consists of the consideration of four basic questions, *in the following order:*

1. *Why* should instruction in this content area be included?
2. *What* topics from this content area should be selected?
3. *When* should this instruction be included?
4. *How* should this instruction be carried out?

Most of the work on new curricula in geometry has been directed at answering *one* much more limited question: Given that (*a*) most instruction in geometry will take place in a one-year course following one or two years of instruction in algebra, (*b*) the essential content of this course will be some subset of the geometric content of Euclid's *Elements* (4), and (*c*) the content will be developed in the spirit of Euclid's *Elements* as a unified mathematical system whose "objects" are abstractions and with most of the facts about the system being deduced from a limited number of undefined terms and assumptions, how can such a course be taught so as to be meaningful and useful to the wide range of students who will take it? Note that this question assumes answers to questions 2 and 3 presented earlier. It ignores question 1, leaving only question 4 to be answered.

It would be unfair to imply that no attention has been given to when geometric topics should be taught. The increase in the geometric content of textbooks for elementary school (an increase not always reflected in the amount of geometric instruction given) has been dramatic. Most secondary programs include some analytic geometry in both the first and the second courses in algebra. Geometric illustrations of important properties of functions, such as continuity, periodicity, and bounded variation, accompany the increased emphasis on these properties in trigonometry courses. Some programs go much further in integrating geometric content with the other content of school mathematics. (See chapter 9 for one such program.)

Still, the geometric experience of most students is largely concentrated in a one-year "formal" course in geometry. It is for this course that most new materials in geometry have been written. Although the variation in these materials is broad, they differ primarily only in how the instruction is to be conveyed—synthetically, numerically, using coordinates, using vectors, using transformations, and so on. They are remarkably similar in content and organization. The content is essentially some subset of the geometric content of Euclid's *Elements*. Variations in content and organization are largely those necessary for the development of the particular mathematical tools to be used, together with some enrichment topics. The organization of the content is that of a single deductive system, with most results being derived from one of several lists of undefined terms and postulates in a more or less rigorous manner. Thus both the content (Euclidean) and the spirit (deductive) of traditional geometry courses remain largely intact.

The impact of this formal course in geometry is felt throughout the curriculum. In the absence of well-established alternate objectives, "getting ready for geometry in high school" seems the de facto objective of most elementary school geometry instruction. This tends to determine both the content of elementary programs and the spirit in which they are taught.

This formal course in school geometry also influences the geometric education of teachers. The limited time assigned to instruction in geometry must be spent preparing them to teach a deductively oriented course in the geometry of Euclid.

Since the one-year course in high school geometry has such influence throughout the curriculum, it seems to deserve a more careful examination than it has been given. This examination should not tacitly assume the appropriateness of traditional content organization or the validity of traditional (often unstated) objectives. In fact, the examination should begin with the consideration of *why* geometry and proceed to *what* geometry, and only then consider the questions of *when* and *how* geometry should be taught. One attempt at such an examination is given in the sections that follow.

First, we shall list and comment on some of the commonly accepted objectives for a formal course in geometry. Next, we shall suggest geometric topics—and approaches to these topics—that seem more likely to produce the achievement of those objectives than any approach to traditional Euclidean content. Finally, we shall indicate how many of the advantages claimed for this new approach can be realized within the framework of existing courses with existing textbooks.

Objectives

There are many objectives for a formal course in geometry—some stated, some tacitly assumed, some vaguely hoped for. From recommendations that have been made and materials that have been written, it is reasonable to infer that the *primary* objective of the course is a content objective: To develop the entire body of traditional content as a unified, abstract mathematical system—a geometry of ideas—based on undefined terms and reasonable assumptions from which the remaining information is obtained by proper application of rules of inference. It is the basic premise of this chapter that this objective is *not* an appropriate one.

Many observations support this premise. First, consider the fact that much of the content to be developed formally is not new to the student. He has had extensive exposure to it and has participated in convincing demonstrations of its validity within the context of a "geometry of pictures." It takes more mathematical maturity than most students possess to view results that they "know" and "believe" as merely conjectures that require formal proof. It demands considerable intellectual discipline to separate unusable observations from usable facts while continuing to use these intuitively appealing observations to give meaning to these facts. (We all *know* that congruence is an undefined term, but we continue to think "same size and shape" to give it meaning.) In short, as students say, "It's hard to pretend you don't know it and prove it." Abstraction, in mathematics or otherwise, is not easy. It is doubly hard when too much is known about one particular physical model for the abstraction—especially when that model is the all-pervasive "real world" of pictures and objects.

Although no one would argue against familiarity on an intuitive level as a necessary aid to formal development, this is not the situation with most students. They are familiar with the results, and furthermore, they believe them in the sense that they have participated in demonstrations of them—demonstrations that to them may be more convincing than formal proofs. It is not surprising that an introduction to formalism in this environment leads to the common student definition of proof as "a formal argument establishing something you already know to be true."

Not only does the student know much of the content of the traditional program, but his exposure to geometry has probably been limited to that content. This makes it particularly difficult to motivate him to consider in depth the basic ideas that are necessary in a formal approach. It is not easy to understand the importance of a concept when no alternatives to it have been encountered. For example, would the commutative property of addition be of interest if all operations on all numbers were commutative? Similarly, is separation on a line an interesting concept until situations are known in which other types of separation apply, such as on a circle? Thus, not only an excessive familiarity with the content to be formalized but also a lack of experience with alternatives to that content make much of the introduction to traditional formal geometry "much ado about nothing" to many students.

Further weaknesses of an attempt at a one-year course in formal geometry as a student's first exposure to an abstract mathematical system are intrinsic within the content itself. It is a system with a large number of undefined terms and a large number of assumptions. In modern treatments of this content it is common practice to introduce these terms and assumptions as needed, that is, to develop the overall system as a hierarchy of related subsystems, such as incidence, order, congruence, and so on. Whereas this practice reduces the impact of the massiveness of the overall structure, it does not alter the fact that the student is engaged in a yearlong task of the construction of an enormous, complicated structure as his first try at coping with an abstract system. He is being asked to "run hard and long" as his first attempt at walking!

An in-depth look at the various subsystems of which Euclidean geometry is composed points up additional problems. First, the nature of the content of these subsystems is quite different. The results concerning incidence are far removed from those involving mensuration or parallels, and these results, in turn, differ greatly from those having to do with congruence. Since these results are essentially different, it should not be surprising that any one approach—synthetic, vector, or coordinate, for example—may produce a very simple development of one subsystem while being far from optimal for the development of another. These essential differences in the subsystems raise the question of the appropriateness of any approach to the traditional content as one abstract system.

Furthermore, the various subsystems differ greatly in the level of sophistication of the concepts they include. This would not be a serious problem if it were possible to study them in the order of their increasing sophistication. However, these subsystems rest upon each other in a relatively inflexible sequence with some of considerable sophistication coming near the beginning (e.g., separation). This necessary, logical ordering precludes a pedagogically acceptable sequencing from easy, to hard, to harder, to hardest. Maintaining a uniform level of rigor throughout produces a broad range of levels of difficulty for students and produces these in a far from optimal sequence. Allowing the level of rigor to fluctuate so that levels of difficulty are controlled tends to induce even more student confusion about when he is to accept something and when he must prove it.

This heterogeneous nature of the content of the traditional program is not surprising. It has been developed as an abstraction from the real world, which is itself quite complex. One should not fault the system. However, it is appropriate to ask why a student's initial contact with the concept of producing and working with abstract systems should be with one of such breadth and complexity.

That Euclidean geometry is a good abstract system in that it fits the real world "in the small" cannot be denied. In fact, in a sense it is too good—or at least too complete—to illustrate to students the efficiency of working with abstractions. A primary motivation for work with abstract systems is that a result established once *in general* can be used with *essentially different* models for the system. Unfortunately, this is not so with Euclidean geometry. The structure of this system is so extensive that it is categorical. That is, any two models of the system are isomorphic— *essentially the same*. This is a satisfying logical result inasmuch as it shows that the system provides a definitive description of the physical model from which it was abstracted—the real world "in the small." However, it is a "so what" result to students. Would not the value of working with abstract systems be much better illustrated for them by the consideration of simpler systems for which there exist nonisomorphic models, for example, vector spaces, fields, projective geometries, or models for proper subsets of the complete set of Euclidean postulates?

As noted earlier, instruction in geometry throughout the curriculum— from elementary school through the education of teachers—is dominated by this one-year course in formal geometry. More precisely, the curriculum is dominated by the primary objective (geometry as a unified abstract system) of that course. In particular, attempts to achieve this primary objective set the environment within which other worthwhile objectives for the course must be achieved. Let us consider the following list of such objectives (including the primary one) and comment on the positive and negative contributions that the attempts to achieve the primary objective make toward the achievement of the others:

1. (Primary objective) To develop the entire body of traditional content as a unified, abstract mathematical system

2. To develop within students an appreciation for both the intellectual strength and the intrinsic beauty of working with abstractions

3. To provide students with the opportunity for original investigation

and the construction of valid arguments within the context of geometric content

4. To extend the students' facility in dealing with real-world problems involving geometric content

5. To exhibit the unity of mathematical ideas through an integration of arithmetic, algebraic, and geometric concepts

6. To build the students' geometric intuition so that geometric models can be used in further mathematics instruction

7. To introduce those mainstream mathematical ideas that arise most naturally within the context of geometry

Even a casual reading of this admittedly limited list of objectives raises the question of the existence of any single approach to any unified body of content that will maximize the probability of the achievement of such diverse objectives, considering the wide variety of students served by a formal course in geometry. If these objectives are not contradictory, they are at least not complementary! Some are very difficult to achieve within any context. Let us consider some of the particular problems that accompany attempts to achieve these objectives within the context of a unified approach to traditional content, regardless of the mediating devices used.

Objective two—appreciation for the value of using abstractions—is an objective in the affective domain. It is even more difficult to determine how attitudes are formed than it is to describe how cognitive skills are attained. Still, it seems safe to say that many of the observations made earlier about a single–abstract-system approach to all of traditional content would support the assertion that such an approach is not conducive to the achievement of this attitudinal objective. The excessive familiarity with, and belief in, the content, the complexity, the massiveness, and the heterogeneity of the system and the lack of payoff inherent in a categorical system seem more likely to "turn off" than "turn on" students. The author (and probably many readers) can produce empirical evidence of this result.

Within a program whose primary objective is the development of traditional content as a single abstract system, the achievement of *objective three*—providing opportunities for original investigation—is usually assigned to those problems often referred to as "originals." Unfortunately, most interesting originals are based on considerable structure and therefore appear too late in the course to salvage the interest and creative spirit of many students. Furthermore, many are of a "so what" nature even to those students who can cope with them. Yet many teachers report work with these problems as the high point of the course from the standpoint of student interest and subsequent effort. Surely, original investigation is important both in itself and for the contribution it can make to developing appreciation for the mathematical method. Consequently, several important questions about the program in geometry are raised. Is it possible to provide students with more such opportunities for original investigation, but at an earlier time and in a less complicated geometric environment? Is this possible within the environment of the single–abstract-system approach? Is it easier to achieve with some other approach? What should be the relative values of the single-system and original-investigation objectives?

Objective four—extending the students' facility in dealing with real-world applications of geometry—is an important one. Its connection to the achievement of the objectives of developing appreciation and providing for original investigation is obvious. It is also closely connected to objective five—integrating mathematical concepts. Objective four is certainly not incompatible with a single–abstract-system approach to geometry. However, such an approach is in no sense necessary for the achievement of this objective, nor is a consideration of real-world applications a necessary part of such an approach. In fact, doing all that is necessary within a formal approach often leaves little time to emphasize applications. It is certainly safe to assert that the deletion or modification of the single–abstract-system objective would not preclude, and might well aid, the achievement of objective four.

The effect of the primary single–abstract-system objective on the achievement of *objective five*—unifying mathematical ideas—depends to some extent on the approach used in developing the overall system of geometry. Surely, Birkhoff's postulates (2), which include number from the beginning, contribute more to achieving objective five than Hilbert's (8). Most treatments using coordinates, vectors, or transformations make even larger contributions. Still, it is questionable whether this important objective can be optimally (or even adequately) achieved through *any* approach as a by-product to the achievement of the *primary* objective.

At first thought, *objective six*—the building of geometric intuition—seems directly contradictory to the abstract-system objective. How can a student's intuition be developed while his reliance on that intuition is simultaneously reduced by a demand for rigorous proofs of what he often considers intuitively obvious? However, any continuing, meaningful experience with concepts should contribute to intuition concerning those concepts. Whether the abstract-system approach makes a maximum contribution to the achievement of this objective is open to serious question. Also, the scope of traditional content, as broad as it is, does not provide experience with all aspects of geometry with respect to which intuition is important (e.g., the important relationships between algebraic and geometric constraints).

One reason given to support the use of coordinates, vectors, or transformations in the development of formal geometry is the contribution such an approach makes to the achievement of *objective seven*—the introduction of mainstream ideas within a geometric context. Other reasons are that such use "makes geometry easier" and contributes to achieving the objective of unifying mathematical ideas. As noted earlier, the heterogeneous nature of traditional content makes the selection of *the* most appropriate tool difficult, for what works best with one subsystem may be far from optimal for another. Why not develop facility with a wide variety of tools and use each where most appropriate? There simply is not time enough for this if one is to achieve the primary objective of presenting all traditional content in a formal manner. If one is willing to modify the single-system objective, it should be possible to provide opportunities to work with the wide variety of interesting mathematical tools whose introduction is most natural within the context of geometry. This seems, perhaps, the most compelling argument for considering the abandonment or modification of the single–abstract-system objective.

It is easy to see that the primary objective—the development of all traditional content as a single, abstract mathematical system—influences, in fact dominates, all other objectives. As stated earlier, it is the premise of the author that this primary objective is an inappropriate one. It is too limiting in its effect on other worthwhile objectives, in the narrow view it gives of geometry "in the large," and in its contribution to the overall mathematical development of the students. This is not to say that experience with an abstract system (or systems) is not an important objective of instruction in geometry, but rather that for the reasons stated earlier, *all traditional, Euclidean geometry* is not the best system with which to begin this experience.

In the sections that follow, we shall use a modified version of objective one:

1*. To use geometric content to provide students with the opportunity to work with abstract mathematical systems

Within the broader context of this set of objectives, we shall make some recommendations for an improved program in geometry.

An Eclectic Approach to Geometry

It is traditional to build a curriculum in geometry by beginning with the primary single–abstract-system content objective, next determining how to conduct the instruction so that this objective can be achieved, and finally doing as much as possible within this limited context to achieve other worthwhile objectives. We shall take a different approach. Having discarded this all-pervasive primary objective in favor of a more limited one, we are now free to look at *all* the content of geometry—not just that contained in the traditional program—and select those geometric concepts that will make maximum contributions to the achievement of *all* our instructional objectives. In this new atmosphere of freedom, one is immediately impressed with the wide choice of "good things to do."

In recognition of the broad spectrum of content within geometry, the diverse nature of the objectives to be achieved, and the wide range of interests and abilities of the students who take this one-year course in geometry, one would surely opt for a "unitized" program with many of the units largely self-contained. It seems unwise to surrender our newly acquired freedom from one massive single-structure approach to *any* alternative unified treatment, which is likely to have all the disadvantages of the traditional program with, perhaps, even fewer advantages. Rather, we shall choose to construct units that will have both logical and pedagogical reasons for their existence. That is, units that are natural entities, not merely collections of bits and pieces of geometry. Each unit will be constructed to make particular contributions to the achievement of one or more of the objectives (1* through 7) of the course.

In the remainder of this section, we shall give descriptions of some such units. This list of units is intended to be representative, but in no sense exhaustive. Which of these units a given teacher might select for a given class depends on the relative value the teacher puts on the various objectives, on the backgrounds, interests, and abilities of the students, on the students' future educational plans, and on many other local variables. Whereas some units should be part of a common background of all students, other units could be assigned to subsets of the class on the basis of particular interests. Perhaps members of each such subset could present an overview report of their work as a class project.

The numbering of the units is largely arbitrary. When an interdependence of units exists, it is noted. With each unit are listed the course objectives relevant to it and the references in which background information can be found.

Unit 1—review, organization, and extension of traditional content (two-dimensional)

This unit contains the important basic terms, facts, results, and formulas of the traditional program. The objectives are to develop the ability to recall, recognize, and use all of these. Minor emphasis would be placed on foundations; the abstract nature of the subject would *not* be stressed.

However, limitations of pictures should be pointed out. As students progress through the unit, there should be a gradual shift toward more formal argument as a basis for establishing results. No claim of rigor should be made.

The content of this unit includes the basic terms, facts, results, and formulas concerning incidence, order, congruence, similarity, parallelism, and mensuration. Regular reference should be made to what is being assumed and to which assumptions are essential and which could be established. Alternatives to the observations about separation, parallelism, the ordinary distance function, and so on, should be mentioned as possibilities for later investigation. Informal observations about the infinite extent, completeness, Archimedean property, and so on, of the line and possible alternatives to these should be inserted and, if interest develops, expanded. Informal treatment of the limit concept should accompany discussions of the circumference and area formulas for circles.

This unit should be studied by all students early, but perhaps not first, in the course. It may be better to begin the course with a shorter unit assured to be new and of high interest to all, such as unit 4 or unit 7 of this section.

In terms of the course objectives, unit 1 makes a major contribution to the achievement of objectives three through six—original investigation, real-world applications, integration of previous learnings, and developing intuition.

The content, but *not* the spirit, of many existing texts could serve as material for this unit. However, this content should be approached in a developmental and experimental, as opposed to a formal, spirit. For a sample of geometric material written in this spirit, see (6).

On completion of this unit, the student would take all this basic content as "known and usable," with no distinction between assumptions and derived results. With this large amount of background, much interesting geometry can be done.

Unit 2—modern topics in Euclidean geometry

Until the emphasis on foundations of geometry in the last fifteen years, a course in modern geometry or college geometry was part of the undergraduate education of most teachers. Such courses were usually based on the Euclidean postulates but presented results not contained in Euclid's *Elements*. (Here, "modern" means post-300 B.C.!) Many of these results are less sophisticated than some in the traditional program. Furthermore, many will be new and unexpected for most students. Some such results in the plane follow.

MENELAUS'S THEOREM. *Given triangle ABC and points X, Y, and Z on $\overleftrightarrow{AB}$,*

*$\overleftrightarrow{BC}$, and $\overleftrightarrow{CA}$, respectively. (See fig. 10.1.) Then X, Y, and Z are collinear
if and only if*

$$\frac{AX}{XB} \cdot \frac{BY}{YC} \cdot \frac{CZ}{ZA} = -1.$$

Here AX, BY, and so on, denote the *directed* distances from the first point to
the second. Why cannot exactly one of the collinear points be between vertices
of the triangle?

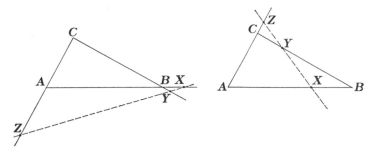

Fig. 10.1. Two cases of Menelaus's theorem

DESARGUES'S THEOREM. *Given triangles ABC and A'B'C' such that $\overleftrightarrow{AA'}$, $\overleftrightarrow{BB'}$,*

*and $\overleftrightarrow{CC'}$ intersect at point P, then the points of intersection of the lines con-
taining corresponding sides are collinear. (See figs. 10.2–10.4.)*

There may be three (fig. 10.2), two (fig. 10.3), or no (fig. 10.4) points of
intersection of corresponding sides. Why not one?

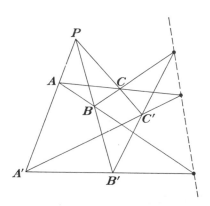

Fig. 10.2. Desargues's theorem—three points of intersection

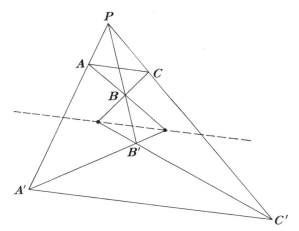

Fig. 10.3. Desargues's theorem—two points of intersection

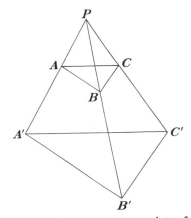

Fig. 10.4. Desargues's theorem—zero points of intersection

PASCAL'S THEOREM. *Given simple hexagon ABCDEF inscribed in a circle, then the points of intersection of the lines containing opposite sides are collinear.* (See figs. 10.5–10.7.)

There may be three (fig. 10.5), two (fig. 10.6), or no (fig. 10.7) points of intersection of pairs of opposite sides. Why not one?

QUADRANGLE THEOREM. *Given the simple quadrangle PQRS whose opposite sides intersect at points X and Y and in which $\overleftrightarrow{PR}$ and $\overleftrightarrow{QS}$ intersect $\overleftrightarrow{XY}$ at W and Z (see fig. 10.8), then*

$$\frac{XW}{WY} \div \frac{XZ}{ZY} = -1.$$

XW, WY, and so on, denote directed distances from the first point to the second.

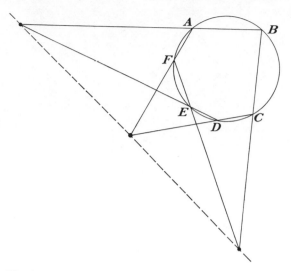

Fig. 10.5. Pascal's theorem—three points of intersection

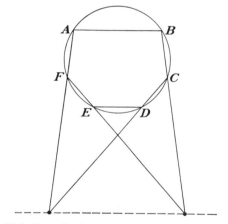

Fig. 10.6. Pascal's theorem—two points of intersection

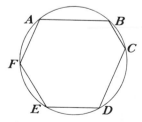

Fig. 10.7. Pascal's theorem—zero points of intersection

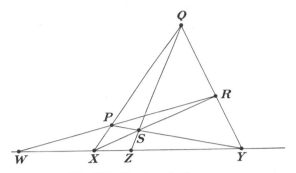

Fig. 10.8. Quadrangle theorem

All these, and many other theorems from modern geometry, are easy to prove. Some (especially Desargues's theorem) give rise to many interesting straightedge constructions. The quadrangle theorem introduces the concepts of double ratio and harmonic sets of points, which are closely related to harmonic progressions of numbers. Pascal's theorem holds for other conics also and forms the basis for straightedge constructions of points of a conic and tangents to conics (see unit 10, later in this section).

A unit such as this one can contribute to the achievement of the objectives on original investigation, application, and (since many such theorems involve numerical results) integration of previous learning.

Source material for such a unit can be found in any text on modern or college geometry, (3) or (14), for example.

Unit 3—coordinate systems

Students entering a geometry course will be familiar with Cartesian coordinates for the plane from first-year algebra. Following are some topics that can form, or be included in, this unit.

1. The distance formula, concept of slope, formula for the coordinates of midpoints of segments, and slope condition for perpendicularity. Their use can be illustrated by establishing some of the "known" results from unit 1. This should include discussions of selecting coordinate systems so that no generality is lost but the algebra is "easy." For example a proof using coordinates is given in figure 10.9.

2. A simplified version of the work of Descartes with points, lines, and between defined in terms of real number arithmetic and algebraic forms, with the subsequent establishment of the basic assumptions of unit 1 as theorems (5, pp. 96–102).

3. Locus problems, including locus definitions of conics.

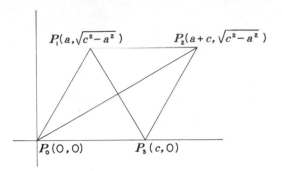

THEOREM. *The diagonals of a rhombus intersect at their midpoints and are perpendicular.*

Proof. Given rhombus $P_0P_1P_2P_3$, the coordinates of its vertices may be selected as shown in the accompanying figure. By the midpoint formula,

$$\text{the midpoint of } \overline{P_0P_2} = \left(\frac{a+c}{2}, \frac{\sqrt{c^2-a^2}}{2}\right)$$

and

$$\text{the midpoint of } \overline{P_1P_3} = \left(\frac{a+c}{2}, \frac{\sqrt{c^2-a^2}}{2}\right).$$

Since these midpoints are the same point, the diagonals intersect at their midpoints. By the slope formula,

$$\text{the slope of } \overline{P_0P_2} = \frac{\sqrt{c^2-a^2}}{a+c}$$

and

$$\text{the slope of } \overline{P_1P_3} = \frac{-\sqrt{c^2-a^2}}{c-a}.$$

Since

$$\frac{\sqrt{c^2-a^2}}{a+c} \cdot \frac{-\sqrt{c^2-a^2}}{c-a} = \frac{-(c^2-a^2)}{c^2-a^2} = -1,$$

the diagonals are perpendicular, by the slope condition for perpendicularity.

Fig. 10.9. A proof using coordinates

4. One- and two-parameter families of lines with emphasis on the relationship between algebraic and geometric constraints.

5. Parametric equations of lines and other curves, orientation of curves, motion along a curve. For example, suppose (x, y) is the position in a plane of particle P at time t (with $-1 \le t \le 1$) and

$$x = \sqrt{1-t^2}$$
$$y = t.$$

Describe the motion of the particle (see fig. 10.10). Write parametric equations for the motion of a particle that moves along the same curve from $(-1, 0)$ at $t = -1$ to $(1, 0)$ at $t = 1$ (two answers). See figure 10.11.

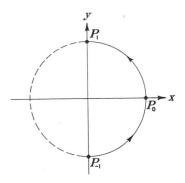

Path: $x^2 + y^2 = (\sqrt{1 - t^2})^2 + t^2 = 1$
$-1 \leq t \leq 1$ gives $0 \leq x \leq 1$
$-1 \leq y \leq 1$
$P_t =$ location of P at time t.

Fig. 10.10

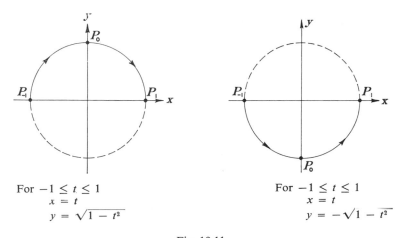

For $-1 \leq t \leq 1$
$x = t$
$y = \sqrt{1 - t^2}$

For $-1 \leq t \leq 1$
$x = t$
$y = -\sqrt{1 - t^2}$

Fig. 10.11

6. Other coordinate systems for a plane: polar coordinates, nonrectangular versions of Cartesian coordinates (nonperpendicular axes), algebraic and geometric consequences.

A unit containing some of these, or comparable, topics would make major contributions to the achievement of the objectives dealing with original investigation, applications, integration of previous learning, development of intuition, and mainstream ideas. A rather formal treatment of topic 2 provides worthwhile experience with an abstract mathematical system.

Source material for these six topics can be found in most texts on analytic geometry, for example, (11) and (13).

Unit 4—mathematical systems

In selecting topics to aid in the achievement of objective one—work with abstract systems—we can choose any one of several subsets, such as incidence, of the overall structure of Euclidean geometry and treat it in a more or less traditional, or formal, manner. However, a better choice would be finite projective geometry because here formalism is forced on us and intuition is no help at all. Furthermore, this system has minimal structure—four axioms (16, pp. 34–36). These four axioms exhibit a symmetry (duality) in terms of points and lines absent from Euclidean geometry. (Why? Each pair of points determines a line. Some pairs of lines determine no point.) It is a remarkable result of this system that specifying that there are to be n points on one line assures exactly n points on every line, n lines on every point, and $n^2 - n + 1$ points in the plane.

The following axioms give a seven-point projective geometry (see fig. 10.12):

AXIOM 1. *There exist a point and a line that are not incident.*

AXIOM 2. *Every line contains at least three points.*

AXIOM 3. *Any two distinct points are joined by exactly one line.*

AXIOM 4. *Any two distinct lines intersect in exactly one point.*

Model.   Points: A_1, A_2, A_3, A_4, A_5, A_6, A_7.

               Lines: $\{A_1, A_2, A_3\}$, $\{A_1, A_4, A_7\}$, $\{A_1, A_5, A_6\}$,
                      $\{A_2, A_5, A_7\}$, $\{A_2, A_4, A_6\}$,
                      $\{A_3, A_6, A_7\}$, $\{A_3, A_4, A_5\}$.

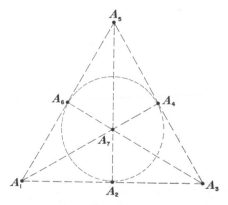

Fig. 10.12

Another interesting system for study is rational-point geometry, a geometry whose points are those points (x,y) of a coordinatized Euclidean plane with both x and y rational numbers. Two such points determine a line $ax + by + c = 0$ with a, b, and c rational numbers. Are such lines "full" of points in the Euclidean sense? Although the circles $x^2 + y^2 = 4$ and $(x - 3)^2 + y^2 = 4$ have common interior points, they have no points of intersection. Can this happen in Euclidean geometry? Why or why not? Questions like these concerning these two geometries can form the basis for a discussion of the need for some of the less intuitively appealing Euclidean postulates.

The main purpose of this unit is the achievement of the objective concerning work with abstract systems. However, it can also contribute to the objective of producing appreciation for the mathematical method and can aid in strengthening intuition.

Unit 5—other geometries

The geometric systems discussed in unit 4 exhibit alternatives to the incidence and completeness results of Euclidean geometry. It is also interesting to consider alternatives to other basic ideas of Euclidean geometry and the consequences of these alternatives. We might ask the following question:

> *What if* on each line there were one "new" point that is the point of intersection of that line with all lines that *were* parallel to it and all these "new" points lie on one "new" line in the plane?

What results of Euclidean geometry (besides the parallel postulate) would be changed? What about betweenness on these new lines? Compare with betweenness on a circle (16, pp. 47–50; 12, chap. 1). What about a coordinate system for this "augmented" Euclidean plane? (See 16, p. 60.) Would such a geometry have any applications? (See 10.)

> *What if* the set of points remained the same as in ordinary geometry but there were more than one parallel to a given line through a given point?

What else would be different from ordinary geometry? What would be the same? (See 16, chap. 1; 17.)

> *What if* we consider "distance between points" as we would in a city with rectangular blocks, that is, distance along the streets?

What properties does ordinary "straight line" distance share with this new concept of distance? Are there other ideas of distance that also have these properties?

Each of these suggestions could be a unit in itself. In addition to the references given, any book on classical projective geometry will give background material for the first suggestion. A source for the second is any book on classical non-Euclidean geometry. Information related to the third suggestion can be found in a text by Eves and Newsom (5, pp. 247–49).

Such a unit (or units) can contribute to the achievement of the objectives of working with abstract systems, developing appreciation, providing opportunities for original investigation, and providing new applications of geometry.

Unit 6—transformations

Proposals and textbooks for the use of transformations as the method for presenting a unified program of more or less traditional content exist (see chapter 6). This is not what is meant here; rather, the purpose of this unit is to teach about transformations using the basic facts of geometry from unit 1 as a mediating device. One can view a particular geometry as the study of those properties of figures which are invariant, or unchanged, under a certain type of transformation. This is the view taken by Klein in his *Erlanger Programm* (9; 16, p. 65). Such an approach can be carried out at any level of formalism, from the very intuitive to the very rigorous. It can be done with or without using coordinates. Such sets (groups) of transformations vary from the quite restrictive "rigid motions" of Euclidean geometry through homothetic transformations (dilations and translations), similarity transformations, affine transformations, inversions, and so on. Of course, a detailed study of any of these can become quite deep; however, with a minimum amount of complexity much can be learned. A basic result concerning dilations is shown in figure 10.13.

In particular, many of the results covered in unit 1 can be established easily with very little "machinery." It is only when one demands that the entire traditional program yield to a transformation approach that sophisticated facts about transformations are necessary.

If coordinates are used, matrices can be introduced to represent transformations, thus introducing students to a powerful mathematical tool in a completely natural setting.

In addition to the references already noted, any textbook using transformations could be used as source material for this unit. (See chapter 6 for references.) As with the use of existing texts for unit 1, the use of these texts will probably require deviation from the unified-system spirit in which they are written. A particularly interesting unit could be based on the ideas of Giles (7).

A unit like this one can make major contributions toward the achievement of the objectives concerning original investigation, integration of

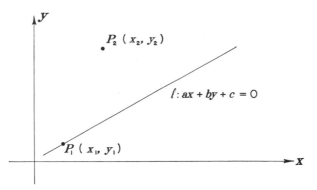

THEOREM. *Under the dilation $x' = kx$, $y' = ky$ for k any nonzero real number, the image of any line is a line parallel to it.*

Proof. For any point P_1 (x_1, y_1), there is a unique image point P'_1 (kx_1, ky_1). For any point P_2 (x_2, y_2), there is a unique point $P_3 \left(\dfrac{x_2}{k}, \dfrac{y_2}{k}\right)$ that has P_2 as its image point. For $P_1 \in l$, $ax_1 + by_1 + c = 0$, so

$$k(ax_1 + by_1 + c) = a \cdot (kx_1) + b \cdot (ky_1) + kc = 0.$$

Therefore

$$P'_1 \in l': ax + by + kc = 0,$$

and l' is parallel to l. (See the accompanying figure.) Also, for $P_2 \in l'$, $ax_2 + by_2 + kc = 0$, so

$$\frac{1}{k}\left(ax_2 + by_2 + kc\right) = a \cdot \frac{x_2}{k} + b \cdot \frac{y_2}{k} + c = 0.$$

Therefore $P_3 \in l$ and l' is the image of l. Note that for $k = 1$, $l' = l$ for all l; for all l such that $c = 0$, $l' = l$ for all k. That is, all lines containing $(0, 0)$ are invariant under all such dilations; all lines are invariant under the identity transformation.

Fig. 10.13.   A basic result concerning dilations

learning, intuition, and mainstream ideas, with special emphasis on the last of these.

Unit 7—elementary topology

This unit leads to the consideration of how geometric objects fit together in a plane and on other surfaces. Attention is focused on such properties as incidence, connectedness, and separation—properties independent of "size and shape." More precisely, those properties that are invariant under very general transformations (homeomorphisms, which carry points "close together" into points "close together" and which have inverses that do the same) are studied. An appealing model is that of perfectly elastic figures (as opposed to the rigid figures of Euclidean geometry) that can be stretched at will and can be separated and put back together as long as points originally close together are put back close together.

With size and shape removed from consideration, the students' first impression may be that there is nothing left to study. However, they can be led to discover and prove the Euler formula

$$V + R - A = 2,$$

which relates the number of vertices, V, and arcs, A, of a connected plane network and the number of disconnected regions, R, into which the network partitions the plane. See figure 10.14.

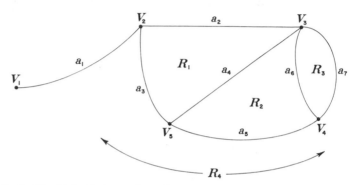

Fig. 10.14. The Euler-Poincare formula for a plane network. Since $V = 5$, $R = 4$, and $A = 7$, $V + R - A = (5 + 4) - 7 = 2$.

The establishment of a comparable result on a sphere and the observation of the equivalence of a network on a sphere with the faces (regions), sides (arcs), and vertices of a polyhedron provide a simple proof of the existence of exactly five regular polyhedra. Consideration of a comparable formula on a torus (doughnut) leads both to a discussion of simply connected regions and to the discovery of an important method of classifying surfaces.

Other paths of exploration lead naturally to the traversibility theorems, orientable curves, and the famous map-coloring problems. In few areas of mathematics can results and questions so close to those being considered by active research mathematicians be presented in such a meaningful manner to secondary school students.

This unit contributes to the achievement of the objectives on appreciation, original investigation, integration, intuition, and mainstream ideas.

Material for a unit such as this one is contained in (1) and (15).

Unit 8—vectors

As was true with transformations in unit 6, the emphasis here is not on using vectors to teach geometry but on using the facts of geometry (unit 1) to teach about vectors. A free-vector approach with or without coordinates is recommended. With coordinates, one associates the ordered pair of real numbers (X,Y) with all directed line segments $\overline{P_iP_j}$ for which

$$x_j - x_i = X \quad \text{and} \quad y_j - y_i = Y$$

with the convention that $\overline{P_iP_i} = P_i$. Now, each ordered pair (X,Y) is a *vector*. Geometrically, the relation "is associated with a given ordered pair of real numbers" is an equivalence relation on the set of directed line segments (including points). Any member of a given equivalence class is a *representation* of the associated vector. The arithmetic of vectors is defined by

$$(X,Y) + (Z,W) = (X + Z, Y + W)$$
$$k(X,Y) = (kX, kY),$$

for k any real number.

From these definitions the triangle law and the existence of collinear representations of (X,Y) and $k(X,Y)$, equidirected for $k > 0$ and opposite-directed for $k < 0$, follow immediately. The parallelogram law is an immediate consequence of the triangle law and the free-vector concept. See figures 10.15 and 10.16.

It is an easy exercise to show that $(0, 0)$ is the additive identity, that $(-X,-Y)$ is the additive inverse of (X,Y), and that $-1 \cdot (X,Y) = -(X,Y)$. The distance formula for a coordiniatized plane yields

$$\text{length of } (X,Y) = \sqrt{X^2 + Y^2}$$
$$\text{length of } k(X,Y) = |k| \cdot \text{length of } (X,Y).$$

With this structure it is easy to establish that

(X,Y) and (Z,W) have collinear representations if and only if there is a real number k such that $(X,Y) = k(Z,W)$.

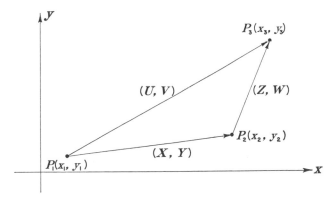

Fig. 10.15

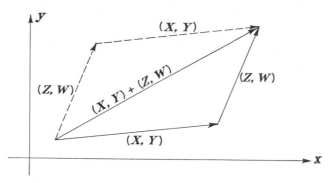

$$X = x_2 - x_1 \qquad Z = x_3 - x_2 \qquad U = x_3 - x_1$$
$$Y = y_2 - y_1 \qquad W = y_3 - y_2 \qquad V = y_3 - y_1$$
$$(X, Y) + (Z, W) = (X + Z, Y + W)$$
$$= (x_3 - x_1, y_3 - y_1)$$
$$= (U, V)$$

Fig. 10.16

This result leads to a discussion of the very important mathematical concept of linear dependence and independence, which is basic to the justification of techniques used to solve systems of equations as well as to many other important ideas.

This minimal structure is sufficient for the proof of many of the results of unit 1. A proof using vectors is shown in figure 10.17.

There are several directions we can go from here. The slope condition for perpendicularity can be used to establish that

(X,Y) and (Z,W) are perpendicular if and only if $XZ + YW = 0$.

With this concept many more results from unit 1 are accessible.

An alternate direction is to increase the number of dimensions under consideration to three or more and consider previously discussed concepts in this new setting. Another possibility is to introduce the concept of a basis for the two-dimensional (or higher) spaces and present transformations in terms of changes of bases.

It is also interesting to develop the entire concept of vectors in the plane without using coordinates. In this case a free vector becomes an equivalence class of directed line segments, with all structure developed completely geometrically. It is an interesting challenge to define *equivalent directed line segments* (congruent *and equidirected*) using only the ideas of synthetic geometry.

Chapter 8 of this yearbook, together with references listed there, contains background material for a unit such as this one (with, again, a possible

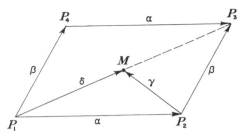

THEOREM. *The diagonals of a parallelogram intersect at their midpoints.*

Proof. Let M denote the midpoint of $\overline{P_1P_3}$, as shown in the accompanying figure. (To simplify notation, vectors are denoted by Greek letters.) The directed line segments $\overline{P_1P_3}$ and $\overline{P_2P_4}$ represent the vectors $\alpha + \beta$ and $\beta - \alpha$, respectively, by the triangle law. Since δ is collinear with, equidirected with, and one-half the length of $\alpha + \beta$, we have

$$\delta = \frac{1}{2}\left(\alpha + \beta\right).$$

By the triangle law,

$$\alpha + \gamma = \delta = \frac{1}{2}\left(\alpha + \beta\right);$$

so

$$\gamma = \frac{1}{2}\left(\beta - \alpha\right).$$

Therefore γ is collinear with, equidirected with, and one-half the length of $\beta - \alpha$. Hence, M is a point of $\overline{P_2P_4}$ and is, in fact, the midpoint of $\overline{P_2P_4}$, as was to be proved.

Fig. 10.17.   A proof using vectors

change in spirit). Such material is also contained in many analytic geometry texts and in the introductory parts of college texts on linear algebra and vector analysis.

This unit makes a major contribution to the achievement of the objectives dealing with abstract systems, original investigations, applications, integration of learning, intuition, and mainstream ideas.

As with most of these units, this one has the advantage that it can be presented at many different levels for different classes or for different students within a class.

Unit 9—Euclidean geometry in three dimensions

The spirit of this unit parallels that of unit 1. Little emphasis would be placed on a formal development, but rather the major stress would be on developing intuition and perceptive abilities, with considerable work with mensuration being included. Study would not be limited to the solids of Euclidean geometry but should include surfaces of revolution, ellipsoids, paraboloids, and so forth. The ability to recognize and sketch curves of intersection should be developed. If this unit follows those on coordinate systems, transformations, or vectors, then generalizations of the methods used there should be included. Material from traditional solid geometry texts can be used if the spirit of the presentation is modified. Most analytic geometry texts contain useful sections on three-dimensional geometry.

This unit can contribute to the achievement of the objectives of original investigation, applications, integration, and intuition.

Unit 10—conics

In unit 3, the introduction of conics as loci was recommended, and in unit 2, Pascal's theorem for a circle was presented. Many other interesting facts about conics are also accessible. These include the following:

1. Consideration of conics as plane sections of a right circular cone (after unit 9)

2. Consideration of conics in the "new" augmented plane of unit 5, their equations in the "new" coordinates mentioned there, and their relationship to the "new" line of the "new" plane

3. Discussion and proof of the standard paper-folding constructions of conics. Paper folding yields the set of tangent lines of the conics when, in all three cases, the point P is folded onto points T_i. The fold lines are the lines X_iM_i of the figures in figure 10.18, and the points X_i are the points of tangency of these tangent lines.

4. Extension of Pascal's theorem to conics and use of it to construct additional points of and tangents to conics. Observations that five independent conditions determine a conic and that there are five independent parameters in a general equation of second degree in two variables provide considerable insight concerning relationships between algebraic and geometric constraints—a logical extension of the situation for first-degree equations in topic 4 of unit 3. A problem involving the construction of conics is given in figure 10.19.

Material for this unit is contained in many analytic geometry and classical projective geometry texts (11; 12, chap. 8).

This unit contributes to the achievement of the objectives of original investigation, applications, integration, intuition, and mainstream ideas.

This list of units is far from exhaustive. It is, however, representative of the types of things that can be done. No doubt each reader has thought of other possible units or of other concepts that could be included in the units described here. If so, the purpose for which this chapter was written has been achieved. It has been our objective to present a new way of thinking about a geometry program rather than to outline in detail what

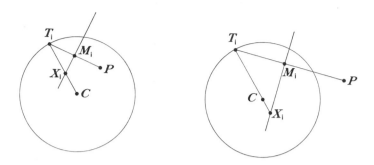

THEOREM. *Let $\mathcal{C}$ denote a circle with center C and radius of length r, $P \neq C$ a fixed point, T_i any point of $\mathcal{C}$, and X_i the point of intersection of $\overleftrightarrow{CT_i}$ and the perpendicular bisector of $\overline{PT_i}$, then—*

    (a) *if P is a point of the interior of $\mathcal{C}$, the locus of all X_i for $T_i \in \mathcal{C}$ is an ellipse with foci C and P;*

    (b) *if P is a point of the exterior of $\mathcal{C}$, the locus of all X_i for $T_i \in \mathcal{C}$ is a hyperbola with foci C and P.*

Proof. For both (a) and (b), $\triangle\, X_i T_i M_i \cong \triangle\, X_i P M_i$ by SAS. Thus in both cases $X_i P_i = X_i T_i$. In (a), $r = X_i C + X_i T_i = X_i C + X_i P$. Therefore, for each i, the sum of the distances of X_i from C and P is the constant r, which is the locus definition of an ellipse with foci C and P. In (b), $r = X_i T_i - X_i C = X_i P - X_i C$. Therefore, for each i, the difference of the distances of X_i from C and P is the constant r, which is the locus definition of a hyperbola with foci C and P. (Since $X_i = C$ for $P \in \mathcal{C}$ the betweenness relations $T_i - X_i - C$ and $T_i - C - X_i$ used above can be justified.)

THEOREM. *Let l denote a line, $P \notin l$ a fixed point, T_i any point of l, and X_i the point of intersection of the perpendicular bisector of $\overline{PT_i}$ and the perpendicular to l at T_i, then the locus of all X_i for $T_i \in l$ is a parabola with focus P and directrix l.*

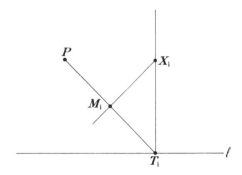

Proof. $\triangle PM_i X_i \cong \triangle T_i M_i X_i$ by SAS. Therefore, $X_i P = X_i T_i$ and for each i, X_i is equidistant from P and l, which is the locus definition of a parabola with focus P and directrix l.

Fig. 10.18. Paper-folding constructions of conics

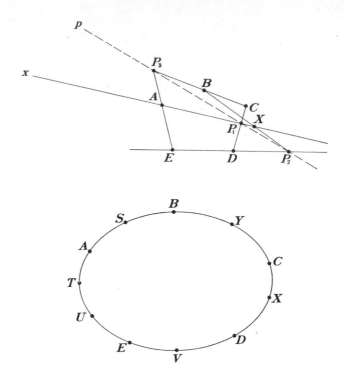

PROBLEM. *Given points A, B, C, D, E of a conic, construct additional points of the conic.*

Solution. Let x be any line on one of the given points, say A. We shall determine an additional point $X \in x$ that is also a point of the conic (if such a point exists). Consider the six points A, X (as yet unknown), B, C, D, E and the hexagon $AXBCDE$ determined by them in the order given. By Pascal's theorem for conics, lines containing opposite sides of this hexagon intersect in collinear points. Thus $\overleftrightarrow{AX} \cap \overleftrightarrow{CD} = P_1$, $\overleftrightarrow{XB} \cap \overleftrightarrow{DE} = P_2$, $\overleftrightarrow{BC} \cap \overleftrightarrow{EA} = P_3$. Since A and X lie on x, $\overleftrightarrow{AX} = x$. Therefore P_1 and P_3 are determined. Thus, the Pascal line p is given by $p = \overleftrightarrow{P_1P_3}$. Then $\overleftrightarrow{DE} \cap p = P_2$ and $P_2B \cap x = X$. (In the Euclidean plane some pairs of sides may be parallel. If, for example, x and $\overleftrightarrow{CD}$ are parallel, p is the line through P_3 parallel to these lines, that is, P_1 is the "ideal point" on these three parallel lines.) The first figure below shows the construction of point X as described above. The second figure shows several additional points obtained by repeating this construction using the same five given points.

Fig. 10.19. Constructions of conics using Pascal's theorem

such a program should be. One should think first of students and of a set of reasonable objectives for them to achieve within the context of geometric content and then choose the appropriate content for those students and those objectives.

Each of the units discussed in this chapter is intended to contribute to the achievement of several objectives. Which units are selected and how their content is approached would, of course, depend on such things as the background and interests of the teacher, the availability of appropriate materials, and the ability and interests of the students. Here again, one is impressed with the flexibility possible when the single–abstract-system approach to geometry is abandoned. For a class whose members have career goals related to science or mathematics, the units and approaches to units that would be most valuable for them might be selected. Perhaps a few units, in considerable depth, would be most appropriate. For a class of students taking geometry only for college entrance, a much different approach might be taken—perhaps more units with less depth of coverage and with special emphasis on the objectives of intuition, application, and appreciation. In fact, it is possible to select different units for different subsets of one class or different approaches to, and emphases within, a given unit for different students in one class.

As stated early in this section, when freed from the limiting "all traditional content as a unified, abstract system" objective, one is immediately impressed with the number of good things that can be done—and with the variety of ways of doing them—to provide a geometry program of sufficient flexibility to meet the mathematical needs of the wide variety of students who take geometry.

What about Today?

Some persons view a change of textbook as equivalent to a change in curriculum; however, this is not necessarily so. In fact, a change of text is neither necessary nor sufficient for a change in curriculum. Textbook changes occur at various points on the "time axis." Curriculum change is a gradual, continuing thing that can be speeded or slowed by textbook change. The curriculum is "what happens in the classroom" through the interaction of students, teacher, and instructional materials. The major components of the curriculum are determined by the teacher. These are the objectives (stated or tacitly understood) of the instruction and the strategies selected to effect the achievement of these objectives.

In the previous section we suggested a variety of sources of material for the units recommended. To expect any busy teacher or group of teachers to develop a complete course from a variety of sources in their spare time or in a short curriculum workshop is, of course, ridiculous. It is equally ridiculous to expect an instant change to a curriculum such as that recommended here even if a textbook containing such units were available. (Curriculum guides, standardized tests, and teachers trained in, and oriented to, the traditional content in geometry do exist!) Unless the teachers involved are committed to this new view of geometry, the use of such a text would no doubt produce frustration and an inferior program in geometry. If the teachers involved *are* committed to this view, then significant, though gradual, change can begin *today*—in *any* class with *any* text.

Let us suggest how this can be done. It does not require abandoning the content of the text, but it does require a change in viewpoint concerning this content. It requires an attitude of flexibility, a willingness to sacrifice consistent rigor and abstract unity to gain breadth and depth of understanding. This attitude can be described as one of "never proving the obvious, proving things only when proof contributes to understanding, and being willing to fall back on intuitive arguments when rigor is impeding understanding." Since Euclidean content consists of strongly interrelated subsystems, it is dangerous to omit a subsystem. However, some such "minisystems"[1] (e.g., incidence) can be treated quite rigorously while others (e.g., separation) can be treated on an intuitive level, with the results of both equally accessible for future work. The tendency of most books to designate these subsystems by periodic introduction of axioms makes such changes in level easy. It seems best to maintain one approach, formal or intuitive, throughout a subsystem to avoid confusing students in regard to their teacher's expectations of them. Of course, the overall plan should be explained to them.

This change in strategy is an important one. Not only will it smooth out the difficulty level of the course, but also it can introduce a vital spirit of freedom within which additional content, alternatives to Euclidean content, and alternative methods of proof are possible. Moreover, the time saved by discussing informally both the obvious and the very difficult can be used to try small excursions into some topics recommended in the previous section without deleting any traditional content. For example, it takes very little time to provide sufficient background with coordinates, transformations, and vectors so that these tools are available for use when most appropriate (contrasted with the rather extensive background necessary if any one of these methods is to be used as the basis for *all* instruction in a formal course). Many topics suggested in the previous section provide the basis for informal discussions to establish important ideas in the text (e.g., separation on a circle or on a torus to develop an understanding of the significance of "ordinary" separation).

How much can (or should) be done outside the traditional content the

---

1. For this appealing term, as well as for reinforcement and expansion of his own views on geometry, the author is indebted to Irvin Vance of New Mexico State University.

first time this approach is tried depends on many factors—the ability of the class, the background of the teacher, the availability of reference materials, and so on. Perhaps little more than the adoption of flexible objectives in regard to textbook content and the encouragement of some student projects can be accomplished at the beginning. Still, this is an important step, for such a change will prove rewarding to both teacher and students, and this will, in turn, produce further change, more rewards, more change, ... It is by such iteration that valid and lasting improvements in curriculum are made.

Implications for Teacher Education

If any proposal for curriculum change is to be adopted successfully, it must not require that teachers do things for which their education and experience has not prepared them. It should require neither massive re-education of practicing teachers nor major modification of the quantity of preservice education of prospective teachers. The proposal made in this chapter with the iterative implementations of it recommended in the previous section satisfies these criteria. However, changes in preservice and in-service courses for teachers could aid and speed this implementation.

What instruction in geometry has been (and is being) included in the preparation of most secondary teachers? Few undergraduate programs require more than one advanced course in geometry. This may be the "modern" geometry described earlier—post-300 B.C. topics in Euclidean geometry. It is more likely to be a study of the foundations of Euclidean geometry using the postulates of Hilbert or Birkhoff. It may be a survey course in geometries, giving students a brief view of Euclidean, projective, classical non-Euclidean, and several other geometries, with attempts at comparisons of these.

In addition, it may be possible for students to choose as electives courses in projective geometry, classical non-Euclidean geometries, or topology. Consideration of vector spaces is usually included in a required course in algebra, but geometric aspects or applications are seldom stressed. Transformations may also be included in this course, but again with little reference to geometry.

It seems fair to say that the geometric education required of teachers is minimal and what is available for election is not extensive. Still, to demand that three or four courses be required is unrealistic. Therefore, we must consider carefully the limited time we have for instruction in geometry and seek to maximize the contribution this instruction makes to what we would like a teacher to know and do.

Courses in modern geometry usually give minimal consideration to the foundations of Euclidean geometry. They take school geometry as "given" and proceed to proving and applying theorems not contained in Euclid's *Elements*. As such, they provide some delimitation of the students' views of geometry. However, their primary objective seems to be to develop within prospective teachers the theorem-proving skills that these teachers should develop with their students.

The purpose of most foundations courses is very simple. After a brief exhibition of the weaknesses of Euclid's postulates and, perhaps, some consideration of alternatives to them (Euclidean and non-Euclidean), one set of postulates is chosen (Hilbert or Birkhoff, in most instances) and Euclidean geometry is developed as a unified abstract system. In general, the postulate set is weaker, the detail greater, and the level of rigor higher than in the course the prospective teacher will teach. The overall strategy is to "teach them what they will teach—only on a higher level." Such a course is more likely to train a teacher to teach *one* particular approach to *one* particular geometry than to produce a teacher with a broadened viewpoint of geometry.

In theory, a "survey of geometries" course provides students both information about Euclidean and other geometries and the perspective from which to view the broad field of geometry. In practice, such courses, like survey courses in algebra, often attempt too much and become quick trips through the "mathematical zoo" with fleeting glances at, but no real familiarity with, the "animals."

The major objective of each of these three types of courses is of unquestioned value. Teachers should be skillful in making original proofs, they should know the foundations of Euclidean geometry, and they should have a view of geometry in the large. Must we opt for one or two of these, excluding the rest? Should theorem-proving skills usually be limited to synthetic or numerical work with Euclidean geometry? Is an exhaustive development using one set of postulates the most efficient way to teach the foundations of Euclidean geometry? Is a shallow exposure to many geometries the best way to develop geometric maturity? It is the thesis of this author that the answer to each of these questions is no.

Modern geometry entered the curriculum at a time when emphasis on structure and proof was essentially absent from other undergraduate courses. Such is no longer true.

Much of the precision brought to Euclidean geometry by Hilbert and Birkhoff is very difficult to appreciate except within a broader historical or conceptual context than most undergraduate students have. Furthermore, they are faced with a situation much like the "pretend you don't know it and prove it" difficulty of secondary school students. Only they must pretend they don't know how to prove it and prove it more rigorously.

This both places strong demands on maturity and tends to produce a "so what" attitude.

The development of breadth of viewpoint through comparison and contrast requires that the student have sufficient knowledge of what things are being considered to realize in what ways they are alike or different. In regard to geometries, this amount of knowledge is significant, although not massive. To attempt comparisons based on inadequate knowledge of what are being compared is more likely to produce frustration than maturity.

The prospective teacher knows considerable Euclidean geometry. Certainly this knowledge could be better used and extended in mathematics courses other than geometry courses than now usually occurs. Coordinate proofs of more "high school theorems" should be included in analytic geometry. Geometric characterizations and applications of vectors and transformations should be included in algebra courses both to extend geometric knowledge and for use in instruction in these topics.

In the first of, it is hoped, two courses in geometry for a prospective teacher, three things could be done:

1. Consider the foundations of Euclidean geometry through comparison and contrast of several postulate sets for Euclidean geometry. Develop thoroughly two or three parts of the overall structure that best illustrate differences arising from adopting different postulate sets.

2. Present some non-Euclidean concepts within the "what if?" context of unit 5 recommended in this chapter.

3. Introduce a "high school level" treatment of one geometry other than Euclidean, with major results compared and contrasted with those in Euclidean geometry.

The second course could be either a rigorous treatment of one geometry (perhaps the geometry indicated in the third recommendation above) or, with a prerequisite of exposure to the algebraic concept of groups, the study of geometries as recommended by Klein (16, p. 65).

The first course in this sequence, together with an increased emphasis on the geometric aspects of topics in other courses, should enable a teacher to teach any unified-system approach to Euclidean geometry. It should also provide him with sufficient knowledge of alternatives to Euclidean structure to produce examples for his students that show the necessity for the extensive structure of modern approaches to traditional content. Finally, the third part of the first course and the second course together should provide him the information necessary to introduce in-depth units of non-Euclidean content if he wishes to do this. Most important of all, such a program should help the teacher see geometry as an interesting and important part of mathematics—an attitude that he will, in turn, transmit to his students.

REFERENCES

1. Arnold, Bradford H. *Intuitive Concepts in Elementary Topology*. Englewood Cliffs, N.J.: Prentice-Hall, 1962.
2. Birkhoff, George D. "A Set of Postulates for Plane Geometry, Based on Scale and Protractor." *Annals of Mathematics* 33 (1932): 329–45.
3. Daus, Paul H. *College Geometry*. New York: Prentice-Hall, 1941.
4. Euclid. *The Thirteen Books of Euclid's Elements*. 2d ed. Translated by T. L. Heath. 1926. Reprint. New York: Dover Publications, 1956.
5. Eves, Howard, and Carroll V. Newsom. *An Introduction to the Foundations and Fundamental Concepts of Mathematics*. New York: Rinehart & Co., 1958.
6. Forbes, Jack E., and R. E. Eicholz. *Mathematics for Elementary Teachers*, pp. 413–73. Reading, Mass.: Addison-Wesley Publishing Co., 1971.
7. Giles, G. *Mathematical Reflections*, pp. 139–66. Association of Teachers of Mathematics. Cambridge: At the University Press, 1970.
8. Hilbert, David. *The Foundations of Geometry*. Translated by E. J. Townsend. 1902. Reprint. La Salle, Ill.: Open Court Publishing Co., 1965.
9. Klein, Felix. "A Comparative View of Recent Researches in Geometry." Translated by M. W. Haskell. *Bulletin of the New York Mathematical Society* 2 (1892–93): 215–49.
10. Kline, Morris. *Mathematics for Liberal Arts*. Reading, Mass.: Addison-Wesley Publishing Co., 1967.
11. Mason, Thomas E., and C. T. Hazard. *Brief Analytic Geometry*. 2d ed. Boston: Ginn & Co., 1947.
12. Patterson, Boyd C. *Projective Geometry*. New York: John Wiley & Sons, 1937.
13. Protter, Murray H., and Charles B. Morrey, Jr. *Analytic Geometry*. Reading, Mass.: Addison-Wesley Publishing Co., 1966.
14. Shively, Levi Stephen. *An Introduction to Modern Geometry*. New York: John Wiley & Sons, 1939.
15. Stein, Sherman K. *Mathematics: The Man Made Universe*. 2d ed. San Francisco: W. H. Freeman & Co., 1969.
16. Tuller, Annita. *A Modern Introduction to Geometries*. Princeton, N.J.: D. Van Nostrand Co., 1967.
17. Wolfe, Harold E. *Introduction to Non-Euclidean Geometry*. New York: Dryden Press, 1945.

PART

Contemporary Views of Geometry

Geometry as a Secondary School Subject

HOWARD F. FEHR

THE STUDY of geometry in the secondary school has been one of the most controversial issues debated by mathematicians and educators during the entire twentieth century. In this debate—and especially during the last fifteen years—two distinguishable positions have become apparent. One group has devoted its efforts mainly to preserving Euclid's synthetic development as a year or more of consecutive study by modifying the axioms. At recent conferences, some speakers have proposed various sets of axioms that could serve to save Euclidean synthetic geometry (1). The other group, looking at all the developments in geometry during the last 150 years, has seen a need for a completely new approach to the study of geometry at the secondary level. One of the chief advocates of a modern approach, Jean Dieudonné, has said, "Euclid must go!" (7, p. 35).

If taken out of context, this quotation may lead to some misunderstanding. What Dieudonné was complaining about is the year or two of formal study in the high school of Euclid's *synthetic* geometry, in which congruence of triangles, circles, and similarity, as there developed, have little to do with the further study of mathematics or its application. Other, more efficient procedures are now available for the study of *Euclidean space*.

From 325 B.C. to A.D. 1827 only one geometry, that of Euclid, existed as a means of the study of space. The only important question to the mathematicians of this period appeared to be, Is Euclid's parallel postulate independent of his other axioms or is it provable from them? That a space

369

different from that described by Euclid could exist was not intuitively evident. From 1827 on, however, various other geometries, some of which have assumed an eminent role in science, have been created. Today the second group of educators mentioned earlier believes that the developments in geometry since 1827 must be considered in answering the question of what to teach in the secondary school. Another reason advanced for a change is to have the subject taught so as to reflect the spirit of the contemporary conception of mathematics as a unified, interlocking set of structures breaking down the traditional barriers separating algebra and geometry.

An important reason for the survival of Euclid's geometry rested on the assumption that it was the only subject available at the secondary school level that introduced the student to an axiomatic development of mathematics. This was indeed true a century ago; however, recent advances in algebra, probability theory, and analysis have made it possible to consider using these topics, in an elementary manner, to introduce axiomatic structure in the secondary school. If we assume that reform in mathematical education is to be carried out in a broader context than just in the field of geometry, it is no longer necessary for geometry to carry the burden, or even the larger part of the burden, of introducing the student to axiomatics.

In arriving at a conception of geometry and geometric instruction from a contemporary viewpoint, it is useful to review the historical background and development that led to this conception.

The Euclidean Era

The history of the origins of geometry is similar to the development of the other classical branches of mathematics. Arising out of practical activity and man's need to describe his surroundings, geometric forms were slowly conceptualized until they took on an abstract meaning of their own. Thus from a practical theory of earth measure, there developed a growing set of relations or theorems that culminated in Euclid's *Elements,* the collection, synthesis, and elaboration of all this knowledge. This axiomatic presentation of a store of geometric knowledge represents one of the greatest achievements in the history of mankind. In fact, for more than two thousand years nothing was added to this geometry that necessitated a change in its foundations.

It is well known, however, that the geometry presented in *The Elements* was neither complete nor flawless, and as a first effort one would not expect it to be so. Among the lack of essentials was his failure to use primitive, undefined terms in order to avoid a noncircular system and an incompleteness in his set of axioms. The most controversial axiom was the parallel postulate. Down through the two millenia from 325 B.C. to A.D. 1827, many mathematicians had a feeling that this assumption was not independent of the others and could be proved. The theory of parallel lines became a focal point of the energies of such mathematicians as Wallis, Saccheri, Lambert, Legendre, Gauss, Bolyai, Lobachevsky, and Riemann. This research paved the way to the first major development since Euclid's time. It is a tribute to Euclid's genius that he included his statement as a postulate and made no attempt to justify it.

Non-Euclidean Spaces

As we now know, if the postulate concerning parallel lines is replaced by the statement "Through a point not on a line there exists more than one line containing that point and parallel to the given line," then a perfectly rigorous, logical synthetic non-Euclidean geometry (hyperbolic) may be developed. This was the contribution of Bolyai and Lobachevsky, who used this non-Euclidean axiom along with Euclid's other axioms. Since they recognized that no contradiction arose, they concluded—

1. that the parallel postulate of Euclid is not provable from the other axioms;
2. that a geometry exists that appears to contradict our intuition but is logically consistent.

Thus the obvious implication is that *there is more than one geometry.*

The mathematical world did not immediately accept the conclusions of these men. It was not until Bernhard Riemann's publication "The Hypotheses Which Lie at the Foundation of Geometry" (originally given as a lecture in 1854 but published posthumously in 1868) that mathematicians gave serious recognition to spaces other than Euclidean. In his text, Riemann not only generalized the concept of space by considering various n-dimensional spaces with metrics, but he allowed for the creation of another non-Euclidean synthetic geometry by the replacement of the parallel postulate with the statement "Any two straight lines in a plane intersect." Again a perfectly logical geometry (elliptic) was developed. Further, he pointed out the logical flaws in Euclid's development and as a result initiated a forty-year era of the perfection of Euclid.

The immediate result of Riemann's publication was a burst of activity with emphasis at first on the development of different types of geometries. A new light was thrown on these different geometries by Felix Klein in 1872. In his *Erlanger Programm* he showed quite clearly that one of the criteria that may distinguish one geometry from another is the particular group of transformations. Different geometries are viewed as those possessing particular properties of space that are preserved under particular groups of transformations. A geometry is determined by a group; every group determines a geometry. (However, there are geometries that do not possess a group structure.) In particular, similitudes and isometries lead to affine and Euclidean spaces.

Moreover, Riemann extended the growing subject of differential geometry from a study of curves and surfaces in three-dimensional Euclidean space to a study of quadratic forms with n coordinates. The story of the advance from Riemann to the present-day global differential *geometry* and differential *topology* is well known to researchers in this field (10, pp. 216–24). Today the development of geometry and its counterpart, topology, is proceeding in all directions. The geometries being studied include projective space; Euclidean space; Hilbert and Banach spaces; four-, n-, and infinite-dimensional spaces; convex spaces; metric spaces; topological spaces; and so on. These theories are finding applications in *and* outside mathematics—in the relativistic space of the physics of time and gravity and in the quantum theory of nuclear physics, for example. From all this activity it is plainly evident that geometry today has a tremendously different aspect from the geometry that is still prevalent in today's high school program.

The Perfection of Euclid

Beginning a century ago with the revelation of Euclid's flaws, a movement developed among the outstanding mathematicians to really clear Euclid of all blemishes. The search was for a minimal complete set of independent axioms that would place Euclid's synthetic geometry on a perfect, logical foundation. All inconsistencies, fallacies, and hidden or unmentioned assumptions were to be eliminated. The task was first completed by Moritz Pasch in 1882. It was followed by others, namely, Peano and Pieri (members of the "Formulaire" group, a forerunner of the "Bourbaki" group), and culminated in 1899 with the publication of *Grundlagen der Geometrie* by David Hilbert. The problem of perfecting Euclid was solved for the mathematicians. However, the solution was far too complicated and abstract to be used as a high school subject on axiomatics and proof.

There then followed a period of sixty years of sporadic efforts to do something about geometry as a secondary school subject. Euclid must be saved! The first significant modification of Hilbert's axioms was given in 1929 by G. D. Birkhoff, who effected a great economy in the number of axioms to be admitted by substituting the order and completeness properties of the real numbers. This modification was the first to use the "real ruler" and "real protractor" axioms. All other modifications in current use follow this same procedure.

Geometry Today

Today, geometry must be defined as a *study of spaces*. Each geometry is an ordered pair, namely, (set, structure), where the elements of the set are called *points* and the elements of the structure are *axioms* (and definitions) that relate the points and the important subsets of them. It is useless to attempt to list all these geometries, synthetic or nonsynthetic, since to do so would certainly result in unintentional omissions. Riemann has seen to this. This new definition has evolved slowly as a result of two phenomena. The first was purely mathematical by the discovery and description of non-Euclidean geometries and "spaces," such as topological, vector, Banach, Hilbert, and metric. The second phenomenon, and more influential, occurred as a result of advances in science and technology. The advent of relativity was most significant. After Einstein showed that the existence of matter in a space/time relationship is actually described by a fourth-dimensional model of Riemannian space, other spaces found application in physics, astronomy, biology, and economics. Euclid's geometry is just one of many, and to imply otherwise would be to deny all that has happened in mathematics and science during the last 100 years.

Geometry for Secondary Schools

Despite the introduction of metric axioms, all the efforts to save Euclid bear too strong an impression of Euclid's synthetic brand to satisfy the present trend of mathematical thought. Euclid's synthetic treatment of his "space" is simply out of the mainstream of current mathematical thinking and is seldom used by pure or applied mathematicians. As a year of formal mathematical study, it must depart. Geometry as a study of space must in due time be related to the structures of algebra, and thus it must be developed so as to permit and exhibit the use of algebraic techniques. This is the spirit of the times. One of the goals at the secondary school level should be the achievement of a basic understanding of vector spaces and linear algebra.

The basis for a contemporary study of geometry in the secondary school was established in the Dubrovnik seminar of 1960, at which the following goals were given (8, p. 189):

1. Develop the concept of space as a "set" with special subsets having structures which are linked to each other; especially affine, euclidean and vector space.

2. Develop knowledge of the precise relationships between the "line" and the set of real numbers.

3. Develop an understanding of the principal transformations applicable to different geometries, and groups of transformations, especially with respect to affine and euclidean geometry.

4. Develop an understanding of an axiomatic structure through this type of study: the affine line, the affine plane, affine space, euclidean metric space, vector space.

5. Develop skill in applying the several methods of geometric development to the solution of original problems—both mathematical and applied.

The use of the word *set* in the ordered pair (set, structure) is not in the physical sense, in which it is correctly used in a preformal study of geometry, but in its strict mathematical sense, that is, a collection of beings or abstract entities that are undefined but that take on meaning only by the structure that is placed on them. What is required for this purpose is a little formalization of elementary set theory to the extent that one understands the operations of union, intersection, and complementation; the nature of subsets and partitioning; the cross product of sets; the power set; and the interrelations of these ideas (6, pp. 1–15).

With these objectives available, a number of subsequent conferences on geometry and the teaching of geometry have been held in Europe and the United States in which the "saving of Euclid's *Elements*" has gradually lost some of its strength to a more powerful argument for the teaching of geometry from a contemporary viewpoint. A number of ways, all valid, can be used for approaching the study of space. One can proceed synthetically, choosing a convenient set of axioms in the manner of Euclid, or one can coordinatize space and make use of the real numbers and a distance function as Descartes indicated it could be done. One can follow the *Erlanger Programm* of Klein, making a group of transformations the basic idea in the study, or one can develop an algebra of points (*n*-tuples) and develop an affine and Euclidean vector space. It appears that a contemporary geometry program must contain all these approaches.

At Dubrovnik in 1960, a group of seventeen mathematicians and educators delineated programs in geometry and algebra which supplement each other. The proposed program in geometry incorporated the following content (8, pp. 189–90):

1. Groups of Transformations (Mappings)
   a) Line symmetry
   b) Point symmetry
   c) Translations

  d) Reflections
  e) Rotations
  f) Glide reflections
  g) Shearing
  h) Dilations
Isometries and homotheties and their relation to congruence and similarity evolve from this study.

2. Affine Geometry
  a) Synthetic affine geometry
  b) Finite geometries
  c) Real numbers and the line
  d) Coordinates
  e) Analytic geometry
  f) Vectors and vector space

3. Euclidean Geometry
  a) Perpendicularity
  b) Inner product
  c) Vector spaces, norm
  d) Trigonometry

4. Conics
  a) Geometric loci
  b) Affine transformations
  c) Projective properties; projective and descriptive geometry
  d) Quadratic forms, parametric representation

The Advice of Some Mathematicians

The entrenchment of Euclid's synthetic geometry and the desire on the part of teachers and some university professors to keep it as a year of study can be well understood. This is the only subject they recall that had an axiomatic treatment, and the proof of theorems and original exercises gave them satisfaction. Although debate may continue to occur, we present here the advice of great mathematicians who gave deep consideration to the way geometric instruction should turn.

As early as 1901, Felix Klein made the following observations with regard to geometric instruction (4, p. 191):

> Modern mathematics has advanced beyond that of the Greeks. One of the important differences is that the Greeks possessed no *independent* arithmetic or *analysis,* neither decimal fractions which lighten numerical calculation, no general use of letters in reckoning. . . . As a substitute

the Greeks had only *a calculus* in geometric form, . . . a process much more cumbersome than in our arithmetic.

In 1934, Oswald Veblen, the great American geometer, made the following observation in a talk on "The Modern Approach to Elementary Geometry" (9, pp. 214):

> If you ask a modern mathematician or physicist what is a Euclidean space, the chances are that he will answer: It is a set of objects called points which are capable of being named by ordered sets of three real numbers (x, y, z) in such a way that any two points (x_1, y_1, z_1) and (x_2, y_2, z_2) determine a number called their "distance" given by the formula,
>
> $$\sqrt{(x_1 - x_2)^2 + (y_1 - y_2)^2 + (z_1 - z_2)^2}\,.$$
>
> With a little refinement of logic this answer is a perfectly good set of axioms. The undefined terms are "points" and certain undefined correspondences between points and ordered sets of numbers (x, y, z) which we call "preferred coordinate systems." The axioms state that the transformations between preferred coordinate systems preserve ratios of distances.

After advocating the introduction of analytic methods in elementary geometry and pointing out the obvious thing—that geometry has to do with physical reality, which means that algebra and geometry must be used, and hence taught, together—he concludes:

> The thesis I have tried to advance is that although we can now see Euclidean geometry more clearly than ever before as a distinct subject capable of being treated without reference to analysis by its own peculiar and highly elegant methods, nevertheless the spirit of our time requires that we should present geometry as an organic part of (mathematical) science as a whole. This requires analytic methods from the first . . . and a non-dogmatic attitude both as to physical and as to logical questions. [9, p. 221]

More recently, a conference held in Bologna, Italy, in 1963 under the auspices of the International Commission on Mathematical Education considered the teaching of geometry in view of the general reform taking place in secondary school instruction. The opening address was given by Emil Artin, in which he outlined several possible approaches (2, pp. 1–4):

> The third extreme possibility is to abandon every axiomatic presentation of geometry. After having introduced the notions of coordinates in a preparatory course (in a manner more or less intuitive), one can define Cartesian space as a vector space of two dimensions furnished with a bilinear form. [p. 2]

After listing some of the difficulties encountered in this third extreme viewpoint, Artin proceeded:

> Contemporary teaching of algebra uses the geometry of vectorial space which is indispensable in the study of linear algebra. The introduction of this space in algebra simplifies all the proofs, and renders intuitive all that made rigorous.
>
> This third point of view does not exclude axiomatics. But it proposes to limit the use of axiomatics to algebra. The notion of group, field, vector space can only be introduced by an axiomatic description. But these axioms can be easily used and students have only a little difficulty in assimilating them. On the other hand, these axioms are not arbitrary and are also useful in other domains of mathematics. Another advantage of this point of view is that it leads to the notion of space of arbitrary dimensions and thus the students obtain ideas that will be useful to them throughout all their university studies. . . . Personally I do not lean to a pure-axiomatic method. I, myself, prefer the third point of view. [p. 3]

The final speaker at the Bologna conference was Henri Cartan of France. In his summary, Cartan spoke of past conferences and of those who search for axiomatic systems to preserve Euclid (3):

> I am reminded of the personage of Claudel, who, in "The Satin Shoe" demanded at all costs, for the new, but on one condition that "the new must resemble exactly the old."
>
> Notably some have attempted to recommend the continuance of ways of exposing Euclid for geometry study. I ask myself if Euclid today would be on the side of those who take his defense; I myself rather imagine that his character would carry him to recast the whole of geometry on new foundations and take into account all the ideas which have acquired a right of entry into the city of mathematics during the last twenty centuries. . . .
>
> We are all in agreement, I believe, of the urgent necessity of a change in mathematical instruction (and especially in geometry) at the secondary school level. This instruction must be pertinent to the education of the young people who later will follow scientific study at the university. But it is also necessary to think of those who no longer study mathematics after leaving the secondary school. . . .
>
> Above all, I join entirely the opinion that the notion of vector space (and that of affine space which one deduces forthwith) must be one of the fundamental notions of the new teaching of mathematics. This notion, with that of scalar and inner product, permits the foundation (or, if one prefers, the reconstruction) of all of Euclidean geometry. But the notion of vector space evidently cannot be parachuted arbitrarily; and to release it for use by the students, it is necessary that they will previously have seen and manipulated vector spaces (without knowing them as such). . . .
>
> In effect, once these results are obtained, it will be possible

1) to initiate the students to the indispensable techniques of analytic geometry.

2) to found, on a solid base, (recapturing all things from the start) the essential notions of geometry without any future necessity to resort to axiomatics. One can, in effect, proceed then by means of explicit definitions of an algebraic character. . . .

These algebraic *explanations* do not exclude the geometric language. Such language is justified. It does not exclude, moreover, the solution of problems by a geometric manner. It will always be of interest that that which is the same problem should be treated in two ways, by a geometric way, and by an analytic way. [3, pp. 84–90]

The essence of these comments is, first, that the classical separation of mathematics into separate, watertight compartments is no longer a way of conceiving of our subject and, secondly, that we must no longer conceive of geometry as a separate subject. As Willmore succinctly put it, "What is important is a geometrical way of looking at a mathematical situation— *geometry is essentially a way of life*" (10, p. 221).

Summary

Geometry today as a junior-senior high school program should enter every year of the study of mathematics. It should grow in depth and complexity until it becomes embedded in all the other areas of mathematical study—linear algebra, analysis, and applied mathematics—*until it becomes a way of thinking.* This study should be made from many approaches:

1. *Physical-informal*—working with drawings, paper folding, and apparatus to gain an intuitive feeling for geometric figures in Euclidean two- and three-space.
2. *Synthetic-axiomatic*—considering the affine plane with a minimum of axioms and with finite as well as infinite models (7). (With a few more axioms, the axiomatic study can be extended to three-space, but the value of this extended formality is doubtful.)
3. *Coordinate-analytic*—assigning coordinates to the affine plane, and subsequently to Euclidean two- and three-space, and using the usual algebraic techniques.
4. *Transformations*—studying mappings of the plane and three-space and eventually relating the study to group structure and the algebra of matrices.
5. *Vectors*—examining the algebra of points in two- and three–affine space and introducing the inner product for the study of perpendicularity and Euclidean space.

6. *Vector spaces and linear algebra*—building an axiomatic *n*-dimensional vector space and its linear algebra. (All the above approaches serve as foundation work for this.)

Each of these six approaches is to be made, not as an isolated bit of study, but in a spiral ascent in which one returns to each procedure at a higher level of abstraction and with a deeper insight into its nature. Into this spiral ascent must also be interwoven those elements of the algebraic structures of group, ring, field, and vector space that permit the student to see the interdependence of geometry and algebra. There is no single axiomatic treatment of all of geometry, but several small axiomatic structures may be introduced for a number of the approaches.

Thus physical-informal approaches are made to the study of figures in the affine plane and the Euclidean plane, first without coordinates, then with coordinates. This study involves the use of transformations in order to arrive, by induction, at properties of figures in the plane. At a higher level, the small set of axioms of plane synthetic affine geometry opens a new world of finite geometries as well as the geometry of the continuous plane. If one cares to do so, the axiom list may be extended to develop the three-dimensional affine space, but this is not a required action, since today this type of synthetic geometry does not have as much value as an analytic approach.

If one desires a formal approach to coordinate geometry, the axioms are available. However, coordinatization can take place in an informal, but correct, manner and thus permit a development of the analytic study of the plane and three-space—a study so essential in the subsequent study of analysis and linear algebra. Concomitantly with the study of coordinate geometry, one studies a new type of algebra, the algebra of points in the plane and in three-space. The climax occurs when all this study, including that of matrices, results in the vector space of two and three dimensions. Then a linear algebra permits the introduction of Euclidean three-space and a complete unification of algebraic and geometric concepts, through transformations, for use in all parts of subsequent mathematical study—trigonometry, probability, the calculus, and differential equations, for example.

The geometry program as depicted in this brief outline is an eclectic one that is an actuality in many schools in Europe, notably in England, Belgium, and the Nordic countries. Of all the developed countries of the world, the only country that retains a year sequence of a modified study of Euclid's synthetic geometry is the United States. We must immediately give serious consideration to presenting our high school youth with a mathematical education that will not leave them anachronistic when they enter the university or enter the life of adult society.

REFERENCES

1. Aarhus Conference. *Lectures on Modern Teaching of Geometry.* Aarhus, Denmark: University of Aarhus, 1965.

2. Artin, Emil. "Les Points de vue extrèmes sur l'enseignement de la géométrie." *Enseignement mathématique* 9 (1963): 1–4.

3. Cartan, Henri. "Réflexions sur les rapports d'Aarhus et de Dubrovnik." *Enseignement mathématique* 9 (1963): 84–90.

4. Klein, Felix. *Elementary Mathematics from an Advanced Standpoint: Geometry.* Translated from 3d German edition by E. R. Hedrick and C. A. Noble. 1901. New York: Macmillan Co., 1939. Reprint. New York: Dover Publications, n.d.

5. Menger, Karl. "The Geometry Relevant to Modern Education." Paper read at the Comprehensive School Mathematics Project Conference on Teaching of Geometry, 1970, at Carbondale, Ill. Mimeographed.

6. Moore, John T. *Elements of Abstract Algebra.* New York: Macmillan Co., 1962.

7. Organization for Economic Cooperation and Development. *New Thinking in School Mathematics.* Edited by Howard F. Fehr. Report of the Royaumont seminar. Paris: OECD, 1960.

8. ———. *Synopsis for Modern Secondary School Mathematics.* Edited by Howard F. Fehr. The Dubrovnik report. Paris: OECD, 1961.

9. Veblen, Oswald. "The Modern Approach to Elementary Geometry." *Rice Institute Pamphlet* 21 (October 1934): 209–21.

10. Willmore, T. J. "Whither Geometry?" *Mathematical Gazette* 54 (October 1970): 216–24.

12

An Evolutionary View

SEYMOUR SCHUSTER

ON ACCEPTANCE of the responsibility to contribute to this section of the yearbook, the authors were given directions to (1) say what they thought geometry to be and (2) discuss the implications of these thoughts for the teaching of geometry. Clearly, these instructions called for an essay—an expression of opinion—in which each author was to present and justify his own conception of geometry.

That there are different conceptions of geometry, even at the same point in history, is perhaps an obvious point. But it is a point worth mentioning because if not borne in mind by the mathematics teacher, the contrary notion, namely, that there is only one properly held view, may rise to dominate. If teachers are not aware of different conceptions of a subject, their notions tend toward rigidity and their curricula tend toward a static state. The ill effects of a single viewpoint, rigid presentations, and an unchanging and unreflected-upon curriculum are many. They hardly need belaboring here, save for one that is more pervasive, and hence more culturally damaging, than the rest—namely, that of minimizing the students' chances of appreciating the vitality that characterizes mathematics.

It behooves the teacher to be aware of the existence of different conceptions of geometry and, moreover, to be aware that each of these conceptions has certain implications for the content and development of

NOTE. Part of the writing of this chapter was undertaken while the author was supported by NSF grant GE-3848.

school geometry. Consequently, three different viewpoints are offered in the three chapters of this section in response to the two assignments mentioned in the first paragraph.

A Conception of Geometry

The first task, namely, that of saying what geometry is, calls for a definition in some sense. Anyone who has grappled with the problem of defining "geometry" surely recognizes what a tall order this turns out to be. An Aristotelian definition, with a *genus* and *differentiae,* would provide criteria for the classification of mathematical knowledge (such as concepts, theorems, and questions) into geometry or nongeometry. Such definitions are clearly desirable in science and mathematics: the classification usually leads to additional insights, the precision admits the possibility of using the definition in logical deduction, and these consequences provide a degree of aesthetic pleasure. However, an Aristotelian definition of geometry that would be acceptable to any significant portion of the mathematics world is no longer attainable. Even thirty years ago the Harvard geometer J. L. Coolidge wrote that "it is not possible to distinguish perfectly between, let us say, geometry and analysis, . . ." (5, p. viii). A few of the difficulties can be appreciated by pondering the following questions:

1. Is the real number line a geometric object? Or is it analytic? Or is it, perhaps, topological? Could the answer be yes to all three questions?

2. Euclidean space can be defined as a real inner-product vector space (4; 10). Is then Euclidean space an algebraic (vis-à-vis geometric) entity?

3. The famous four-color conjecture states that every map on the sphere (or plane) can be colored using only four colors. Is this a geometric statement? Or is it a topological statement? Perhaps it would be preferable to classify the conjecture as part of combinatorial mathematics. (See Coxeter [6] and Ore [18].)

4. What is the essential difference between an axiom system that is generally accepted as geometric (e.g., projective geometry) and one that is generally accepted as algebraic (e.g., group)? (While pondering this question, it should be noted that an axiom system for projective geometry can be phrased in purely set-theoretic terms with no reference to points or lines. O'Hara and Ward [17, pp. 17–18] provide an interesting illustration of such phrasing.)

5. Is the study of finite planes geometry? Or is it algebra, or combinatorics? (See Albert and Sandler [1] or Hall [13, pp. 283–86].)

6. Among the most profound results of projective geometry is the

innocuous-sounding theorem of Pappus: *If alternate vertices of a hexagon lie on two sides, then the three pairs of opposite sides meet in three collinear points.* The only known proof of Pappus's theorem in finite projective planes is its equivalence to the celebrated Wedderburn theorem: *Every finite corpus (or skew-field or division ring) is a field.* Is Pappus's theorem, then, a theorem of algebra?

7. The impossibility of trisecting a general angle, duplicating the cube, and squaring the circle by ruler and compass constructions in a finite number of steps can be proved only by studying the roots of polynominals and invoking group theory. (In addition, squaring the circle requires sophisticated analysis to show that π is transcendental.) Should the "impossibility theorems" be classified under geometry even though proofs are legitimately given only in courses in algebra?

Even if we were successful in obtaining an Aristotelian definition of geometry, one wonders whether it would serve our purposes, namely, the purposes of teachers. Administrators require neat classifications. They find it convenient to report that n hours of the syllabus are devoted to subject S and course A is clearly in a category that permits it to be counted toward a minor, whereas course B is not in that category and must therefore be counted as an elective. However, teachers need less than a neat classification—and yet more. Even if we have only vague notions of the classification of our knowledge, we can know—and indeed we *need* to know —about its historical roots, its evolution, its analytical techniques, and its relationship to other subject matter and to the culture. In fact, a genus-differentiae classification might hinder our purposes by providing a pigeonhole for the dreaded isolation of the subject matter.

Geometry is ubiquitous. It is intertwined with so much of mathematics and science that it is beyond extrication and isolation. The famous mathematicians Veblen and Whitehead were sufficiently moved by this fact to write, "Any objective definition of geometry would probably include the whole of mathematics" (22, p. 17). As partial evidence, consider the following lists of titles, which come from library cards. They are either book titles or chapter titles.

I	II
Euclidean Geometry	Geometric Algebra
Hyperbolic Geometry	Geometric Topology
Elliptic Geometry	Geometric Number Theory
Non-Riemannian Geometry	Geometric Measure Theory
Projective Geometry	Geometric Integration
Combinatorial Geometry	Geometric Optics
Analytic Geometry	Geometric Inequalities

I	II
Differential Geometry	Geometric Invariant Theory
Hilbert Geometry	Geometric Function Theory
Algebraic Geometry	Geometric Mechanics
Minkowskian Geometry	Geometry of Quantum Theory
Continuous Geometry	Geometry of Relativity
Integral Geometry	Geometric Orderings
Transformation Geometry	Geometry of Art
Vector Geometry	Geometry of Sandstone Bodies
Linear Geometry	Geometry of International Trade
Distance Geometry	Geometry and the Imagination

Among the major appeals of geometry is the pervasiveness attested to in these lists, and its power in the curriculum can be drawn from this characteristic. Thus, sharp boundary lines between geometry and other subjects might very well contribute to the sterility of the subjects concerned in contrast to the enrichment that may be gained from the freedom offered by vague or flexible boundaries.

How, then, can we define geometry and still come away with something of consequence for teaching? It has been held that a historical or evolutionary point of view is worthwhile in providing a description (20). Etymology tells us that geometry means *earth measure,* and so the origins of the subject most certainly lie embedded in attempts to solve problems of physical science and engineering. The Greeks, who seem to have been the first to develop geometry into a mathematical science, took two major steps to achieve this credit. First, they took the subject out of the physical sciences by abstracting or idealizing the basic notions: points are without length, lines are without width, and surfaces have no thickness. It is quite understandable that the Greeks thought their geometry to be truly a science of space when we reflect that this was the age that gave birth to Platonic idealism. Physical space, in Plato's terms, was a "shadow" of the "reality" with which geometry dealt. Secondly, the Greeks made the study deductive and developed it axiomatically. (The historical details and a philosophical analysis of the origin of geometry are not especially necessary for our purposes here. Readers who are interested in such material can easily find a large number of good sources; however, it is particularly recommended that teachers begin with Wilder's *Evolution of Mathematical Concepts* [24] and also see the first chapter of Whittaker's *From Euclid to Eddington* [23].)

Thus, from the first geometers we understand that *geometry is the axiomatic or deductive study of the ideas that arise as abstractions of the concepts that are encountered in analyzing physical space.* For two millenia the first part of this description was considered foremost. Geometry was inextricably bound to axiomatics. Spinoza was said to have developed a geometric theory of ethics when he attempted to develop his thoughts axiomatically, and in the *Caine Mutiny* Captain Queeg, defending himself at his court-martial, stated that he proved himself right by geometric logic. This phraseology is satisfactory for poetry or figurative description, but any twentieth-century student of mathematics realizes that axiomatic developments are not peculiar to geometry. In fact, it is now the intent of mathematicians to develop every branch of mathematics axiomatically.

What we have left from the Greek point of view, then, is *some* notion of the content of geometry. We are on relatively safe ground because the content includes the "ideas that *arise* as abstractions." This admits the possibility of an ever-increasing body of concepts and the possibility that the word "geometry" can change in meaning, as it has most certainly done. Hence, the discussion will be shortened by making a somewhat vague but flexible statement that will allow inferences to be drawn for the curriculum and for teaching the curriculum:

> Geometry should be regarded as a body of knowledge that had its origins in the study of physical space and physical objects but concerned itself with the abstractions that derived from such study. Hence, early geometry dealt with concepts such as points, lines, curves, surfaces, distance, area, and volume. Over the centuries, the imagination and creativity of mathematicians (influenced considerably by the changing ideas in physics) have produced many extensions of this study. They have developed higher levels of abstractions, variations of axiom systems, and many different techniques for the analysis of geometric problems. Thus we have different geometries: Euclidean and non-Euclidean geometries, n-dimensional and infinite-dimensional geometries, projective geometries, and a host of others, as in list I above. And we have different analytical techniques that are indicated by some of the following familiar labels: analytic geometry, vector geometry, transformation geometry, differential geometry, algebraic geometry, combinatorial geometry, and still others, as in list II.

With this outlook one can easily understand the claim that geometry is pervasive. It appears in courses that come under the rubrics of analysis, topology, and probability. Algebra is not an exception, in spite of the fact that some algebraists, much to the detriment of their students and their subject, write and teach in a manner that avoids geometry as well as other insight-giving devices. (Fortunately, this tendency among algebraists is now waning somewhat, having been attacked by several leaders, among whom was the virtuoso teacher, Emil Artin. See [2, p. 13] for an interesting discussion of how geometric considerations clarify ideas in linear algebra.)

Implications for Teaching

Axiomatics

Geometry should not be the principal vehicle for teaching the axiomatic method, formal structure, and formal proof.

The arguments behind this inference are several, and they are being heard in many quarters (3; 12; 14; 15; 20). Two such arguments will be stated with brevity: (1) Since axiomatic developments are now present in all branches of mathematics, there are many choices of subjects for introducing formal structure; and (2) axiomatic development of Euclidean geometry is difficult—it should be reserved for more advanced study than school mathematics. This latter conclusion is tempered somewhat by an alternative proposal that Freudenthal and Schuster have independently called *locally axiomatic* teaching (12; 20). It suggests that as they grow in mathematical maturity, students would benefit from studying smaller axiomatic systems within the subject of Euclidean geometry. This point of view might be regarded as a request to teach mathematics the way it existed among the Greeks prior to Euclid. The theory of congruence no doubt existed as a set of interrelated propositions; the theory of ratio, proportion, and similarity was developed as a unit by Eudoxus. Locally axiomatic developments of units on transversals, circles and angle measure, and parallelograms are certainly feasible and could be offered earlier than grade 10 precisely because of their simplicity.

Empirical geometric experience

What was earlier referred to as the pervasiveness of geometry leads to the belief that teaching should capitalize on this characteristic of geometry from the earliest stages of education.

Arithmetic studies using the number line are now common practice, but this is just a beginning to measurement studies that would enrich the students' grasp of both arithmetic and geometry.

The physical study of polyhedra would be both useful and fun at early levels. The need for nomenclature arises; combinatorial questions can be asked, and these have exciting answers; symmetry, coloring, existence, and group-theoretic questions can all be studied at an elementary level in such a way as to contribute to the curriculum as a whole.

The dissection of two- and three-dimensional figures can be another worthwhile experience at the elementary and secondary levels. This kind of meaningful play can be tied to questions of congruence, measure (additivity of area and volume functions), combinatorial questions, and motions. The tie-up with motions in this context perhaps deserves another few sentences. Students know intuitively, even if they have not seen a formal development, that comparison of the area measure of polygonal regions can be reduced to *equidecomposability;* that is, if P and Q are polygonal regions, then the area measure of P is equal to the area measure of Q if there exist dissections of P and Q into polygonal subregions such that there is a one-to-one correspondence between the subregions of P and the subregions of Q in which corresponding subregions are congruent. Symbolically, let m be the measure function. Then the $m(P) = m(Q)$ if there are dissections of P and Q,

$$P = \bigcup_{i=1}^{n} P_i, \qquad P_j \cap P_k = \phi \Leftrightarrow j \neq k,$$

$$Q = \bigcup_{i=1}^{n} Q_i, \qquad Q_j \cap Q_k = \phi \Leftrightarrow j \neq k,$$

such that P_i is congruent to Q_i, for all i. Several questions then arise: Is the converse true? Is there a systematic method for producing dissections that exhibit the equidecomposability? And getting back to motions, what are the motions that exhibit the congruences $P_i = Q_i$? What is the minimum number of isometries required? Can the motions be restricted to orientation-preserving motions, or is it necessary to employ reflections?

Serious play with mirrors is a fruitful geometric experience for students and one that is gaining favor. The idea of reflection is basic to an understanding of geometry from a classical, as well as a modern, point of view. A single mirror produces a visualization of this basic isometry, namely, reflection and gives rise to questions concerning orientation. Pairs of mirrors produce dihedral kaleidoscopes, which at different levels tie up with studies of symmetry, group theory, angle-measure, rotations, and translations (if the mirrors are parallel), and can be united with matters in the sciences (e.g., crystals, molecular structure, and morphology in biological studies) and the arts (e.g., painting, sculpture, and architecture). Extensions using three mirrors are difficult except for a few special cases, but experience with even one special case (three mutually perpendicular mirrors, for example) can be instructive and motivating.

The list of empirical geometric experiences that would be worthwhile introducing can be extended. The work of Dienes (9), Paul Rosenbloom at the Minnesota School Mathematics and Science Center, the Comprehensive School Mathematics Project in Carbondale, Illinois, and the Educational Development Corporation in Newton, Massachusetts, to name just a few, give many more possibilities.

Geometrization and applications

Geometry had its roots in the study of a physical science. As a mathematical study it solved problems in that science (surveying) and proved useful in all the sciences, both pure and applied. It is precisely because of its usefulness that it has survived as a necessary part of the education of all engineers and scientists. For various reasons, historical and otherwise, the usefulness of geometry per se has been slighted—even ignored (11; 14; 15; 20). The definition that has been given in this chapter implies that this state of affairs should be changed, and changed in several ways:

1. Geometry should be related more to the physical world. More attention must be given to the contributions geometry can make toward solving physical problems, and this also means discussing the limitations of (mathematical) geometry. To many people this suggests introducing applications. Certainly applications should appear both as introductory material to serve as motivation for the study of a topic and as concluding material to exhibit the power of a given study; however, even prior to struggling for applications, another change in philosophy and curriculum material must be instituted—three-dimensional geometry should be promoted from its current secondary role to a primary role!

The rationale that three-dimensional concepts are essential for applications is certainly sufficient argument for this assertion, but another reason should be added. If students at the elementary and secondary level are challenged only by plane geometry, they are being robbed at a crucial time in their lives of the practice of reasoning about space and developing the intuition necessary for the analysis of higher-dimensional problems of science and mathematics. Nurturing and developing mathematical intuition should certainly be a principal concern of the mathematics curriculum.

2. *Geometrizing,* the process of constructing geometric conceptualizations, is the basic ingredient in applying geometry to other branches of mathematics as well as to the problems of science. The fact that so varied a class of concepts and problems admit geometric interpretations is another important rationale for the presence of geometry in the curriculum. Hence, geometry itself should be learned as an instrument for interpreting the concepts and problems that arise in other branches of knowledge. Examples in physics abound. At the early stages, the play with mirrors provides golden opportunities for geometric interpretations of problems of optics and mechanics and for abstractions of the physical concepts. Problems involving the mechanics of constant forces and, later, velocities are more easily understood once the geometric formulations are given. The interpretation of work as an area is likewise illuminating. On the not so familiar side, geometric interpretations of the laws of chance and

the study of errors in measurement might be undertaken, and even graph-theoretic interpretations of elementary combinatorial problems.

3. Closely associated with geometrizing is the question of building a formal structure, that is, an axiomatic system. The present manner of teaching this aspect of mathematics rarely teaches students where the primitive frame (the undefined terms and unproved propositions) comes from, that is, *how* and *why* a mathematician creates his deductive system the way he does. If teaching at the early stages would give students experience—both physical and logical—with the concepts, then there would be a much better basis for teaching the important matters in foundations. For example, it is important that students see *where* the definitions come from; the study of interrelationships of propositions will turn up necessary and sufficient conditions that can often be used to define a concept in a convenient manner.

4. Another matter that should be made more concrete is the process of generalizing—How is this done? With a view toward what? Most important are the questions of how to abstract, how to choose axioms, how to mathematize a subject that we already have some knowledge about. This sequence of considerations would be closer to real mathematical development than that which exists in the current curriculum because students would then learn that the axiomatic treatment of a subject is usually the last stage in the creation of a mathematical subject. Moreover, there would be more training in mathematical creativity than could possibly come from merely studying the finished and highly polished product and completing exercises that textbook authors select to fill the gaps.

Mathematical geometric experience

All of the suggestions put forth in the preceding section could have come under the heading of this section. They deserved separate treatment simply because applications are so underemphasized in the current mathematics curriculum.

The major point of this section, however, draws from the final portion of the description of geometry—that a variety of techniques is used to attack geometric problems. Understanding this, we see the clear inference that none of the techniques should be slighted and that each should be appreciated for its own special power; each has virtues and, also, short-comings.

We all know about problems that appear to be difficult until they are examined in "just the right way." Geometry abounds in such problems. There are problems solved by the simple and direct synthetic methods of Euclid. There are others that do not yield to these methods but require analytical methods. Vector techniques may be very elegant for some problems but not for others. Transformation methods, which may now be coming into a stronger position in teaching in this hemisphere, have great philosophical as well as other advantages, enabling one to see underlying invariants that may be playing key roles; yet there are questions for which a transformational study would be quite artificial.

Hence, recommendations like those of Freudenthal (11) and Klamkin (14; 15)—that there be a reorientation of the study of geometry toward problem solving—should be supported. Problems, or classes of problems, should be designed to exhibit the power and beauty of one geometric method at a time. These problems would then prompt the study of that method *qua* geometric method. Other methods come to the fore by exhibiting new problems for which the old methods are inadequate.

Finally—and here is where the sharpest criticism is expected—among the techniques for solving geometric problems should be that of interpreting a geometric problem in some other context, a technique sometimes called "transforming" the problem. For example, mechanical considerations can be used to solve geometric problems (21), electrical networks can often accomplish such results, optics (the mirrors once again) can be used to solve geometric inequalities, and probability (combinatorial analysis) can solve some geometric problems, too. In short, if concept X has a geometric interpretation, then it is very likely that concept X can be used to solve some geometric problems just as well as geometry can be used to solve problems involving concept X.

Particularly because of the generality of the foregoing, but also for other reasons, some special attention should be given to the transformation method. To begin the discussion, let us state that *invariance* is one of the most important ideas in all of mathematics and that geometry is unquestionably the most natural subject for the demonstration and use of this idea. Geometry, therefore, can serve well the entire curriculum by maintaining the invariance notion as one of its themes and using it as an instrument for problem solving as well as for the usual purpose of classifying geometries.

In a sense, this notion should be brought into a child's early mathematical experience. Some of Piaget's astounding experiments show that small children do not realize that the size (cardinality) of a set is invariant under physical rearrangement of the elements (19). Thus arithmetic concepts cannot possibly have meaning until this hurdle is passed, that is, until an invariance is recognized. True geometric experiences with invariance are easy to generate, beginning with the mirror reflections mentioned earlier.

Before proceeding further along these lines, let us get the horse before the cart. One can talk about invariance only *after* transformations. That is, a set or structure is invariant relative to, or under, a transformation. Hence, the subject of *geometric transformations* is prior to, and therefore essential for, invariance. The presence of transformations as a topic in the curriculum would have many advantages: they serve as a natural tie-up with algebra and the function concept, the Euclidean transformations can be taught concretely with mirrors or with physical translations and rotations, translations serve as a basis for an early study of vectors, and the notion of isometry gives the simplest formulation of the idea of congruence. Incidentally, isometry is an excellent example of the mathematizing of a physical concept, for an isometric transformation is a mathematical abstraction (and generalization) of Euclid's idea of superposition.

Once transformations are available, the *symmetry* concept can be defined with precision: A structure (such as a geometric figure or an algebraic equation) possesses symmetry if it is invariant under some transformation other than the identity. There are many virtues to reaching this subject early in the curriculum—important properties of figures can be gleaned from their symmetries, symmetry concepts are fundamental in science from the most elementary work on optics to the most sophisticated work in elementary particle theory, symmetry principles constitute an important tool for problem solving, and finally, with the modern view of symmetries as transformations that preserve some structure (2; 25), all geometries are simply studies of symmetries.

After students feel at home with the basic ideas of transformations, it would be appropriate to go beyond the Euclidean isometries. The most natural next step would be to introduce first magnification and contraction and then point reflections (all examples of *central dilatations*). This is the beginning of the study of *affine transformations,* which also relate directly to physical concepts and should therefore be applied to physical problems.

After these concepts are explored, it would be appropriate to study *parallel projection* and, in particular, *orthogonal projection* by means of shadows and diagrams. The question "What are the invariants?" should be asked. The answer helps to define affine symmetry, which incidentally explains why spirals are symmetric in a technical sense, and opens up a world of applications in the biological sciences, for example, *phyllotaxis,* the phenomenon of order of leaf arrangement on a stem and other analogous phenomena of order in botany, and *conchyliometry,* the study of the geometric forms of shells. See Coxeter (7, pp. 169–72 and 326–27).

Since it has been asserted that invariance should be an ever-present theme in a geometry curriculum, it is perhaps advisable to present an example of how invariance becomes a tool for problem solving and mathematical creativity apart from its philosophical (or foundational) use in classifying geometries. (Further interesting examples are given by Klamkin and Newman [16].) Consider the problem of determining whether it is possible to inscribe an ellipse in a given triangle so that the ellipse touches the triangle at the midpoints of its three sides. (This problem was given to college calculus students, most of whom failed to obtain any worthwhile insight in spite of their relatively good knowledge of analytic geometry.) A small amount of information about affine transformations provides the key:

1. An affine transformation, f, can be used to transform the given triangle T into an equilateral triangle E:

$$f: T \to E$$

2. A circle C inscribed in E touches the sides of E at their midpoints.

3. Since the affine transformation f^{-1} *preserves midpoints,* the image of C under f^{-1} is an ellipse that satisfies the desired conditions.

All the properties of affine transformations used in the solution of this problem are among those that can be taught to junior high school students by means of parallel projections:

1'. Three given noncollinear points of a plane can be mapped onto any other three given noncollinear points of the plane.

2'. Affine images of circles are ellipses. (Perhaps an ellipse would be defined this way at the junior high school level.)

3'. Ratios of distances, and hence midpoints, are invariants of affine transformations.

A possible next step in the physical motivation (i.e., providing an empirical geometric experience) of geometry is *central projection,* which can be demonstrated by casting shadows using a light bulb. Again, this leads to more mathematizing of physics when the light bulb becomes a mathematical point, the projection screen becomes a line or plane extending without bound, and the notion of shadow is extended and abstracted in terms of geometric incidences. Of course this is *projective geometry,* which Lehner, Whitehead, and Coxeter, in historical order, have indicated as possible to treat in secondary schools. (See Coxeter [8, pp. vii–viii].) Coxeter's text *Projective Geometry* was an attempt to make the subject accessible to advanced high school students. It would seem advisable to introduce some of the notions of projective geometry even earlier because they are so intuitive and they offer so much.

The projective equivalence (invariance!) of conic sections is seen immediately when the center of the perspectivity is taken as the vertex of a cone. This not only will serve to delight students but will be another problem-solving tool, as it no doubt was for Pascal when he proved his famous theorem on hexagons inscribed in conics. Even the invariance of cross ratio can be derived from physical considerations.

In all the transformations mentioned thus far, collinearity has been an invariant. One can imagine enrichment units that would introduce more general transformations leading to topological and combinatorial invariants. Combinatorical considerations in particular are important to inject into school geometry. For example, Euler's formula relating the vertices, edges, and faces of a polyhedron should be observed as a combinatorial result (another invariance relation). This would be one of the by-products of using graph theory, as suggested earlier in the section "Geometrization and Applications."

"Globally axiomatic" development of Euclidean geometry

Questions concerning the full-fledged axiomatic treatment of Euclidean geometry have been assiduously avoided. Although the *when* and the *how* constitute a major controversy in the world of mathematical education, other matters seem more important, and that is why these "other matters" were argued first.

On the question of *when,* the heretical answer is offered first: perhaps *never* for the majority of school geometry students. "Locally axiomatic" developments of the subject matter, as proposed in the section "Axiomatics," not only are sufficient but are to be preferred for the general population. There is no reasonably rigorous treatment of Euclidean geometry simple enough for the current student population, that is, no treatment that would not entail a serious sacrifice of things more relevant to the educational needs of students. In any event, no matter what population is to receive it, the question of when to teach an axiomatic development should be begged, since the answer can only be this: After students are sufficiently prepared. Referring to earlier remarks, this means after they have had (empirical) experience with the subject matter and after they understand the basic notions of the subject, have some feel for the propositions that might hold, and have discovered some interrelationships and interdependences between the propositions. In short, axiomatic developments are in order only after the pedagogical groundwork has been laid.

The question of *how* to axiomatize Euclidean geometry should also be begged, since the decision must be made on the basis of ease of development, economy, elegance, and usefulness of the approach in the students' future mathematical training. There is no "best" decision that can be made with assurance, for any decision must be based on the mathematical preparation given to the students prior to their "globally axiomatic" study. For example, if economy is a major concern, it appears at this point in history that the most economical definition of Euclidean geometry comes from linear algebra: *A Euclidean space is a real inner-product vector space*. However, this "economy" of words hides a great deal; it compresses as many as eleven algebraic postulates, to say nothing of their geometric interpretations. Hence, the basic groundwork for this approach requires studying the geometric basis for the axioms of a vector space and giving attention to the importance of the concept of perpendicularity to Euclidean geometry. The latter point must be appreciated before the notion of inner product can be meaningful.

Summary

In conclusion, let us reiterate: (1) The definition of a subject for the purposes of teachers should take cognizance of the roots and historical evolution of the concepts; (2) the definition is best left somewhat vague in order to tolerate different viewpoints and to retain a dynamic quality for the subject; and (3) the implications of such a definition contribute to a flexible and dynamic curriculum that is sensitive to the culture role of the subject.

The particular definition of geometry offered in this chapter has the following implications: (1) There should be a radical change in the relationship between the teaching of geometry and axiomatics—less emphasis should be placed on axiomatic structure and formal proof per se. Axiomatic developments of portions of the subject matter spread out over several years are to be preferred to a full-fledged formal development. More attention should be given to the motivation and origin of the primitive frame of undefined terms and unproved propositions. (2) Geometric experiences of all sorts—empirical and mathematical—should be injected into the curriculum from the earliest grades. (3) Many more applications should be studied with a view toward providing training in geometrizing, namely, providing a geometric conceptualization of a problem. (4) More emphasis must be given to problem solving. (5) The relatively complete rigorous development of Euclidean geometry should appear as a culmination uniting the "locally axiomatic" developments of the preceding years in contrast to current programs that attempt a complete "globally axiomatic" treatment the first time students encounter the subject. Moreover, the "globally axiomatic" development, whenever it comes, must be based on axioms and definitions that are plausible and acceptable to students, and plausibility and acceptability can be expected only if students are provided with a thorough experience with the ideas involved. If the students have previously dealt with the technical concepts (entities, properties, and relations) contained in the axioms and if the axioms annunciate familiar facts concerning the concepts, then the hurdle of motivation for the axiom system presents no obstacle. Students will then—and only then—have a chance of appreciating the creative process involved in developing a branch of mathematics deductively.

REFERENCES

1. Albert, Adrian A., and R. Sandler. *An Introduction to Finite Projective Planes.* New York: Holt, Rinehart & Winston, 1968.

2. Artin, Emil. *Geometric Algebra.* New York: Interscience, 1957.

3. Blank, Albert A. "The Use and Abuse of the Axiomatic Methods in High School Teaching." In *The Role of Axiomatics and Problem Solving in Mathematics,* pp. 13–19. Boston: Ginn & Co., 1966.

4. Choquet, Gustave. *Geometry in a Modern Setting.* New York: Houghton Mifflin Co., 1969.

5. Coolidge, Julian L. *A History of Geometrical Methods.* 1940. Reprint. New York: Dover Publications, 1963.

6. Coxeter, H. S. M. "The Four-Color Map Problem, 1840–1890." *Mathematics Teacher* 52 (April 1959): 283–89.

7. ———. *Introduction to Geometry.* New York: John Wiley & Sons, 1961.

8. ———. *Projective Geometry.* Waltham, Mass.: Blaisdell Publishing Co., 1964.

9. Dienes, Zoltan P. *An Experimental Study of Mathematics Learning.* London: Hutchinson's University Library, 1963.

10. Dieudonné, Jean. *Linear Algebra and Geometry.* New York: Houghton Mifflin Co., 1969.

11. Freudenthal, Hans. "Why to Teach Mathematics So As to Be Useful." *Educational Studies in Mathematics* 1 (May 1968): 3–8.

12. ———. "Geometry between the Devil and the Deep Sea." *Educational Studies in Mathematics* 3 (June 1971): 413–35.

13. Hall, Marshall. *The Theory of Groups.* New York: Macmillan Co., 1959.

14. Klamkin, Murray S. "On the Ideal Role of an Industrial Mathematician and Its Educational Implications." *American Mathematical Monthly* 78 (January 1971): 53–76.

15. ———. "On the Teaching of Mathematics So As to Be Useful." *Educational Studies in Mathematics* 1 (May 1968): 126–60.

16. Klamkin, Murray S., and Donald J. Newman. "The Philosophy and Applications of Transform Theory." *SIAM Review* 3 (January 1961): 10–36.

17. O'Hara, Charles W., and D. R. B. Ward. *An Introduction to Projective Geometry.* New York: Oxford University Press, 1937.

18. Ore, Oystein. *Graphs and Their Uses.* SMSG Monograph series, no. 10, New Mathematical Library. New York: Random House, L. W. Singer Co., 1963.

19. Piaget, Jean, Barbel Inhelder, and Alina Szeminska. *The Child's Conception of Geometry.* New York: Basic Books, 1960.

20. Schuster, Seymour. "On the Teaching of Geometry—a Potpourri." *Educational Studies in Mathematics* 4 (June 1971): 76–86.

21. Uspenskii, V. A. *Some Applications of Mechanics to Mathematics.* Waltham, Mass.: Blaisdell Publishing Co., 1961.

22. Veblen, Oswald, and J. H. C. Whitehead. *The Foundations of Differential Geometry.* Cambridge: At the University Press, 1932.
23. Whittaker, E. T. *From Euclid to Eddington.* Cambridge: At the University Press, 1949.
24. Wilder, Raymond L. *Evolution of Mathematical Concepts: An Elementary Study.* New York: John Wiley & Sons, 1968.
25. Yale, Paul B. *Geometry and Symmetry.* New York: Holden-Day, 1968.

13

A Quick Trip through Modern Geometry, with Implications for School Curricula

ROSS L. FINNEY

ASSUMING that he has not already decided, just how does one go about determining what geometry is? He can ask mathematicians reputed to be geometers exactly what kind of mathematics they do; he can ask all available mathematicians what geometry they have done; and he can ask all these mathematicians what geometry others have done. In return, he will receive much advice, some of it conflicting, and many beautiful theorems, some of which will be couched in astonishing language.

A mathematician who is asked what geometry is may spend a certain amount of time asking his colleagues what they think it is, but sooner or later he must answer in one of two ways. Either he can reply that geometry is what geometers do (a perfect response, of course, but one that does not describe exactly what they do) or, better, he can give examples of the work geometers do.

The latter choice has been made for this chapter, and from these examples and their variety, implications for school curricula will be drawn and predictions about the role geometry will play in the schools of the future will be made, for the mathematical and pedagogical discoveries of this century are proving to be powerful forces for change. The effects of these discoveries have already been felt through the appearance of a variety of new curricula—curricula that on the whole seem to be improvements over the more traditional ones that they have replaced—and it is likely that they will produce even greater changes in the years ahead.

Since the examples of geometry given in this chapter are fairly independent of each other, it should be possible for the reader to skim them or skip about in them as he pleases and still make sense of what follows. In any event, the future role of geometry in our schools, discussed in the section "Implications for School Curricula," can be read independently of the survey of geometry that precedes it. Among the conclusions discussed in the last two sections of the chapter is the following: Euclid is not dead, that is, Euclidean geometry is still a valid geometry for study; rather, it is geometry in the *style* of Euclid that is of waning interest and importance.

Euclidean Spaces

One of the powerful mathematical forces at work in education today is the explosion of mathematical knowledge that has occurred during this century, particularly in the last twenty years. To illustrate this explosion, a brief list of current mathematical fields that are all different but nonetheless intimately concerned with geometry can be made. (The list as it stands is by no means complete, but it did come up in a natural way several years ago when the mathematicians at the University of Illinois at Urbana-Champaign decided to revise the geometry curriculum. Everyone who felt a professional interest in what geometry our students should learn was invited to participate, and the group that convened found itself representing many different fields.) This list includes dynamical systems and control theory, measure-theoretic number theory, differential geometry, differential topology, combinatorial topology, partial differential equations, algebraic and analytic homotopy theory, three-dimensional topology, general and algebraic topology, category theory, geometric topology, point-set topology, projective geometry, and differentiable manifolds.

The group all agreed that our undergraduate course in set theory and metric spaces did not give a preparation broad enough for graduate work in geometry and topology, and all except the projective geometer felt that an undergraduate's exposure to projective geometry should be limited to what came naturally into the undergraduate course on linear algebra. We then designed a year's course to introduce undergraduate mathematics majors to geometry, a yearlong answer to the question, What is geometry? A description of the course, now being given for the fourth consecutive year, follows.

The first section, and one of the shortest, is a review of some of the Euclidean geometry that appears in our courses on linear algebra and elementary real variables. A *point* or *vector* x of *Euclidean n-space* R^n is defined to be an *n*-tuple $(x_1, x_2, \ldots, x_n)$ of real numbers. Thus R^2, whose points are pairs of real numbers, is the familiar Cartesian plane, and R^3 is the usual Euclidean three-space. The space R^1 is the set of real numbers itself.

The *scalar multiple* tx of a vector x is the vector $(tx_1, tx_2, \ldots, tx_n)$. The *scalar product* $(x|y)$ of the vectors x and y is the real number $x_1y_1 + x_2y_2 + \ldots + x_ny_n$. Vectors x and y are defined to be *perpendicular* whenever $(x|y) = 0$. The *length* of the vector x is the number $\sqrt{(x|x)}$ and is denoted by $|x|$ for short. See, for example, the vectors in figure 13.1. The length $|(2,-1)|$ of the vector $(2,-1)$ is $\sqrt{5}$ because

$$(2,-1)|(2,-1) = 4 + 1 = 5.$$

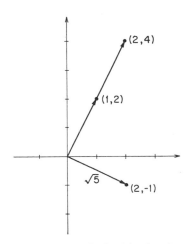

Fig. 13.1   Vectors (points) in the plane R²

The vector $(2,4)$ is twice the vector $(1,2)$ and is perpendicular to the vector $(2,-1)$ because

$$((2,4)|(2,-1)) = 4 - 4 = 0.$$

The course then defines the *distance* $d(x,y)$ between points x and y of R^n to be the length $|x - y|$ of the difference vector $x - y$. This definition shows that for all points x, y, and z—

1. $d(x,y) = d(y,x)$;
2. $d(x,y) \geq 0$;
3. $d(x,y) = 0$ if and only if $x = y$;
4. $d(x,y) + d(y,z) \geq d(x,z)$.

The formula $|x - y|$, when written out in full, is the usual square-root formula for distance. In figure 13.1, the distance between $(2,4)$ and $(2,-1)$ is

$$|(2,4) - (2,-1)| = |(2 - 2, 4 + 1)| = |(0,5)| = \sqrt{25} = 5.$$

It is then possible to define a *motion* to be a distance-preserving function from R^n to R^n. That is, f is such a function if and only if $d(f(x),f(y)) = d(x,y)$ for every pair of points x and y. Motions, of course, are the transformations that many of us were taught to call *rigid motions:* translations, rotations, reflections, and the like. The main theorem of this section describes motions in terms of matrix multiplication, and the course then spends a certain amount of time looking at matrices to see what kinds of motions they do or do not produce.

Finally, *Euclidean geometry* is defined to be the study of the properties of the subsets of R^n that are invariant under motions. Collinearity, incidence, length, volume, angle, separation, and convexity are all examples of such properties.

Euclidean Spaces Are Metric Spaces

One's knowledge of Euclidean spaces may be broadened considerably by studying them as metric spaces, and the next section of the course introduces the appropriate ideas.

A *metric space* is a set X equipped with a function $d(x,y)$ that is defined for all points x and y of X and that satisfies the four properties listed above for the distance function in R^n. The function d is called the *distance function* of X, or the *metric* of X, whence the name "metric space." The Euclidean space R^n, with its distance function $d(x,y) = |x - y|$, is thus a metric space and can be studied with all the tools that have been developed over the years for studying metric spaces.

For a given set there are usually a great variety of metrics. For the points of the Euclidean plane, we have the usual Euclidean metric d given by the rule

$$d(x,y) = \sqrt{(x_1 - y_1)^2 + (x_2 - y_2)^2}.$$

The points of the plane with this metric make up the space called R^2. However, we also have the *city-block metric* d' given by the rule

$$d'(x,y) = |x_1 - y_1| + |x_2 - y_2|,$$

where the vertical bars now mean absolute value.

It is always interesting to compare conic sections in these two metrics. The unit circles (the circles with radius 1 and center at the origin) look like those in figure 13.2. The parabolas with focus $(1,0)$ and directrix the vertical axis look like those in figure 13.3.

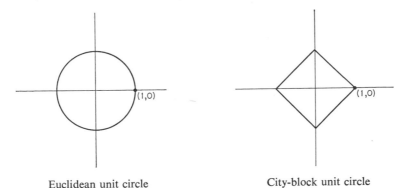

Euclidean unit circle   City-block unit circle

Fig. 13.2

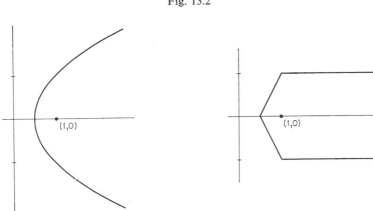

A parabola in the Euclidean metric   A parabola in the city-block metric

Fig. 13.3.

Whenever x is a point of a metric space X and r is a positive real number, the *open ball with center x and radius r* is the set of points of X whose distance from x is less than r.

The name *ball* comes from R^3, where open balls look exactly like balls whose outer-bounding spheres have been stripped off. In R^2, open balls are disks minus their bounding circles. In R^1, open balls are simply open intervals. (See fig. 13.4.) For higher dimensions, we still find ourselves drawing pictures just like those in figure 13.4.

Open balls come in all sizes, depending on the radius r, and in each R^n each point x is the center of infinitely many of them.

A subset of a metric space X is said to be *open* if it is a union of open balls. There is a trick here: the empty set is open because it is an empty union of open balls, and each open ball is open. The space X, itself the union of all its open balls, is open.

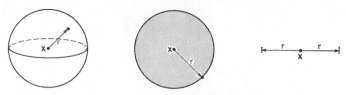

Fig. 13.4.   Open balls in R^3, R^2, and R^1

In R^1 the closed unit interval $[0,1]$ is not open because neither of the endpoints 0 or 1 lies in an open interval that also lies in $[0,1]$. If the endpoints are discarded, however, the result is the open unit interval $(0,1)$, which is open.

In the plane R^2 the set of all points with both coordinates positive is open. The set of all points with nonnegative coordinates, however, is not open because no point of either axis lies in an open ball that lies in the set. The *closed unit ball* E^2, consisting of all points of the plane with distance no more than 1 from the origin, is not open. The *open unit ball,* consisting of all points with distance less than 1 from the origin, is open. The *unit sphere* S^1, consisting of all points with distance exactly 1 from the origin, is not open.

In Euclidean three-space R^3, the open unit ball is again open, but the closed unit ball E^3 and the unit sphere S^2 are not.

Whenever X is a subset of R^n, we may define a metric for X by defining $d(x,x')$ to be the R^n-distance between the points x and x'. Open balls in X need not look round anymore, but they are called open balls nonetheless.

In figure 13.5, the set X is a comb with three teeth. The open ball in X about x of radius r consists of a piece from each tooth.

A function f from a metric space X to a metric space Y is *continuous* if the set $f^{-1}V$ is open in X whenever the set V is open in Y. This requirement may look backwards, but it is equivalent to the calculus definition that f is continuous if $f(x_n) \to f(x)$ whenever $x_n \to x$.

A continuous function is a *homeomorphism* whenever it has a continuous inverse. Each motion of R^n is a homeomorphism, but most homeomorphisms from R^n to R^n are not motions. Spaces X and Y are called *homeomorphic* if there is a homeomorphism from one onto the other.

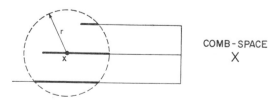

COMB - SPACE

X

Fig. 13.5.   An open ball in a metric space need not look round.

The *topology of Euclidean space* is the study of the properties of the subsets of R^n that are unchanged by homeomorphisms.

In the plane R^2 the subsets homeomorphic to the unit circle S^1 are the simple (i.e., do not cross themselves), closed (i.e., start and stop at the same point) paths. For example, in figure 13.6 the plane sets are homeomorphic to S^1, but in figure 13.7 they are not.

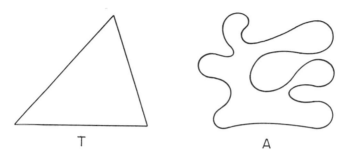

T   A

Fig. 13.6.   These plane sets are both homeomorphic to S^1.

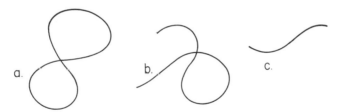

a.   b.   c.

Fig. 13.7.   These plane sets are not homeomorphic to S^1.

To illustrate the kind of role homeomorphisms can play in geometry, we shall digress from the course outline for a moment. In figure 13.6, the triangle T and the outline of the amoeba A differ in many respects: they do not enclose equal areas, only one is convex, and only one has angles. No motion takes one to the other because motions preserve area, convexity, and angle. But T and A are alike in one respect already mentioned: they are both homeomorphic to the unit circle S^1. Moreover, the property of being homeomorphic to S^1 is preserved by motions; so being homeomorphic to S^1 is a property of Euclidean geometry.

To convince us that this geometric property is really worth something, we have the following theorem (14, p. 94):

SCHOENFLIES THEOREM.   *Whenever two subsets of the plane are both homeomorphic to S^1, there is a homeomorphism from the plane onto the plane that carries one subset onto the other.*

From this single geometric property shared by the triangle and the amoeba, we conclude that although there is no motion of the plane carrying one to the other, there is still a homeomorphism of the plane that does so.

There are two natural ways to rephrase the Schoenflies theorem for Euclidean three-space, and both are false. They are false, however, for reasons that are worth investigating and that have led to research that is by no means over.

FALSE STATEMENT 1. *If two subsets of R^3 are homeomorphic to the unit circle S^1, then there is a homeomorphism of R^3 onto R^3 that carries one set onto the other.*

A counterexample is given by a knotted circle K and a circle C that lies in a plane (fig. 13.8). Each of K and C is homeomorphic to the unit circle, but no homeomorphism of R^3 onto R^3 superimposes K on C or C on K. Homeomorphisms of R^3 do not untie knots. (This assertion involves some group theory that we will sketch later.)

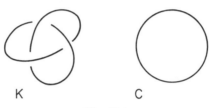

K   C

Fig. 13.8

On reexamining false statement 1, we might decide that it was not a very good analogy because the dimension of the Euclidean space went from two to three while that of the circle S^1 remained unchanged. However, if we put the unit 2-sphere S^2 in place of the circle S^1 in false statement 1, we get the following:

FALSE STATEMENT 2. *If two subsets of R^3 are homeomorphic to the unit sphere S^2, then there is a homeomorphism of R^3 onto R^3 that carries one set onto the other.*

To get a feeling for why this statement is false, see figures 13.9 and 13.10. Figure 13.9 shows the unit 2-sphere in R^3. Figure 13.10 shows a very wiggly sphere, called the Fox-Artin (FA) sphere. The point is not that S^2 and FA fail to be homeomorphic; they *are* homeomorphic (in spite of the wiggles). The point is that there is no homeomorphism of R^3 onto R^3 that superimposes the two. The proof, by Fox and Artin, involves a delicate geometric construction of a particular group (7, p. 989). This proof will be discussed later in the chapter.

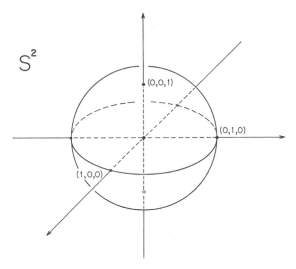

Fig. 13.9   The unit 2-sphere in R^3

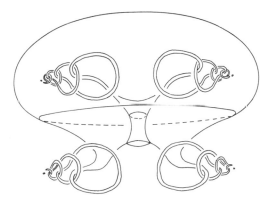

Fig. 13.10.   The Fox-Artin 2-sphere, FA

In contrast to the Fox-Artin example, it is possible to find a homeomorphism of R^3 onto R^3 that superimposes S^2 and the rather lumpy sphere shown in figure 13.11. In fact, false statement 2 can be made true by adding restrictions that keep the sets involved from being too wiggly. The course, to which we now return, provides the necessary language.

Paths and Calculus in Euclidean Spaces

After the geometry and topology of Euclidean space is defined, the course describes conditions that can be put on curves and surfaces in R^3 to produce a manageable geometric theory.

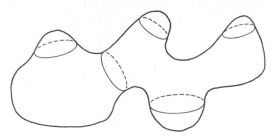

Fig. 13.11.   Lumpy 2-sphere

A *path* in R^3 is any continuous function p from an interval $I = [0,t_1]$ of real numbers into R^3. For each value of t, the point $p(t)$ in R^3 is a triple $(x(t), y(t), z(t))$ of real numbers. This notation suggests that we may think of p as a triple of real-valued functions defined on I, and that is precisely what we do.

One reason for thinking of p in this way is that we can then tell when a candidate for a path is continuous: p is continuous if and only if each of the coordinate functions x, y, and z is continuous. Another reason is that it lets us define paths easily. There are, in fact, many examples from calculus, but before looking at some of them, it will help to have a notation for the phrase

function f from A to B.

The notation is

$$f : A \to B.$$

With this notation, the phrase

function f from R^n to R^n is continuous

can be shortened to

$$f : R^n \to R^n \text{ is continuous.}$$

An example from calculus is the path $p_1 : [0,2\pi] \to R^3$, given by the following formulas:

$$x(t) = \cos t$$
$$y(t) = \sin t$$
$$z(t) = t.$$

The path starts at $(1,0,0)$ when $t = 0$, winds in a helix around the z-axis, and stops when $t = 2\pi$ at $(1,0,2\pi)$. (See fig. 13.12.)

A third reason for defining paths to be functions is that we can bring our knowledge of calculus to bear on the coordinate functions x, y, and z.

Unfortunately, however, if a path is a function, it is no longer merely a geometric object in R^3. In fact, the path $p_2 : [0,1] \to R^3$, given by the correspondence $t \to (\cos 2\pi t, \sin 2\pi t, 2\pi t)$, traces out the same helix as

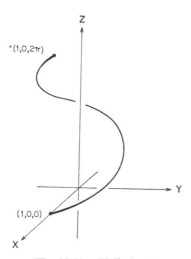

Fig. 13.12.   Helix in R^3

the path p_1. To complicate things further, the path $p_3: [0,2\pi] \to R^3$, given by the correspondence $t \to (\cos(2\pi - t), \sin(2\pi - t), 2\pi - t)$, traces out the helix again, but in the opposite direction. It starts at $(1,0,2\pi)$ when $t = 0$ and stops at $(1,0,0)$ when $t = 2\pi$. If our study of these paths is to be a study of the helix itself, we need a way to say that the paths are essentially the same.

A *parameter change* between intervals I and I' is a continuous function from I onto I' that is either strictly increasing or strictly decreasing. For example, the function $f_1: [0,1] \to [0,2\pi]$, given by $t \to 2\pi t$, is strictly increasing. As t goes up, so does $2\pi t$. The function $f_2: [0,2\pi] \to [0,2\pi]$, given by $t \to (2\pi - t)$, is strictly decreasing. As t goes up, $(2\pi - t)$ goes down.

We say that paths $p: I \to R^3$ and $p': I' \to R^3$ are *equivalent* if there is a parameter change $f: I \to I'$ such that $p(t) = p'(f(t))$ for every t. The three helix paths are equivalent because for each t, we have $p_2(t) = p_1(f_1(t))$ and $p_3(t) = p_1(f_2(t))$.

Equivalence of paths is an equivalence relation, and so the set of paths falls naturally into equivalence classes, which we now call *curves*. The paths p_1, p_2, and p_3, being equivalent, all represent the same curve.

Tracing out the helix in opposite directions, however, shows a distinction between the paths p_1 and p_3 that is too useful to obliterate by lumping them into the same curve. The observation that the parameter change that gives their equivalence is decreasing is the clue for what to do. If we call paths equivalent only if they differ by an increasing parameter change, then each of the former equivalence classes that we called a curve splits

The function $|\ddot{p}|$ is thus a measure of how much the curve represented by p deviates from being a straight line. If we call $|\ddot{p}(s)|$ the *curvature* of the curve at s, we may rephrase the theorem to say that *a regular curve is a straight line if and only if it has curvature zero.* This rephrasing should generate a certain amount of faith in the theory (or at least in the nomenclature). As further evidence that $|\ddot{p}(s)|$ deserves its name, let us add that for a circle of radius r, the number $|\ddot{p}(s)|$ is $1/r$ for every value of s, a result coinciding with the common prejudice that circles should have constant curvature equal to the reciprocal of their radii.

If a curve lies in the xy plane, which we may think of as R^2, its curvature tells us even more.

THEOREM. *Suppose that two oriented regular curves have arc-length paths defined on the same interval, $p,p':I \rightarrow R^2$. Then there exists an orientation-preserving (the determinant of its matrix is 1) motion M with $p = M \circ p'$ if and only if the curvature functions $|\ddot{p}|$ and $|\ddot{p}'|$ are identical.*

By $M \circ p'$ is meant the composite function that is p' followed by M. We may indicate, schematically, that $p = M \circ p'$ by drawing a diagram like that of figure 13.14.

Roughly speaking, then, oriented regular plane curves are congruent under an orientation-preserving motion if and only if they have the same length and curvature. In particular, every regular curve with constant curvature $k \neq 0$ is congruent to part (or all) of a circle with radius $1/k$.

Not all curves in R^3 are planar, however, and curvature alone does not tell how planar a given regular curve is. The arc-length path p_4 for the helix has curvature $\frac{1}{2}$:

$$|\ddot{p}_4(s)| = \left| \left(-\frac{1}{2} \cos \frac{s}{\sqrt{2}}, -\frac{1}{2} \sin \frac{s}{\sqrt{2}}, 0 \right) \right|$$

$$= \sqrt{\frac{1}{4} \left(\cos^2 \frac{s}{\sqrt{2}} + \sin^2 \frac{s}{\sqrt{2}} \right)}$$

$$= \frac{1}{2}$$

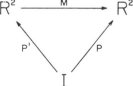

Fig. 13.14. This schematic diagram says that motion M moves the image of path p' to the image of path p. More precisely, it says that $p = M \circ p'$.

But the helix does not lie in a plane, and it certainly is not a circle of radius 2.

One of the pleasant side effects of choosing arc-length paths to represent curves is that the length $|\dot{p}(s)|$ of the vector $(\dot{x}(s), \dot{y}(s), \dot{z}(s))$ of first derivatives with respect to s is always 1. This is because the arc-length formula gives

$$s = \int_0^s \sqrt{\dot{x}^2 + \dot{y}^2 + \dot{z}^2},$$

and so

$$\frac{ds}{ds} = \sqrt{\dot{x}^2 + \dot{y}^2 + \dot{z}^2} = |\dot{p}(s)|.$$

The derivative ds/ds is, of course, 1.

For each value of s, the vector $\dot{p}(s)$ is called the *tangent vector* to the curve represented by p. It is so named because in every known example it actually looks like a tangent vector.

Another pleasant side effect is that for each s the vector $\dot{p}(s)$ is perpendicular to the vector $\ddot{p}(s)$. This is true because

$$0 = \frac{d}{ds}(1) = \frac{d}{ds}(\dot{p}|\dot{p}) = (\dot{p}|\ddot{p}) + (\ddot{p}|\dot{p}) = 2(\dot{p}|\ddot{p});$$

so $(\dot{p}|\ddot{p}) = 0$. This argument, although standard, is convincing only when one knows why the equalities hold. The first holds because the derivative of a constant is always 0. The second holds because $\dot{p}$ has length 1 (so $\dot{p}|\dot{p} = 1$). The third is a routine identity that can be verified by writing out $\dot{p}|\dot{p}$, taking its derivative, and then condensing. The fourth is an instance of the symmetry of scalar products.

The vectors $\dot{p}$ and $\ddot{p}$, being perpendicular, determine a plane whenever $\ddot{p} \neq 0$. Whenever the curve represented by p is a plane curve, the plane of $\dot{p}$ and $\ddot{p}$ is the same as the plane of the curve. Suppose that at each point $p(s)$ of a nonplanar curve we could find a measure of how much the curve twists away from the plane of $\dot{p}(s)$ and $\ddot{p}(s)$. We should then have a measure of how much the curve deviates from being a plane curve at s. Fortunately, to produce such a measure takes only a modicum of patient work. First take the cross product $b = \dot{p} \times \ddot{p}$ of the vectors $\dot{p}$ and $\ddot{p}$; then for each s the vector $\ddot{p}(s)$ is a real scalar multiple $-\tau(s)$ of db/ds.

Although this procedure is certainly not obvious, the function τ, called the *torsion function of the curve*, does serve as the measure we want, because we have the following theorem:

THEOREM. *The curve is planar if and only if τ is identically 0.*

In fact, as the next theorem says, it will not merely serve, it will serve perfectly.

THEOREM. *Let p_1, p_2: $[0,L]$ → R^3 be arc-length representatives of oriented curves. Then there exists an orientation-preserving motion M with $p_1 = M \circ p_2$ if and only if the curvature functions $|\ddot{p}_1|$ and $|\ddot{p}_2|$ are identical and the torsion functions τ_1 and τ_2 are identical.*

So if two paths in R^3 have the same length and if they bend and twist the same way, then they are congruent.

In addition, the theorem says that both curvature and torsion of curves in space are geometric notions; they are unaltered by motions.

It is now possible to see why regularity in R^3 is assumed to mean three continuous derivatives: with three, τ is continuous. In the plane τ is not needed, and two continuous derivatives are enough to make the curvature continuous.

This last theorem has implications for the Schoenflies problem. It gives us a restricted version of false statement 1 that is true: *If two subsets of R^3 are homeomorphic to a circle and if they can be described by arc-length paths that have identical curvature and torsion functions, then there is a rigid motion of R^3 that places one on the other.*

Since each motion is a homeomorphism, this would seem to solve the problem. However, just as our hypotheses were too weak when we first stated false statement 1, now they are too strong. To generalize the Schoenflies theorem, we do not need to produce a motion; any homeomorphism will do. Can we not weaken the hypotheses and still come up with a homeomorphism? (The answer is yes, we can, but not easily.)

The course continues its treatment of regular curves with a study of winding number, vector fields, and the Jordan curve theorem.

JORDAN CURVE THEOREM. *If a subset of the plane is homeomorphic to a circle, then it separates the plane into two connected sets, one bounded and one unbounded, of which it is the common boundary.*

Although this theorem is considerably weaker than the Schoenflies theorem, its proof is accessible at this stage in an undergraduate's education; that of the Schoenflies theorem is not. Moreover, the higher-dimensional generalizations of the Jordan curve theorem are all true: A subset of R^n ($n \geq 2$) homeomorphic to the unit sphere S^{n-1} in R^n separates R^n.

The course then looks at ways to apply calculus to a study of surfaces in R^3. Historically, this kind of study has led to the use of derivatives in still higher dimensions and to applications that have helped to solve geometric problems with stunning success in the last few decades (12).

Building with Line Segments and
Triangles in Euclidean Spaces

It is to *combinatorial topology* that the course turns when it leaves curves and surfaces. Combinatorial topology is the study of the metric spaces that can be made by fitting together, in an orderly fashion, geometric building blocks called *simplexes*. The *standard 0-simplex* is the point 1 in the real line R^1. The *standard 1-simplex* is the line segment spanned by the points $(1,0)$ and $(0,1)$ in R^2. The *standard 2-simplex* is the triangular plate in R^3 whose vertices are $(1,0,0)$, $(0,1,0)$ and $(0,0,1)$. (See fig. 13.15.) In general, the *standard n-simplex* is the smallest convex set in R^{n+1} that contains the tips of the $n + 1$ unit vectors.

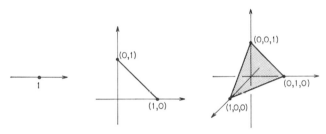

Fig. 13.15.   Standard simplexes of dimensions 0, 1, and 2

A *geometric n-simplex* (we shall just say *n-simplex*) is a space homeomorphic to the standard *n*-simplex. Each of the simplexes in figure 13.16 is a 1-simplex because it is homeomorphic to the standard 1-simplex. Similarly, a 2-simplex and a 3-simplex might look like those in figure 13.17.

The rule for assembling 1-simplexes is that they must meet in common endpoints or not at all. The two pictures in figure 13.18 show what can

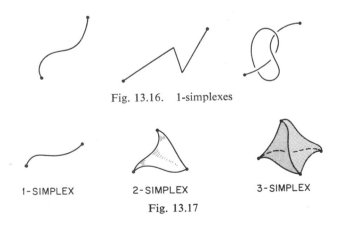

Fig. 13.16.   1-simplexes

1-SIMPLEX   2-SIMPLEX   3-SIMPLEX

Fig. 13.17

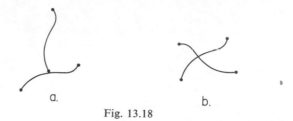

a.   b.

Fig. 13.18

go wrong: the simplexes touch or cross at interior points. To build correctly the space shown in figure 13.18a takes at least three simplexes (see fig. 13.19a); to build correctly the space shown in figure 13.18b takes at least four (fig. 13.19b). We must make certain that all points of contact are vertices.

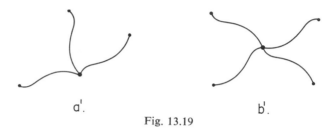

a'.   b'.

Fig. 13.19

When a finite number of 1-simplexes are assembled according to this rule, the resulting metric space is a *1-complex* (pronounced "one-complex" and short for "one-dimensional complex"). Infinite 1-complexes do exist; for example, the unit intervals give a simplicial decomposition of the real line. But there is no need to consider them here.

Famous and entertaining problems about 1-complexes abound. The problem of the seven bridges of Kaliningrad, as Königsberg is now called, is quite well known. Kuratowski's construction theorem, however, is not so familiar, and so a brief discussion of it is in order.

Not all 1-complexes can be assembled in a plane. For example, the complete five-point CFP and the gas-lights-and-water circuit GLW cannot be assembled in a plane, for both require three dimensions for their construction (fig. 13.20).

It might be imagined that a great variety of such recalcitrant 1-complexes exists, but Kuratowski was able to prove in 1930 that a 1-complex that requires R^3 for its construction already contains CFP or GLW somewhere inside it (9, p. 271). These two 1-complexes are entirely responsible for nonconstructability in the plane. Astonishing!

The rule for assembling 2-simplexes is that they must connect either (1) at a vertex or along an entire edge or (2) not at all. (See fig. 13.21.)

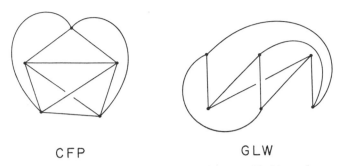

CFP   GLW

Fig. 13.20.   These complexes cannot be assembled in a plane

When a finite number of 2-simplexes are assembled by this rule, the resulting metric space is a *2-complex*.

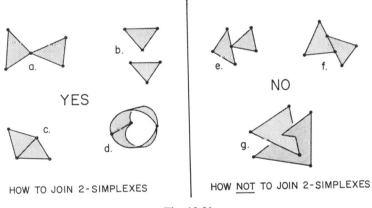

b.

YES   NO

c.   e.   f.

d.   g.

HOW TO JOIN 2-SIMPLEXES   HOW NOT TO JOIN 2-SIMPLEXES

Fig. 13.21

A number of 2-complexes are surfaces. For example, let us begin with four particular 2-simplexes, as shown in figure 13.22.

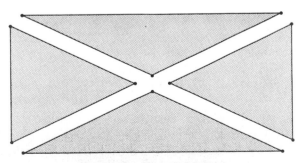

Fig. 13.22.   Four 2-simplexes

1. Connect neighboring edges to get a *rectangle* (fig. 13.23).

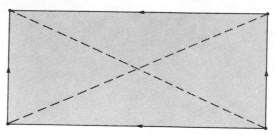

Fig. 13.23.   A rectangle

2. Then connect the top and bottom edges to match the arrows in figure 13.24 to make a *cylinder*.

Fig. 13.24.   A cylinder

3. Finally, connect the right and left ends to match the arrows in figure 13.25 to produce a *torus* (surface of an anchor ring).

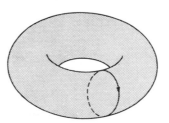

Fig. 13.25.   A torus

If in step 3 we had connected the ends of the cylinder so that the arrows were opposed instead of matched, we would have produced the surface known as the Klein bottle. The Klein bottle cannot be assembled in R^3, however. Its construction requires one more dimension. It can be assembled in R^4.

There is a general theorem, easily proved in linear algebra, that an n-complex (a metric space assembled from n-simplexes according to analogous rules) can always be assembled in R^{2n+1}. In the early 1930s Van Kampen (13, p. 72) and Flores (6, p. 17) were able to show that certain

n-complexes needed the full $2n + 1$ dimensions. They did this by finding a suitable generalization of the 1-complex CFP.

The boundary of a 2-simplex is made of three vertices and the three 1-simplexes that connect the vertices in pairs. The boundary of a 3-simplex consists of four vertices, the six 1-simplexes connecting the vertices in pairs, and the four 2-simplexes connecting the vertices in threes. The boundary of a 4-simplex consists of five vertices, the ten 1-simplexes connecting them in pairs, and so on. (See fig. 13.26.) And here we have the key to Van Kampen's and Flores's discovery: these five vertices and ten 1-simplexes constitute the complete five-point CFP.

THE 3 VERTICES
AND 3 1-SIMPLEXES
IN THE BOUNDARY
OF A 2-SIMPLEX

THE 4 VERTICES
AND 6 1-SIMPLEXES
IN THE BOUNDARY
OF A 3-SIMPLEX

THE 5 VERTICES
AND 10 1-SIMPLEXES
IN THE BOUNDARY
OF A 4-SIMPLEX

Fig. 13.26

The appropriate higher-dimensional analogue, they proved, is the union of all the n-simplexes in the boundary of a $(2n + 2)$-simplex. Such a union cannot be assembled in any Euclidean space of dimension less than $2n + 1$. The complex CFP is the case $n = 1$, the union of all ten 1-simplexes in the boundary of a $(2 \cdot 1 + 2 = 4)$-simplex.

Van Kampen and Flores also discovered that a natural generalization of GLW requires the full $2n + 1$, but seemingly there is no theorem like Kuratowski's to say that every n-complex requiring the full $2n + 1$ contains a generalized CFP or GLW.

Combinatorial topology has been able to solve a number of problems, among them one on identifying spheres posed originally for dimension three by Poincaré—a problem that has been solved in all dimensions except three and four (3; 12). This problem is mentioned here partly because of its long standing—it was first posed at the turn of the century, but the first solutions did not appear until the early 1960s—and partly because solutions in dimensions three and four have not yet been found. The combinatorial solutions of the early 1960s led to a completely unrestricted solution in 1966, but still in dimensions greater than or equal to five.

At the risk of conveying the impression that all geometry is involved with the Schoenflies problem, which it definitely is *not,* some time should be spent on its solution in R^3, since the hypotheses are so purely geometric.

A *2-sphere* is a space homeomorphic to the unit sphere S^2. A 2-sphere in R^3 is *polyhedral* if it is a 2-complex that is made from flat, straight-edged triangular plates. (See fig. 13.27.)

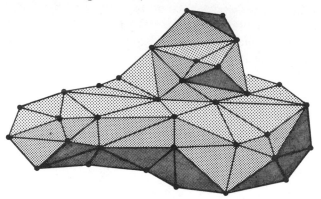

Fig. 13.27.   A polyhedral 2-sphere in R^3

It has been known since Alexander's 1924 paper (1, p. 6) that the bounded component of the complement of a polyhedral 2-sphere in R^3 is always homeomorphic to an open ball (the Fox-Artin 2-sphere is *not* the boundary of a ball in R^3), but it was not until the early 1950s that Moise proved that if A and B are polyhedral spheres in R^3, then there is a homeomorphism of R^3 onto R^3 that moves A onto B (10, p. 172).

Moise's theorem, of course, does not apply to nonpolyhedral spheres. The Fox-Artin 2-sphere in figure 13.10 is not polyhedral and cannot be moved onto S^2 by a homeomorphism of R^3; the lumpy 2-sphere in figure 13.11 is not polyhedral but can be moved onto S^2 by a homeomorphism of R^3. A description of recent work in this direction can be found in an article by Schultz (12).

The undergraduate course restricts its attention more or less to complexes of dimensions one and two and uses its findings to construct *fundamental groups* of surfaces and then to return to the geometry of surfaces. (The fundamental group is the group π_1 that appears in the next section.)

Putting Geometry and Algebra Together

Let us turn for a moment to a kind of problem that pervades all mathematics, and geometry in particular. (It has already appeared here in disguise as the Schoenflies problem.) If A is a subset of a metric space X, then A is itself a metric space, and the distance between two points of A is the X-distance between them. Similarly, if B is a subset of another metric space Y, then B is a metric space, and we may talk of continuous functions from A to B. Given such an f, we may then ask whether there is a continuous function F from X to Y that agrees with f on A. (See fig. 13.28.) When there is, F is called an *extension* of f to X. Sometimes more is asked of F than mere continuity, but in any event determining the existence of F is called an *extension problem*.

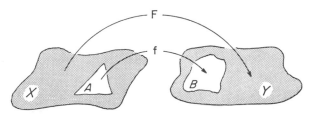

Fig. 13.28. F is an extension of f if it agrees with f on A.

To keep track of what is going on, one usually draws a schematic diagram like the one shown in figure 13.29. Here, i and j are the *inclusion functions* defined by $i(a) = a$ and $j(b) = b$. The horizontal arrow for F is dashed to indicate that its existence is in question.

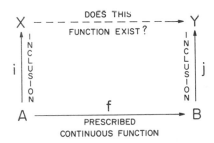

Fig. 13.29.   Schematic diagram for an extension problem

The Schoenflies theorem for the plane says that if A is a subset of the plane that is homeomorphic to the unit circle S^1, then it is possible to find a homeomorphism F of the plane onto itself that moves A onto S^1. But, in fact, even more is true, for if f is any homeomorphism of A onto S^1, then there exists a homeomorphism F of R^2 onto R^2 that agrees with f on A. Thus the solution F of the Schoenflies theorem is provided by the solution of an extension problem, as shown in figure 13.30.

One of the favorite theorems of calculus says that if $f : [0,1] \to [0,1]$ is continuous, then there is some point x in $[0,1]$ for which $f(x) = x$. In

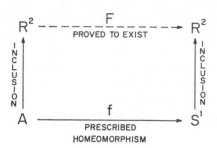

Fig. 13.30.   The Schoenflies problem is solved as an extension problem.

short, every continuous $f: [0,1] \to [0,1]$ has a *fixed point*. The theorem appears in calculus because one of its proofs is a natural consequence of the intermediate-value theorem. However, theorems like it have long been a domain of study of geometers, who ask the question, What geometric objects X have the property that each continuous function from X to X has a fixed point? In short, what geometric objects have the *fixed-point property?* In fact, one of the great remaining problems in the study of the plane R^2 is to find out exactly which of its subsets have the fixed-point property (4, p. 132). Certainly circles do not, for there are small rotations that leave no point fixed. Lines do not, for there are translations that leave no point fixed. Nor do any other Euclidean spaces have the fixed-point property, for the same reason. However, every closed unit ball E^n does have the fixed-point property.

The calculus proof for $[0,1]$ cannot be adapted to higher dimensions, but the statement that the balls E^n have the fixed-point property can be rephrased in terms of an extension problem, which in turn has an algebraic solution. The conversion of a geometric problem to an extension problem and its subsequent solution by algebra is typical of the techniques used by geometers today and is well worth looking into. In order to be specific, we shall prove that the closed unit disk E^2 has the fixed-point property. The pictures that accompany the argument will be those of E^2 and of its bounding circle S^1. The argument, however, will be equally valid for every E^n and bounding sphere S^{n-1}.

The proof that each continuous $f: E^2 \to E^2$ has a fixed point is a contrapositive proof. That is, it begins by supposing that f has no fixed point and then rushes merrily off to a contradiction.

If $f: E^2 \to E^2$ has no fixed point, then $f(x)$ is never the same point as x, and a unique ray starts at $f(x)$ and runs back through x. (See fig. 13.31.) This ray hits the circle S^1 at some point that we call $r(x)$. The correspondence $x \to r(x)$ defines a function from E^2 to S^1, and in our course it becomes a routine exercise to show that r is continuous. If x happens to be a

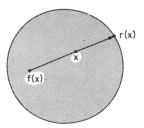

Fig. 13.31. This can be done for each x if $f:E^2 \to E^2$ has no fixed point.

point of E^2 that already lies on S^1, then $r(x) = x$. The function r is then the solution of the following extension problem: Find a continuous function from E^2 to S^1 that agrees with the identity function on S^1. (See fig. 13.32.)

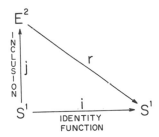

Fig. 13.32. If f has no fixed point, then r solves this extension problem.

We arrive, then, at this point: If f has no fixed point, then this extension problem has a solution. What then? We find that this extension problem *never* has a solution (as will be shown later in this chapter). We conclude that f must have a fixed point after all.

It is in showing that the identity function on S^1 cannot be extended to a continuous function from E^2 to S^1 that algebra comes in. A general theory, *homotopy theory,* assigns to each metric space X a sequence of groups

$$\pi_0(X), \pi_1(X), \pi_2(X), \ldots$$

(pronounced "pi zero of X," "pi one of X," and so on). For a given space it is usually difficult, if not impossible, to compute these groups, but we do know them for some spaces. Roughly speaking, the groups measure the numbers and dimensions of holes in a space. For example, the fact that E^2 has no holes at all is reflected in the fact that every $\pi_n(E^2)$ is the zero group (the group whose only element is the identity element 0). The circle S^1 has one two-dimensional hole, and we find that $\pi_1(S^1)$ is the group Z of integers, a group that can be generated by one nonzero generator. The space S^2 has one three-dimensional hole, and we find that $\pi_2(S^2)$ is Z, whereas $\pi_1(S^2)$ is the zero group, 0.

One of the accomplishments of homotopy theory is that the assignment of groups is not haphazard. Whenever f is a continuous function from metric space X to metric space Y, there correspond to f homomorphisms $f_n : \pi_n(X) \to \pi_n(X)$ of groups. There is one for each natural number n, and the following marvelous things happen:

1. If $f : X \to X$ is the identity function, then f_n is also the identity function.
2. If $f : X \to Y$ and $g : Y \to W$, then the homomorphism $(g \circ f)_n$ assigned to the composite $g \circ f$ is precisely the composite $g_n \circ f_n$ of the homomorphisms assigned to f and g.

The only thing we need to add before we tackle the extension problem for S^1 is a fact from group theory: Whenever $h : G_1 \to G_2$ is a homomorphism of groups, h takes the identity element of G_1 to the identity element of G_2.

Again the proof is contrapositive. Suppose that in figure 13.32 an extension r really did exist. Then homotopy theory would convert figure 13.32 into a diagram of groups and homomorphisms that looks like figure 13.33. In fact, since we know what the groups are, the diagram in figure 13.33 actually looks like figure 13.34.

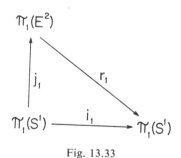

Fig. 13.33

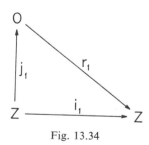

Fig. 13.34

From this diagram we can see that the composite $r_1 \circ j_1$ takes everything from Z on the left side to the identity element 0 of Z on the right side.

This is because j_1 takes everything to 0 and because r_1, being a homomorphism, takes the identity element 0 of the 0-group to the identity element 0 of Z.

According to (1) above, the homomorphism i_1 is the identity function on Z. According to (2) above, the composite $r_1 \circ j_1$ is $(r \circ j)_1$. If r agrees with i on S^1, this says that $r \circ j = i$. Hence $r_1 \circ j_1 = (r \circ j)_1 = i_1 =$ the identity function on Z.

The identity function, however, does not take everything from Z on the left to the identity element of Z on the right. We conclude from this contradiction that the hypothesized extension r does not exist.

To produce a proof of the original fixed-point theorem in higher dimensions, we replace E^2 by E^n and S^1 by S^{n-1}. In the homotopy theory, we then replace π_1 by π_{n-1}. As before, $\pi_{n-1}(E^n) = 0$ and $\pi_{n-1}(S^{n-1}) = Z$. The argument above goes through word for word.

Now that we have the homotopy group $\pi_1(X)$ for each metric space X, we are equipped to go back and look at the assertion that there is no homeomorphism of R^3 onto R^3 that takes the cloverleaf knot K onto any circle C that lies in a plane. (See fig. 13.35.) The key to the proof is a consequence of the rules 1 and 2 above: Whenever a continuous function f is a homeomorphism, the assigned f_n's are actually isomorphisms.

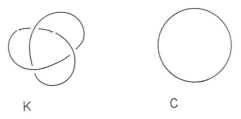

K   C

Fig. 13.35

A homeomorphism h of R^3 onto R^3 that superimposes knot K onto circle C is also a homeomorphism that superimposes the metric space $R^3 - K$ onto the metric space $R^3 - C$. (The notation $R^3 - K$ denotes the the complement of K in R^3, that is, R^3 with K deleted.) It can be computed (but not easily) that the group $\pi_1(R^3 - C)$ is isomorphic to Z, whereas the group $\pi_1(R^3 - K)$ definitely is not. Yet if h existed, the assigned homomorphism h_1 would have to be an isomorphism of these two nonisomorphic groups. We conclude that there is no h.

The proof that there is no homeomorphism of R^3 that superimposes the Fox-Artin sphere FA (see fig. 13.10) and the unit sphere S^2 is accomplished in a similar manner: one produces, with great care and diligence, π_1 of one of the complementary domains of FA and observes that it is not the 0-group, which it would have to be if there were a homeomorphism of R^3 onto R^3 that took FA onto S^2.

A Last Example: Hilbert Space

At the beginning of this geometry survey, it was mentioned that the list of fields given there was not complete, and our last example is taken from a field not on that list—the field of infinite-dimensional geometry. It is particularly appropriate to choose this field because of a recent announcement by Schori and West that they have solved one of the field's long-standing problems (11).

It is simple enough to generalize the notion of Euclidean n-space by letting points be infinite rather than finite sequences of real numbers. But the corresponding formula

$$\sqrt{\sum_{n=1}^{\infty} (x_n - y_n)^2}$$

does not define a distance function. There are sequences x and y for which the sum fails to converge. However, if one throws out all sequences except those for which $\sum x_n^2$ does converge, the formula does define a distance. The resulting metric space, *Hilbert space,* is denoted by H.

We know, of course, that the plane R^2 is the Cartesian product of two copies of R^1, that is $R^2 = R^1 \times R^1$. Likewise, $R^3 = R^1 \times R^1 \times R^1$, and R^n is a product of n copies of R^1. Six years ago Anderson (2, p. 515) solved another long-standing problem of infinite-dimensional geometry by showing that in spite of its special formulation, Hilbert space too is a product of copies of R^1 (this time infinite).

In spite of this similarity between H and the Euclidean spaces, however, a number of exciting differences exist. Among them is Klee's discovery that compact subsets of H always have homeomorphic complements (8, p. 8). For R^n, whose compact sets are the bounded sets that have open complements, such a statement would be entirely false. It would be preposterous to assert, for example, that a point and a sphere in R^n have homeomorphic complements.

Within H is the *Hilbert cube, C,* the subspace of those sequences $(x_1, x_2, \ldots)$ with $|x_n| \leq \dfrac{1}{n}$ for each n. The word *cube* is suggested by the fact that C is homeomorphic to a product of infinitely many copies of the unit interval. What Schori and West have announced is that the cube is homeomorphic to the space of all closed subsets of the closed unit interval $[0,1]$. To describe this space, we first look at $[0,1]$ itself.

The unit interval $[0,1]$ is a metric space, the distance between two points x and y being the usual absolute value $|x - y|$. The open subsets of [0,1] are then the sets that are unions of the open balls given by this distance function (open subintervals, and so on). A subset of [0,1] is *closed* if its complement is open. Single points, closed intervals, and finite unions of single points and closed intervals are all examples of closed subsets of [0,1], as shown in figure 13.36.

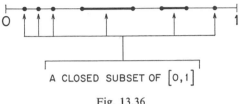

A CLOSED SUBSET OF $[0,1]$

Fig. 13.36

In the *space of closed subsets* of [0,1], the "points" are the closed subsets of [0,1]. The distance $h(A,B)$ between two such points A and B is not their usual distance in [0,1], however, but rather a number that is defined in the following three inelegant steps:

1. Look at all the numbers $d(a,B)$ where $d(a,B)$ is the usual distance in [0,1] from the point a of A to the set B. Take the largest of these numbers as a ranges over A.
2. Do the same for the numbers $d(b,A)$ as the point b ranges over B.
3. Then $h(A,B)$ is the larger of the numbers produced by steps 1 and 2.

For example, let A be the point $\frac{1}{4}$ and let B be the closed interval $\left[\frac{1}{2}, \frac{3}{4}\right]$. Step 1 gives $\frac{1}{4}$, step 2 gives $\frac{1}{2}$, and step 3 says $h(A,B) = \frac{1}{2}$. (See fig. 13.37.)

The assertion, then, of Schori and West is that the metric space whose points are the closed subsets of [0,1] and whose metric is h is actually homeomorphic to the Hilbert cube C.

Although the fields mentioned in the initial list all have special techniques and points of view (and although there are still other fields con-

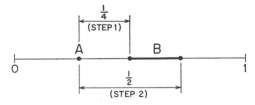

Fig. 13.37.   Here the distance $h(A,B)$ is 1/2.

cerned with geometry that are not in the list), they all have a number of outward features in common: each field is international, each has famous men, each has a profusion of interesting problems, each fills a number of journals every year, each has worthy problems accessible to beginners, each is discussed at mathematical conferences all over the world, and each has grown so large that keeping up with the work of one's colleagues can be a full-time job.

What Has Happened to Traditional Geometries?

It must have been noticed that very little has been said about traditional nondifferentiable, noncombinatorial, nonalgebraic Euclidean and projective geometries. That is not because I do not love them (I do!), but because they constitute only a small part of the geometrical activity that is taking place today. A number of people do work full time in the more traditional geometries and enjoy it thoroughly—and they are contributive. Many other mathematicians work in traditional geometries from time to time. Almost every mathematician likes to see a new theorem about triangles and appreciates deft and elegant proofs of a theorem he already knows.

Yet, although everyone likes them, the feeling of the majority of mathematicians is that the traditional fields have already been worked so hard that they are played out, that they are dead ends. Most graduate students are warned away from traditional geometries, if for no other reason than because whatever they do will almost certainly have been done already. (I once worked for weeks developing a theory about certain kinds of convex bodies in R^n, only to find later that most of what I had done was in Minkowski's collected works. A few years ago I leafed through hundreds of issues of nineteenth-century French and Belgian journals just to make sure that my triangle theorems [5, p. 177] were not there.)

No one disputes that geometry in the style of Euclid (I have called it "traditional" geometry here) was a beautiful beginning. It still is a beautiful beginning, but over the years geometry has developed into such a varied paradise of deep results and powerful tools and such a fertile ground for the imagination that most geometers simply will not give up what it is today to work with it as it used to be.

Implications for School Curricula

As the chairman of a committee that wrote most of the USAID Entebbe Mathematics Program's Secondary-C Geometry, I have had a number of opportunities to discuss the teaching of geometry in schools both here and abroad. It will probably not be surprising to the readers of this chapter that

I have come to a point of view somewhat different from that held by most teachers when I went to school. I believe, for example, that we should no longer spend an entire high school year on the traditional Euclidean geometry of the plane. I also believe that it is wrong to isolate geometry in the traditional way from other mathematics and that a good deal of the geometry that has been reserved for the later years of high school should appear earlier. I also think that it is not a good idea to try to teach geometry to high school students as an exercise in axiomatic foundations.

Among the reasons for my beliefs are the great increase in mathematical knowledge that has taken place in this century and the very productive dependence that each branch of mathematics has been shown to have on the others, and I have tried to illustrate these phenomena by example in the preceding sections. But there are other reasons that I shall bring in as these beliefs are discussed in more detail and as implications for school curricula in the future are drawn.

To start at the beginning, however, let me say that I still think that plane Euclidean geometry should be taught in our schools and, in fact, taught throughout the entire school curriculum. It is beautiful, many people love it, and many of its best theorems and ideas are suggested by things we can all see and draw. In our architectural society it is inescapable and hence a natural subject of study. Like all geometries, it begins as a mathematics of pictures. For example, look at the phrasing of Euclid's first three postulates: to *draw* a straight line from any point to any point, to *produce* a finite straight line continuously in a straight line, to *describe* a circle with any center and distance. Of course, today geometers more often draw other things, but this is a good first step toward understanding what they do.

It is very important for students to have experiences in school that they cannot explain immediately but that are so strongly supported by their intuition that no one has to tell them that they are right. Geometry is unusually well suited for this because it has constructions that, for reasons not immediately obvious, produce results that are not only correct but also remarkably convincing. After drawing enough pictures, a student can come *by himself* to the thought, "Here is a curious thing that I know always happens. Why?" In this context the question "Why?" is not bogus; it is a genuine question about something that in the student's experience is a real event.

Very often plane geometry comes up naturally in other elementary mathematics. If we have learned anything from the mathematics of this century, it is that the most far-reaching research tends to be done by people who apply the tools of more than one discipline. And if we have learned anything about teaching mathematics in the past few decades, it is that it is fruitful and interesting at every level to combine geometry, algebra, and analysis. We have, in fact, learned that it is a good idea to combine mathematics with every kind of measurement, prediction, computation, scientific investigation, visual observation, and data-collecting activity that we possibly can, both in and out of class. In any event, we cannot leave geometry out of school mathematics, and—what is much the same thing— we cannot relegate geometry to a tidy corner by itself in the tenth grade.

The very need to combine geometry with other mathematics should prevent us from trying to teach geometry in high school as an exercise in axiomatic foundations. There are, however, two other reasons for abandoning this attempt: it bores most people, and it does not teach them very much geometry. An extreme example of the latter reason is a book written in the early 1960s. The exposition is elegant, beautiful, careful, and gentle. It is the finest version of Hilbert's geometry that has ever been produced for schools, and the authors may justly be proud. But by the time incidence, betweenness, separation, and congruence have been introduced, there are twenty-five postulates. This beginning limits the second half of the book to theorems the whole class has seen before. When the course ends, it has gone no further than the parallel postulate and some of its consequences.

What kind of geometry, then, should we have in our schools? Again, an example may be given with a certain junior high school program. Although the program is an integrated one, only a description of its geometry will be given here.

This particular program contains, first of all, a generous supply of activities that foster geometric perception and intuition on both two and three dimensions. These activities involve area, shape, shape patterns, size, order, angle, symmetry, right- and left-handedness, inside and outside, motions in the plane, similarity, subdivision, counting, estimation, convexity, length and direction of arrows, parallelism, block piles, perimeter, surface area, inequalities, and equalities. Mathematics deals with all these things, and geometry, too, is associated with them at all levels.

Coordinate systems, both geographical and mathematical, are introduced at an early stage. Functions are graphed whenever possible, and information is deduced about them—including some functions whose rules are not known—from their graphs. Graphs are used a great deal in our society to convey numerical information (be it true or false!), and students are bound to meet them both in and out of class.

Connections between geometry and measurement appear repeatedly in the program. These connections include the usual ones: distance in the plane, units of measure for angles, areas, volumes, and so on. But besides such "practical" connections, there are a number of Euclidean theorems of the geometry of the plane that can be discovered by measuring, and these also appear in the program.

Toward the end of the program—after students are sufficiently experienced—come an introduction to transformation geometry and some elementary applications of linear inequalities. Finally, for those schools fortunate enough to have access to a computer, there are computer-assisted graphing activities.

The level at which the kind of geometry described above can generally be done depends, of course, on the background of the teacher and the class, on the availability of appropriate text materials, and on the creation of suitable activities. But contrary to what might be thought, this program is not written for the upper 20 percent of our junior high school students but for the middle 60 percent (although preliminary testing suggests that the program will suit both groups).

To return to the general role of geometry in school mathematics, the phenomenon on which we must focus is that a great deal of geometry occurs naturally with other mathematics. Some is combined with algebra and analysis—for example, in graphing equalities and inequalities. Some is connected with displaying data and with optimization problems. Still more occurs with estimating, measuring, and counting, and with intrinsically interesting problems in visual perception: shape sorting, volume, area, length, arrows, symmetries, convexity, separation, one-dimensional graphs, cross sections, and geoboard games.

This geometry that occurs naturally leads easily to interesting geometry that is also nonobvious. Nonobvious, however, does not mean difficult or inaccessible. For example, the theory of similarity transformations of the plane is simple and elegant and contains many surprising and beautiful theorems. Ceva's theorem, which itself combines algebra and geometry, requires only the $1/2\ bh$ formula and the unique division-point theorem, and yet it leads immediately to a startling list of concurrence theorems, including three of the standard ones. Ceva's theorem states that *if points X, Y, and Z are points of sides BC, CA, and AB of triangle ABC, then the lines AX, BY, and CZ are concurrent if and only if the product AY/YC · CX/XB · BZ/ZA is equal to* 1. (See fig. 13.38.) For obvious reasons X, Y, and Z are not allowed to coincide with the vertices of the triangle, but they may be exterior points. From the theorem it can be seen immediately that the medians of a triangle are concurrent: each ratio in the product is itself one.

Transformation geometry blends naturally into linear algebra (which should be available at some stage to those going on to college) and thus takes students directly toward the geometries that are the main concern of mathematicians today.

So far this section has dealt with things that have already been done or that can be done in most of our schools today with their present facili-

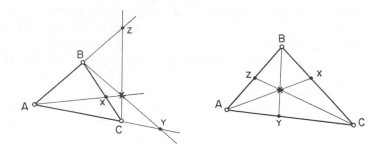

Fig. 13.38.   The lines in Ceva's theorem can meet inside or outside triangle *ABC*.

ties. The section reflects changes that are inevitable consequences of strong mathematical and educational forces: the proliferation of mathematics in this century and the recent educational discoveries about totally integrated programs, about format, and about the ability of children to enjoy and understand at a young age the mathematics that tradition had reserved for their older years.

What has not been mentioned yet is a technological force that is going to make available in the future to most children a kind of mathematics education that simply has not been available: the use of computers, both for instruction and for discovery.

A computer in a school lets students compute at an early age. It makes available to everyone the results of computations that, though simple in theory, are otherwise tedious and prohibitively time consuming. More to the point, since this is an essay about geometry, a computer can plot and display information so rapidly and in so many ways that a great deal of geometrical activity is automatically available to anyone who can use a teletype.

More important still, a computer with a television output can portray dynamic aspects of geometry—aspects whose portrayal is more or less impossible with the standard equipment of blackboard, chalk, and hand computation. What does the back of that block pile look like? Turn it around and see. Or, how does a certain similarity deform that triangle? The computer will show the deformation. What does a four-dimensional cube look like? The computer will show its three-dimensional projections as the cube is moved about. Into what shapes can a given triangle be deformed without changing its area? Watch the screen. What surfaces do certain equations define? The computer will show one, and the changes in the surface can be watched as the parameters in the equation are changed. What happens if certain parameters in an optimization problem are changed? The computer will show how that convex region changes. After all, much of geometry is dynamic, and with a computer that dynamism can be portrayed.

But there is more! Once a student can see things move, can see geometric objects change continuously, he can also develop a fine intuition about the relation—and it is a dynamic one—between elementary algebra, functions, and geometry.

Kenneth Leisenring said that the main problem of high school geometry is to replace the proving of the obvious with mathematics of genuine interest. There is no lack of interesting elementary mathematics, and the ways we can teach are now richer and more varied than ever before. Certainly enough material and technique is available and within our grasp to furnish a large number of excellent school curricula. Perhaps this yearbook can become a guide to the changes that seem to be so clearly on the way and so clearly needed.

REFERENCES

1. Alexander, James W. "On the Subdivision of 3-Space by a Polyhedron." *Proceedings of the National Academy of Science, U.S.A.* 10 (1924): 6–8.

2. Anderson, Richard D. "Hilbert Space Is Homeomorphic to the Countably Infinite Product of Lines." *Bulletin of the American Mathematical Society* 72 (1966): 515–19.

3. Bing, R. H. "Some Aspects of the Topology of 3-Manifolds Related to the Poincaré Conjecture." In *Lectures in Modern Mathematics,* edited by T. L. Saaty, pp. 93–128. New York: John Wiley & Sons, 1964.

4. ———. "The Elusive Fixed Point Property." *American Mathematical Monthly* 76 (February 1969): 119–32.

5. Finney, Ross L. "Dynamic Proofs of Euclidean Theorems." *Mathematics Magazine* 43 (September 1970): 177–85.

6. Flores, A. I. "Uber die Existenz n-dimensiondler Komplexe, die nicht in den R_{2n} topologisch einbettbar sint." *Ergebnisse eines mathematischen Kolloquiums* 5 (1932): 17–24.

7. Fox, Ralph H., and Emil Artin. "Some Wild Cells and Spheres in Three-Dimensional Space." *Annals of Mathematics* 49 (1948): 979–90.

8. Klee, Victor L. "Convex Bodies and Periodic Homeomorphisms in Hilbert Space." *Transactions of the American Mathematical Society* 74 (1953): 8–43.

9. Kuratowski, Casimir. "Sur le problème des courbes gauches en topologie." *Fundamenta Mathematicae* 15 (1930): 271–83.

10. Moise, Edwin E. "Affine Structures in 3-Manifolds, II, Positional Properties of 2-Spheres." *Annals of Mathematics* 55 (1952): 172–76.

11. Schori, Richard, and James E. West.
    Talk given at the Colloquium in Geometrical Topology held 18–20 March 1971 at the University of Utah.

12. Schultz, Reinhard. "Some Recent Results on Topological Manifolds." *American Mathematical Monthly* 78 (November 1971): 941–52.

13. Van Kampen, Egbertus R. "Komplexe in euklideschen Räumen." *Abhandlungen aus dem mathematischen Seminar der Hamburgischen Universität* 9 (1933): 72–78.

14. Wilder, Raymond L. *Topology of Manifolds.* American Mathematical Society Colloquium Publications series, vol. 32. Providence, R.I.: The Society, 1949.

PART 4

The Education of Teachers

14

The Education
of Elementary School Teachers
in Geometry

HENRY VAN ENGEN

O UTLINING a course of study for the geometric education of elementary
school teachers requires that considerable thought be given to certain
basic pedagogical considerations. These considerations naturally involve
questions of mathematical content, but they also involve questions of
method, purpose, symbolism, the relation of the course to classroom
realities (as well as the relation of geometry to reality), and structure. It
will be necessary to dwell briefly on each of these considerations before
discussing the details of the course, since they influence the choice of
content, the sequence of the concepts taught, and the whole spirit of the
course.

Obviously, any such discussion as developed herein will bring forth a
number of different opinions. There is no more unanimity of opinion on
the geometry for the elementary school teacher than there is on the geometry for the teacher in the secondary school or the geometry for the undergraduate classes. The range of opinions will be even broader than usual
because of the pedagogical considerations presented in this chapter.

The introductory part of this chapter presents five basic considerations.
These are not mutually exclusive of each other, for the mathematical structure of a course cannot be discussed without keeping in mind the basic

435

purpose of the course. Nor is method entirely independent of content and purpose. However, to highlight each consideration, it will be convenient to treat each one as a separate and distinct entity.

Five Basic Considerations

The content

The content of a course should be of maximum benefit to the teacher. This is an obvious principle, but it needs interpretation. What benefits are to result from taking the course? Most certainly there should be mathematical benefits; that is, the preservice teacher should achieve a greater degree of mathematical maturity, a knowledge of mathematical principles, and so on. However, one must go beyond these considerations. The course content should have an immediate impact on what the teacher does in the classroom. Too many courses now being proposed for preservice teachers have no such impact. Proving that if three points are on a line than one point must be between the other two points has no payoff for the elementary teacher. However, an intuitive study of isometries can have an immediate payoff.

The organizers of the course

Those who outline a course for elementary teachers should arrive at their recommendations on the basis of long-term experience working with the preservice teacher. The curriculum at all levels of our educational system—graduate, undergraduate, high school, and elementary—is based in large part on the experience knowledgeable people have had trying to teach certain mathematical ideas to people of various ages. After all, one must face the world of reality, and only experience teaches one what the world of the classroom is really like. An apparent violation of this principle can be found in the report of the Cambridge Conference on Teacher Training (1). The participants at this conference consisted of a group of highly knowledgeable mathematicians many of whom have neither taught a class of preservice elementary teachers nor been in an elementary classroom as a teacher. As a result, their recommendations do not meet the test of reality. That their proposals are mathematically desirable in some utopian sense cannot be questioned. However, it is easy to solve difficult problems if one ignores all boundary conditions. For example, the preservice elementary teacher has about two years of high school mathmatics as a mathematical background. Isometries, if treated analytically as suggested by the Cambridge Conference group, are beyond students at this level of mathematical maturity.

The method

The method used in the course for the preservice teacher should approximate that to be used by the teacher. It is difficult to characterize a method of instruction. However, it is easier to say what should not be done than what should be done. The traditional lecture is not satisfactory, particularly if used on all occasions, but teachers must learn to talk about geometric ideas as well as use them on paper. This means that free-discussion, question-and-answer, and guided-discovery techniques are highly preferable to a lecture approach. Whatever instructional techniques are used, the result should be a highly informal classroom as well as a highly informal approach to the subject itself. The "spirit" does not grow if it is subjected to severe restraints.

How delicate the question of method and content can be is amply illustrated by the following comment made in the Cambridge Conference report (1, p. 11):

> The conference membership included three British mathematicians, two of whom had been associated with the School Mathematics Project. For most of the Americans, the presence of these three at the meeting and the display of the SMP books . . . constituted an introduction to what has been done in mathematics reform in the British Isles. While it would be difficult or impossible to pinpoint specific examples, it would not be too much to say that the quite different approach of the British to teaching mathematics, and the wholly different character of their textbooks, exerted a considerable influence on some of the Americans at the meeting.

It is indeed unfortunate for education in America that the few British who attended this conference did not influence more those who attended the conference. One can only conclude that the majority of those at the conference did not really understand the importance of the difference in pedagogical approach between the British and the American programs. The British place more emphasis than Americans on a method of instruction appropriate for the maturity level of the student and less emphasis on the explicit display of the mathematical structure being studied. The British books are informal both in their definitions and in their proofs. For example, an angle is an amount of turning, and a right angle is a quarter-turn. What American textbook author would dare produce this as a definition of angle? And yet it is pedagogically sound.

The language

The language and symbolism used in the course should be precise enough to keep channels of communication clear but not so precise that the language and symbolism become the main objective of the course. One of the more foolish trends in some of the "new" mathematics is a pedantic insistence on the use of an overly prescribed set of symbols. There is little need for special symbols for line, segment, and ray, nor should the course be too pedantic about the measure of an angle or a segment. The criterion should be, *Is communication clear?* If the diagram that usually accompanies a geometric problem does not make it clear that AB is a segment, then it may be necessary to say "segment AB." This practice is highly preferable to devising special symbols for use in the preservice classes for elementary teachers.

Symbolism and terminology has, in some instances, been carried to the extreme. A few illustrations will suffice to illustrate this point. There are instances in which a distinction is made between a "dot" and a "point"! In other instances, the definition of parallel lines is given in terms of symbols for the intersection of sets; that is, in a plane, two lines l_1 and l_2 are parallel if and only if $l_1 \cap l_2 = \phi$. For the elementary teacher, in what sense is this preferable to saying that two lines are parallel if they have no points in common? Others insist that one must not write $\angle A = 90°$, and some even insist that it should be written $m° \angle A = 90$ instead of $m \angle A = 90°$. This is pedantic precision and has no place in the proposed course.

The teacher of the kind of course to be outlined in this chapter should always keep in mind that it is not the objective to teach preservice teachers to write papers that are acceptable for publication in reputable journals. Rather, it is the purpose of the course to introduce these teachers to some important geometric ideas that have their roots in what should be done in the elementary classroom. This means that in addition to the methods of instruction being informal, the language and symbolism will be informal, but *not wrong*. The language and symbolism will not always be "precisely precise," but it must not be misleading.

The structure

To display the structure of a geometry is relatively unimportant for a course of this kind. This does not mean that the course will be a hodge-podge of ideas; it does mean, however, that the explicit statement of the postulates is not too important. Furthermore, the formal proof of theorems is of little importance. The fact that some ideas follow "as naturally as breathing" from others is important. Here the emphasis should be on plausible reasoning and not on formal proof.

The work of the Committee for the Undergraduate Program in Mathematics (CUPM) and the work of the Cambridge Conference may have been harmful to the useful organization of courses for the elementary teacher. This is true partly because these programs say little or nothing about methods of instruction and devote most of their time to a highly organized outline of ideas as the basis for such a course and partly because the writers of textbooks have been overzealous in following *their* interpretations of the recommendations of CUPM. Textbook writers seem to find it necessary to say that their course is based on the recommendations of CUPM and then proceed to give a highly structured course based, for example, on the approach used by the School Mathematics Study Group (SMSG) geometry. Rewriting the SMSG geometry for elementary teachers does not solve the problem and has little to offer toward a solution. Gail Young made the following remarks about this highly controversial problem (3, p. 88):

> They [the courses] should be enjoyable courses; the prospective teacher should learn something and should attain a positive attitude toward mathematics. The courses should not be highly rigorous but not everyone seems to agree with this. Several books have been written which say that they follow the CUPM recommendations. And they do—they follow the letter completely and just as the Bible says, "The letter killeth."

As stated earlier, one of the basic purposes of a course of this kind is to acquaint the teacher with some worthwhile geometric ideas that have, or should have, their roots in the elementary school course of study. In fact, some of the ideas in a geometry course for teachers should be capable of being used almost verbatim in the elementary school. Proving theorem after theorem is not useful—it is not even desirable—because "it killeth" the true spirit of the course.

Although it is true that mathematics can be described as the study of structures, the enthusiasts for structure seem to feel that good mathematics was not developed prior to the mathematical world's discovery that algebra, as well as geometry, could be "structurized." This idea is far from the truth. The study of structure should be played down in the courses for elementary teachers and the emphasis placed on worthwhile ideas.

A Proposed Geometry Course

This course is organized into seven units. Units 2, 3, 4, and 5 should be taught in sequence. Otherwise, the units are relatively independent. Unit 1, of course, is an introductory unit. Unit 6 supplies a background for measurement, and unit 7 is on networks. The last two units are independent of any of the other units.

As seen in the following outline, geometry for elementary teachers is a course making maximum use of the ruler, compass, and squared paper. It is more of an activity course than a course of reading or proving theorems.

Unit 1—an introduction to geometry

This unit serves as a review of some of the simpler ideas of geometry and affords an opportunity to introduce some terminology, but very informally. It is assumed that everybody knows what a line is and that an angle needs no formal definition. Here one might take the position of the School Mathematics Project and simply say that an angle is an amount of turning illustrated in several ways. The following activities might be covered:

1. Making various simple ruler-and-compass constructions
2. Constructing triangles, given certain sets of conditions
3. Constructing tables to find the number of lines that can be drawn through N points, no three of which are collinear
4. Finding Euler's formula, arrived at through making tables and generalizing
5. Working locus problems, culminating in constructing points of an ellipse, parabola, and hyperbola

Unit 2—reflections

This unit is on reflections in a line, all in the plane. The mapping terminology may be used, but it should be used informally; that is, for every original there is one and only one image, and for every image there is one and only one original. Full use should be made of the ruler, compass, and squared paper. The following concepts are developed:

1. Basic properties
   a) Corresponding to every line, there is a reflection of the plane onto itself.
   b) If P' is the image of P, then P is the image of P'.
   c) If P_1 and P_2 are two points in the plane and P_1' and P_2' are their respective images, then $P_1P_2 = P_1'P_2'$. That is, distances are preserved or are invariant.
   d) The image of an angle, a segment, and a line is, respectively, an angle, a segment, and a line.
   e) The size of angles (their measures) is preserved.
   f) Given P_1 and P_2, then there is a line (and only one) such that P_2 is the image of P_1.
   g) The bisector of an angle is a ray with origin at the vertex of the angle and such that each side of the angle is the image of the other side when reflected in the bisector.

h) Betweenness of points is preserved; sense is not preserved.

i) The image of any figure is congruent to the original. (That is, congruency may be defined in terms of reflections.)

j) Some figures map onto themselves if the line of reflection is properly selected (symmetry).

2. Related ideas

   a) Polygons classified on the basis of the number of lines of symmetry

   b) Symmetries of a square, equilateral triangle, and so on (group properties)

Unit 3—rotations and translations

Translations and rotations can be defined in terms of reflections. Hence, reflections play a unique role in transformational geometry. Informal definitions of these ideas continue to be given here because, as previously stated, formal definitions are a delight to the mature mathematical mind, yet very possibly a stumbling block to the immature mind. This unit stresses the following concept:

1. Basic properties

   a) A rotation is the product of two reflections in two intersecting lines.

   b) A translation is the product of two reflections in parallel lines.

   c) Betweenness, distance, sense, and angle measure are preserved by rotations and translations.

   d) If one figure is the image of another figure under a rotation or a translation, then they are congruent.

2. Related ideas

   a) If two figures are congruent, then one can be mapped onto the other by means of rotations and translations, provided they have the same sense.

   b) If two figures have opposite sense, then the one can be mapped onto the other by at most three reflections.

   c) Group properties of rotations and reflections should be studied. (For example, the product of two rotations is not necessarily a rotation.)

Unit 4—glide reflections

On the basis of the previous units, it might seem that the product of any two of the three transformations studied in units 2 and 3 should be a translation, a reflection, or a rotation. That this is not true is immediately obvious when one thinks of moving a triangle parallel to a line (a translation) followed by a reflection in the line. The resultant of these two transformations cannot be expressed as a transformation that was studied in units 2 and 3. Hence glide reflections, a fourth and final transformation in the set of motions, are introduced, and the following concepts presented:

1. Basic properties
   a) Betweenness, distances, and angle size are invariant.
   b) The sense of a figure is not invariant.

2. Related ideas
   a) Characterization of motions in terms of invariant points and lines
   b) Subgroups and group properties of motions
   c) Proofs of simple plane geometry theorems using transformational ideas

Unit 5—congruence and similarity

The informal treatment of geometry is continued in this unit. Although it is possible to give precise definitions of congruence and similarity, it is debatable whether this should be done, since the intuitive feel for "figures that fit" is of more importance for the beginner than the precise statement in mathematical terms. The basic concepts of this unit follow:

1. Basic properties
   a) Congruence is defined in terms of mappings that preserve all distances.
   b) If two triangles are congruent, then all medians, altitudes, and so on, are congruent.
   c) Similarity is defined in terms of mappings in which all distances in the original figure (triangle) are multiplied by a certain constant, k.
   d) If two triangles are similar, then all medians, altitudes, and so on, are proportional.

2. Related ideas
   a) Proof of a simple "theorem," such as: The four points determined by the "floor corners" of a room are congruent to the four points determined by the "ceiling corners" of a room
   b) Proof of a "theorem" such as this: Any two line segments are similar

Unit 6—perimeters, areas, and volumes

No attempt to introduce some of the subtle and difficult mathematical questions that can be raised in considering the topics of this unit should be made. For example, the mathematical problem one confronts in finding the area of a rectangle when the dimensions are irrational has no place in courses for elementary teachers. Included are the following concepts:

1. Basic properties
   a) Areas of simple closed curves can be approximated by overlaying squared paper. Better approximations can be made if finer meshes are used.
   b) The area of a triangle is equal to the product of one-half the base and altitude; the area of a rectangle or a parallelogram is equal to the product of the base and altitude.
   c) Areas and perimeters of some geometric figures can be computed by decomposing the figures into rectangles and triangles.
   d) Volumes of simple closed figures in three-space can be approximated by stacking cubes within the solids. Better approximations can be made by using smaller cubes.

2. Related ideas
   a) Areas with constant perimeters and perimeters with constant areas
   b) Areas with constant base and altitude
   c) Pick's theorem: *The area of a polygonal region on a lattice is* $\dfrac{n + (2k - 2)}{2}$ *where n is the number of lattice points in the boundary and k is the number of lattice points in the interior of the region*

Unit 7—networks

The following concepts are included in unit 7:

1. Basic principles
   a) The number of odd vertices in a connected network is even.
   b) Networks with no odd vertices can be traversed.
   c) Networks with two odd vertices can be traversed if started at one of the odd vertices.
   d) Networks with more than two odd vertices cannot be traversed.

2. Related ideas
   a) The Königsberg bridge problem
   b) The four-color problem and the five-color theorem

Concluding Remarks

Textbooks published in the United States for elementary teachers tend to lean in varying degrees toward the CUPM and Cambridge Conference point of view. However, an instructor of an activity course such as the one envisioned in this recommendation for the training of elementary teachers does not really need a text—ideas for activities and exercises should be sufficient. Hence, a brief annotated list of suitable books for ideas is included here, and all are readily available on the American market.

Algebra and Geometry for Teachers, Charles F. Brumfiel and Irvin E. Vance. Reading, Mass.: Addison-Wesley Publishing Co., 1970.

Geometric Transformations, I. M. Yaglom. Translated by Allen Shields. New York: L. W. Singer Co., 1962.

> This book is a problem book in transformational geometry with a brief exposition of the ideas needed to solve the problems. Some of these problems are suitable for use in the elementary school teacher's course in geometry.

Geometry in the Classroom: New Concepts and Methods, H. A. Elliott, James R. MacLean, and Janet M. Jordan. Toronto: Holt, Rinehart, & Winston of Canada, 1968.

> This book has many activities that are suitable for such a course as proposed in this outline.

Some Lessons in Mathematics—a Handbook on the Teaching of "Modern" Mathematics, T. J. Fletcher, ed. London: Cambridge University Press, 1965.

> A book generally in the British pedagogical tradition—good mathematics without always going back to "two points determine one and only one line."

Transformation Geometry, Max Jeger. New York: American Elsevier Publishing Co., 1969.

> This book is somewhat formal in its approach, but it is a good reference source for an instructional staff.

REFERENCES

1. Cambridge Conference on Teacher Training. *Goals for Mathematical Education of Elementary School Teachers.* Boston: Houghton Mifflin Co., 1967.
2. Committee on the Undergraduate Program in Mathematics of the Mathematical Association of America, Panel on Teacher Training. "Recommendations for the Training of Teachers of Mathematics." *Mathematics Teacher* 53 (December 1960): 632–38, 643.

3. Young, Gail S. "The Training of Elementary Teachers." Report of a conference on teacher training held at Michigan State University in East Lansing, Michigan, in 1967. Mimeographed.

15

The Education
of Secondary School Teachers
in Geometry

F. JOE CROSSWHITE

DRAMATIC new directions for secondary school geometry have been suggested in the earlier chapters of this yearbook, and recent conference reports have amplified the variety of new approaches proposed. New textbooks that embody these recommendations are already available. If the authors of these recommendations are at all prophetic, the coming era will be one of unparalleled diversity in secondary school geometry.

The implications for teacher education should be clear. Existing patterns in the secondary school curriculum cannot provide adequate guidelines for the education of secondary school teachers in geometry and so it seems axiomatic that teacher education should anticipate, as well as reflect, the school curriculum. And yet the past record has not been good—too often it has been the opposite. The mathematical preparation of teachers has lagged behind developments in school mathematics.

Historical Perspectives

Until the recent revolution in school mathematics, no historical parallel for radical change in secondary school geometry existed. Change came slowly in the past. New programs evolved as a restructuring or a redistribution of elements already existing in the curriculum. No serious threat was made to the basic motivation for, or approach to, secondary school geometry. The prospective teacher could reasonably expect to teach in high school the geometry that he had studied in high school.

The National Council of Teachers of Mathematics (NCTM) last published a yearbook entitled *The Teaching of Geometry* in 1930. At that time, geometry had enjoyed a long period of stability in the school curriculum, as Swenson indicated in his chapter in that yearbook:

> The sway which Euclid's *Elements* has held as a textbook for more than two thousand years is without parallel in the history of mathematics. Even the invention of Cartesian geometry in 1637 has not affected the teaching of the so-called Euclidean geometry. An almost unlimited number of textbooks have appeared in modern times but the only way in which they have differed is in the sequence of the theorems. Euclid's treatment has in the main been retained and no modern mathematical methods have been introduced. [20, p. 96]

Likewise, Longley commented in a similar view in the same book:

> Geometry is one of the oldest of educational disciplines. More than any other it has retained its essential character for centuries. Why, then, should this most stable subject of our curriculum be questioned at the present time? Why should we ask: What shall we teach in geometry? [20, p. 29]

Longley's question was rhetorical. The answer was clear.

In 1930, the curriculum in mathematics was under attack. Pressure was being applied to reduce the time devoted to instruction and to the material covered in the courses. The traditional course in solid geometry was a likely casualty. Few colleges continued to require, and many had ceased to offer, a course in solid geometry. The College Entrance Examination Board had already modified its examinations to provide a combined plane and solid geometry option. In fact, the NCTM's Fifth Yearbook was written in direct response to the question of integrating plane and solid geometry. According to its editor, William David Reeve, it was designed to "supplement and assist the National Committee recently appointed . . . in studying the feasibility of a combined one-year course in plane and solid geometry" (20, p. v). Surely there was reason to anticipate what must have then seemed a substantial change in the secondary school curriculum in geometry. Yet nowhere in that yearbook was the potential impact of such a change on teacher education discussed.

This oversight might be discounted on the assumption that the anticipated change was viewed only as a redistribution of content already in the curriculum. If a teacher were adequately trained to teach plane and solid geometry in separate courses, it might have been reasonable to assume he could teach the combined course. But was the teacher of 1930 being adequately prepared to teach geometry? In the judgment of Christofferson, he was not:

> The academic mathematical training of the high school teacher [in 1930], in addition to high school algebra and geometry, consists at best of college algebra, trigonometry, analytic geometry, calculus, and perhaps differential equations, advanced algebra, surveying, mechanics, or astronomy. . . . The weakness of this program of training for geometry teachers is apparent. The student is well trained in algebra because of his wide experience with its symbolism throughout his college course, but his training in geometry consists of very little more, if any, than his high school course. [5, p. 1]

Christofferson was prefacing an argument for the professionalization of subject matter—a concept that was quite new in 1930. The incidental application of geometric concepts in college courses in trigonometry, analytic geometry, and calculus seemed inadequate to him because nowhere was provision made for an emphasis on proving or demonstrating relationships in the way emphasized in high school geometry. This "inadequacy" was viewed as a direct consequence of the fact that colleges of education accepted academic mathematics courses without modification or adequate adjustment to the needs of the prospective teacher.

Professionalized subject matter—the double use of subject matter both to insure the mastery of the content of high school mathematics and to establish a philosophy and a technique of teaching—seemed to Christofferson an appropriate pattern for the college geometry course for prospective teachers. His book, *Geometry Professionalized for Teachers* (5), was a pioneer effort toward integrating subject matter and teaching methodology. It still provides valuable reading for the prospective teacher or teacher educator. His statement that the proposed course was "based on the theory that teachers are more apt to teach geometry as they were or are taught geometry than as they are told to teach it" (5, p. 69) expresses a sentiment widely echoed in recent recommendations for the preparation of mathematics teachers.

These two historical developments—the integration of plane and solid geometry and the professionalization of subject matter—seem to have had some impact on the mathematical preparation of secondary school teachers in geometry in the period from 1930 to 1960. Both developments, however, were slow. Solid geometry suffered a lingering illness in the schools. Professionalized subject matter, especially with respect to geometry, enjoyed at best an incomplete success. It never became the dominant pattern in the mathematical education of secondary school teachers.

Other trends that might have been forecast in 1930 had even less impact on the preparation of secondary teachers. The development of informal geometry in the evolving junior high school was clearly anticipated. However, there was no feeling that either the content or the methods of informal geometry deserved special treatment in college mathematics courses. This material, if treated at all, was reserved for methods courses, and this attitude apparently continues to prevail today.

What eventually proved to be a major development in secondary school geometry was suggested as a potential new direction in the NCTM Fifth Yearbook. Birkhoff and Beatley, in their chapter entitled "A New Approach to Elementary Geometry," developed the framework for a metric approach to plane geometry (3, pp. 86–95). They later wrote a high school text, *Basic Geometry* (2), based on this approach. (It is interesting to note that Birkhoff, who was chairman of the National Committee that the NCTM's Fifth Yearbook was designed to assist in studying the integration of plane and solid geometry, chose instead to address himself to this more fundamental change.) These developments foreshadowed the "ruler and protractor axiom" approach popularized in a School Mathematics Study Group (SMSG) text. Considering that thirty years were to elapse between the initial recommendation and its implementation in the schools, one might argue that no impact on teacher education was warranted; yet appropriate modification in the geometric preparation of teachers might have shortened the period from recommendation to implementation. Surely such modification would have facilitated implementation when it did occur.

The period following 1930—depression, followed by war—can hardly be considered normal. The slow pace of development, both in school mathematics and in mathematics for teacher education, may have been peculiar to the times. For a detailed discussion of this era in mathematics education, the reader may refer to the Thirty-second Yearbook of the NCTM (19). Attention is restricted here to selected recommendations specific to the education of secondary school teachers in geometry.

The report of the Commission on the Training and Utilization of Advanced Students of Mathematics, published in 1935, recommended a "college level treatment of synthetic Euclidean geometry; or possibly descriptive geometry" as part of the minimum training in mathematics for teachers of secondary school mathematics, and projective geometry or solid analytic geometry was included in a listing of "desirable additional work in mathematics" (8, pp. 275–76). A parallel pattern was included in the report of the Joint Commission on the Place of Mathematics in Secondary Education, published in 1940 as the Fifteenth Yearbook of the NCTM. Here the basic course in geometry was described as "a course that examines somewhat critically Euclidean geometry, and gives brief introductions to projective geometry and non-Euclidean geometry, using synthetic methods" (14, p. 201). For a teacher whose full time would be devoted to mathematics in a school providing more advanced courses, "additional work in geometry, such as projective geometry, descriptive geometry, etc." was recommended. The Commission on Post-War Plans in 1945, although not offering detailed recommendations for the mathematical preparation of teachers, did reflect the continuing concern for the professionalization of subject matter in two of its theses (7):

> Thesis 26. *The teacher of mathematics should have a wide background in the subjects he will be called upon to teach.* [p. 217]
>
> Thesis 29. *The courses in mathematical subject matter for the prospective mathematics teacher should be professionalized.* [p. 219]

What trends would these recommendations suggest for the period from 1930 to 1960? It would seem reasonable to expect that the education of secondary school teachers in geometry would include a college-level course in advanced Euclidean geometry, treated synthetically and including topics of solid geometry. Moreover, it would be expected that this course would have been designed for teachers and taught as professionalized subject matter. Was that the situation in 1960?

A survey of 799 teachers of high school mathematics was made in 1959 by the U.S. Office of Education (USOE). This survey indicated that 7.1 percent of the teachers surveyed had taken no college mathematics and that only 61 percent had studied calculus or more advanced courses. The average preparation for those who were teaching one or more courses in mathematics was twenty-three semester hours, much of it at a level below calculus. (4, p. 29.)

A survey of college offerings in 1960/61, also conducted by the USOE, revealed the following pattern of junior- and senior-level courses offered in the 877 institutions surveyed (15, p. 28): college geometry, 42 percent; solid analytic geometry, 19 percent; projective geometry, 19 percent; differential geometry, 8 percent; higher or modern geometry, 8 percent; non-Euclidean geometry, 8 percent.

Other survey evidence tends to support the conclusion that the education of secondary school teachers in geometry had not reached the level anticipated in the recommendations of the 1930s and 1940s (21; 13; 23).

Several factors combined to reduce the probability that the average teacher in 1960 would have completed a college-level course in geometry other than the traditional course in analytic geometry. Geometry was not required in many teacher-education programs. In others, no course was even available. The total number of semester hours required for certification of secondary school teachers was low. The basic mathematics core consistently appearing in recommendations and in teacher-education programs included complete treatments of college algebra, trigonometry, analytic geometry, and differential and integral calculus. This program filled the freshman and sophomore years and frequently carried as much as twenty semester hours of credit. Many institutions offered college credit for high school level courses. As a result, many prospective teachers were able to complete their requirements without ever reaching a junior- or senior-level course. When upper-division courses in geometry were available, they had to compete for student enrollment with courses in advanced or modern algebra, theory of equations, number theory, history of mathematics, mathematics of finance, differential equations, advanced calculus, and so on. If the probability of the average teacher having had an advanced course in geometry was low, the probability of his having had a course especially designed for teachers was lower still.

Recent Developments

Secondary school curriculum developments in the 1950s created renewed concern for teacher education. Conclusions based on studies of the status of teacher education were confirmed by experience with in-service, summer, and academic-year institutes. Teachers were not prepared to teach the newly developing curricula. The concern of professional organizations was reflected in the reports of several committees.

The Commission on Mathematics was appointed by the College Entrance Examination Board (CEEB) to develop detailed proposals for revising the secondary school curriculum in mathematics for college-capable students. Their report, published in 1959, acknowledged that the successful implementation of the proposed nine-point program depended on well-prepared teachers and therefore included recommendations for teacher preparation. The recommended program in geometry for secondary school teachers stressed the foundations of geometry and specifically cautioned against courses dealing primarily with advanced Euclidean geometry. Non-Euclidean geometries, topology, and differential geometry were deemed more appropriate. (6, p. 55.) The report of the Secondary School Curriculum Committee of the NCTM, also published in 1959, suggested that current curriculum demands require that teachers of mathematics in grades 7 through 12 have competence in "geometry—Euclidean and non-Euclidean, metric and projective, synthetic and analytic" (22, pp. 414–15). Guidelines for the preparation and certification of teachers of secondary school mathematics were published in 1960 by the National Association of

State Directors of Teacher Education and Certification (NASDTEC) and the American Association for the Advancement of Science (AAAS). A year course in geometry emphasizing the study of the foundations of geometry, projective and non-Euclidean geometries, and vectors was recommended (26, p. 795).

The most persistent committee effort toward reforming the mathematical preparation of teachers has been that of the Panel on Teacher Training of the Committee on the Undergraduate Program in Mathematics (CUPM). This committee of the Mathematical Association of America (MAA) not only has made and periodically revised recommendations for teacher training but also has conducted regional conferences, developed course guidelines, and publicized its efforts at professional meetings and through journal articles. Its original recommendations (1961) for teachers of high school mathematics suggested a two-semester sequence in geometry that would include elements of plane geometry using synthetic and metric approaches, historical development, parallelism and similarity, area and volume, ruler-and-compass constructions, other geometries, projective geometry, and pure analytic geometry. The two courses were to emphasize a "higher understanding of the geometry of the school curriculum" and were to have a firm axiomatic basis. (9.)

Individual mathematics educators were also making broad proposals for the education of secondary teachers in geometry. The suggestions posed by Meserve to participants in the Mideast Regional State College Conference on Science and Mathematics in 1958 are illustrative. (These proposals were subsequently adopted as recommendations of the group.) Meserve felt that geometry for secondary school teachers should include a discussion of the content and structure of secondary school geometry; coordinate systems; projective geometry as a basis for, or generalization from, Euclidean geometry; a view of geometry as the study of properties invariant under a group of transformations; and topology (17, pp. 439–40). Fehr proposed in 1959 that the axiomatic bases of Euclidean, projective, affine, and non-Euclidean geometries be included in a one-year course required for the certification of teachers of senior high school mathematics (12).

Wong reported in her dissertation study, "Status and Direction of Geometry for Teachers," that the College Geometry Project of the University of Minnesota (under the direction of Seymour Schuster) began as early as 1964 to prepare materials for secondary school teachers on such topics as axiomatics, geometric transformations, vector geometry, convexity and combinatorial geometry, dissection theory, projective geometry, non-Euclidean geometry, geometric constructions, and differential geometry (25, p. 17).

Apparently there was as much diversity in these "early 1960" proposals for teacher education in geometry as in current recommendations for secondary school geometry. Yet there were common elements in the proposals. There seemed to be agreement that the classical advanced Euclidean geometry (or college geometry) was not adequate. Most proposals emphasized a study of the foundations of geometry, the introduction of geometries other than Euclidean (affine, projective, non-Euclidean, differential), and approaches other than the classical synthetic approach (transformational, metric, vector).

Wong's survey of 155 colleges and universities during the 1967/68 school year provides a basis for assessing the impact of these recommendations. For the purposes of her study, she defined "contemporary geometry" as a course primarily concerned with axiomatic structure and a reexamination of elementary Euclidean geometry, "survey of geometry" as a course treating several geometries, and "college geometry" as the traditional advanced Euclidean geometry. The percents of institutions offering each course were as follows: contemporary geometry, 66.1 percent; projective geometry, 39.1 percent; survey of geometry, 33.9 percent; college geometry, 12.4 percent; non-Euclidean geometry, 10.7 percent; and differential geometry, 9.9 percent. (25)

A comparison of these statistics with those reported in the 1960/61 USOE survey reveals the extent of change over a very few years. Wong's data indicate that only 6.5 percent of the institutions surveyed required no course in geometry of prospective secondary school teachers. Fifty percent required a single course, and 35 percent required two courses. Furthermore, each of the four leading texts being used (in a total of 70 percent of the institutions surveyed) was published in 1963 or later. The most frequently used text (in 40 percent of the institutions) was *Elementary Geometry from an Advanced Standpoint* (18). The nature of this and the other leading texts clearly indicates a substantial redirection in the education of secondary school teachers in geometry since 1960. Whether this change was in reaction to, or anticipation of, developments in school mathematics is relatively unimportant. The remarkably accelerated pace of change in the mathematical preparation of teachers has assured that substantially more teachers are now taking a college course in geometry that is at least consistent with recent developments in school mathematics.

Three conclusions resulting from the Wong study are particularly relevant here (25):

> 1. The majority of secondary school mathematics teachers being trained presently at the selected institutions complete at least one course in geometry and are studying Euclidean geometry from the contemporary point of view (p. 123).

2. The majority of university mathematicians and educators involved in the training of prospective mathematics teachers at the selected institutions recommend some change in their preparation in geometry but express no majority opinion on whether the change should be in additional course work, change of emphasis, or both. The majority favor emphasis on transformations from both the algebraic and the synthetic points of view. (p. 125.)

3. The majority of professors of mathematics and of education involved in training prospective mathematics teachers in the selected institutions are opposed to any change regarding formal proof and rigor in the high school program. The majority advocate more coordinate geometry, and less than half indicate approval of vector and transformation approaches in high school geometry. (p. 124.)

Current Status and Recommendations

The Wong study is indicative of the pace of change in geometry for secondary teachers over the past decade. And yet, with the exception of the recommended course in computer science, the two-course sequence in geometry has probably been the least fully implemented of the original CUPM recommendations. Dubisch makes the following point:

> The basic problem in implementing this portion of the CUPM recommendation is that, in contrast to the recommended algebra courses, the recommended geometry courses do not correspond to a standard geometry sequence required of (or even recommended for) non-teaching majors. [11, pp. 295–96]

As Wong's study shows, traditional advanced Euclidean geometry (or "college geometry") has declined in popularity and has been replaced as the most frequently available course by "contemporary geometry." However, where students elect or are required to take two courses in geometry, the second course in the "sequence" is seldom consciously designed to supplement the first. More often, the course is elected from standard offerings and is not designed specifically for prospective teachers.

Several factors work against mathematics departments' providing a unified two-course sequence in geometry appropriate for teachers. In many schools, the number of secondary school mathematics teachers graduated each year may be smaller than the number for which a specially designed course is feasible. Of the institutions surveyed by Wong, less than 19 percent graduated as many as thirty mathematics teachers per year; 62 percent graduated fewer than twenty; and 30 percent graduated fewer than ten (25, p. 43). Dubisch lists additional reasons why a one-year sequence in geometry paralleling the CUPM recommendations is frequently not available:

First of all, geometry is simply not, at the present time, as much in the mainstream of mathematics as are, for example, algebra and analysis. Thus, colleges which must limit their offerings will naturally tend to limit them in the area of geometry rather than in more "essential" areas. Secondly, the kind of geometry that is of maximal current research interest (e.g., convexity and differential geometry) is not the kind of geometry that is as closely related to the present secondary school curriculum as is the geometry of the CUPM recommendations. In particular, the importance of a study of the foundations of Euclidean geometry is near zero except for prospective teachers of Euclidean geometry. [11, p. 296]

In spite of this situation, teacher educators should feel satisfaction in the developments since 1960. Many of the recommendations made at that time have been implemented. Courses approximating the original CUPM guidelines are available to large numbers of prospective teachers. There is little question that the geometric preparation of teachers is beginning to catch up with developments in school mathematics.

Nevertheless, this is no time for complacency. The exciting new directions for secondary school geometry outlined in this yearbook and in recent conference reports will place new demands on the secondary school teacher. Who can be sure today just what sort of geometry tomorrow's teacher will need? Continued change should be anticipated. More than that, it seems increasingly likely that tomorrow's teacher may be asked to choose any one of several geometries or approaches to geometry for the secondary school. Or perhaps even more likely, someone else will make that choice for him.

The obvious need, it seems, is to design a program of preparation in geometry that enables the teacher to teach well existing curricula and to adapt readily to new curricula as they develop. This demands exposure both to the content and methods of current programs in secondary school geometry and to the content and methods appropriate to new programs that may evolve.

Recent guidelines for the preparation of secondary school mathematics teachers have been proposed by the American Association for the Advancement of Science (AAAS) Commission on Science Education and the National Association of State Directors of Teacher Education and Certification (NASDTEC). The potential diversity in secondary school geometry is clearly recognized in their recommendations (1, pp. 23–24):

Geometry is taught using a wide variety of approaches, including coordinates, vectors, and transformations such as isometries, similarities, and affine transformations. A teacher of geometry needs to be able to adapt his teaching to each of these approaches while maintaining flexibility to adapt to new approaches as future geometry courses continue to evolve.

Accordingly, a prospective secondary school mathematics teacher should:

1. Be able to develop many of the usual geometric concepts informally; for example, using paper folding.
2. Know the role of axiomatics in synthetic geometry and the existence of various axiom systems for Euclidean geometry.
3. Be able to use either coordinates or vectors to prove theorems such as the concurrence of the medians of a triangle.
4. Recognize the existence of other geometries by being able to identify a few theorems in at least one other geometry such as a non-Euclidean geometry, projective geometry, or affine geometry.
5. Be able to discuss the role of transformations, at least in Euclidean geometry.
6. Understand thoroughly the interconnections between algebra and geometry.

The most recent CUPM guidelines for the training of teachers also recognize the growing diversity in high school geometry (10, p. 17):

> The nature of high school geometry continues to change. Changes over the past decade have mainly been toward remedying the principal defects in Euclid's *Elements* that are related to the order, separation, and completeness properties of the line, but more recently there has developed an entirely new approach to geometry that links it strongly to algebra. This approach is now finding its way into the high school geometry course. A teacher should be prepared to teach geometry either in the modern Euclidean spirit or from the new algebraic point of view.

For teachers of high school mathematics the CUPM guidelines recommend two courses in geometry at the college level: one emphasizing a traditional approach concentrating on synthetic methods and the study of the foundations of Euclidean geometry, the other strongly linked to linear algebra and including an investigation of the groups of transformations associated with geometry (10, p. 18). Suggested outlines for each kind of course are included in an appendix to the recommendations.

Dubisch makes the following observation in *Mathematics Education,* the 1970 Yearbook of the National Society for the Study of Education (NSSE):

> Regardless . . . of how closely the syllabi for teacher training courses match the most carefully prepared recommendations of the most highly qualified committees, there are still likely to be serious deficiencies in teacher training at all levels. It is becoming increasingly clear that it is not sufficient to recognize the desirability of special courses for teachers; we must also recognize the need for special procedures in the teaching of such courses. [11, p. 302]

Two specific aspects of this problem are identified as follows (11, pp. 302–3):

1. There is a great need to relate more closely what is being studied to problems of teaching elementary and secondary school mathematics.
2. Closely related to the need just described is the need to provide for a discussion of teaching methods along with content, rather than a study of content and methods in separate courses.

These recommendations seem very close to the professionalized subject matter Christofferson described forty years ago. However, the context is different now. Christofferson was defining a course related to an established, stable curriculum in geometry. Today, the professionalization of subject matter must be more broadly conceived to encompass a more dynamic school curriculum.

Many mathematics educators are legitimately concerned with the idea that the prospective secondary school teacher should complete an undergraduate major leading toward graduate study in mathematics. Professionalization of subject matter does not necessarily imply that prospective teachers should be isolated from undergraduate majors in mathematics. The content of many of the courses in the undergraduate major is equally appropriate for the prospective teacher. In these situations, it is appropriate for the two groups to study together. (The prospective teacher's experience might be supplemented by methods seminars directly relating the content of the college course to the school curriculum.) In other areas, and this may be particularly true in the area of geometry, the needs of the prospective teacher do not coincide with those of the undergraduate major, and here it is essential that special courses be designed for teachers.

The CUPM guidelines explicitly recognize a unique role for geometry (and for probability and statistics) in the professional training of teachers (10, p. 16):

> While the program we recommend for prospective teachers will leave the student with a deficiency in analysis and algebra in order to meet the CUPM recommendations for entry to graduate school . . . , the prospective graduate student in mathematics would normally need courses in geometry and probability and statistics to meet our recommendations for teachers. *We regard it as a matter of great importance that a program for teachers should be identical to the one offered to other mathematics majors, except for a few courses peculiarly appropriate to prospective high school teachers.*

The implication in the above passage would seem to be that geometry is an appropriate area for the professionalization of subject matter.

The AAAS/NASDTEC and CUPM guidelines provide a reasonable framework within which to design a sequence of courses in geometry for teachers. Both recognize the dual need to prepare the teacher both to teach contemporary programs and to adapt to new programs as they evolve. The nature of contemporary school mathematics programs imposes natural constraints on a college course designed to meet the first goal. It is less likely that a unique content emphasis is necessary to accomplish the latter task. Any one of several content patterns may be equally effective in developing flexibility and adaptability in the teacher. This yearbook and such recent reports as that of the Comprehensive School Mathematics Program's Second International Conference on the Teaching of Mathematics illustrate a range of topics and treatments that might be appropriate for this purpose. (In fact, these sources in themselves provide appropriate text material for courses in geometry for teachers.) In the final analysis, the specific content or approach chosen may be less important than the spirit in which it is taught and learned.

Many people have recently called (again!) for a combination of subject matter and teaching methods in college mathematics courses for teachers. The basic intent of this recommendation might be accomplished if college instructors of mathematics would simply exhibit appropriate methods *in their own teaching.* The old adage that "teachers are more apt to teach as they are taught than as they are told to teach" has never, to my knowledge, been proved completely invalid. Consciously or unconsciously, college instructors do provide models for prospective teachers. When the models are appropriate, specific reference to teaching methods in the college mathematics course may be unnecessary. Moreover, the methods to which one would like prospective teachers exposed are quite likely appropriate for all undergraduate students.

The above paragraph is not intended to minimize either the need to relate the content of college courses to the secondary school curriculum or the need to integrate instruction in content and methodology. These are critical needs. Where qualified instructors are available, this should be a major focus in mathematics courses designed specifically for teachers. Departments of mathematics and of mathematics education should cooperate extensively in providing methods seminars related to the mathematical content of the college course. Instructors of mathematics might find they could make their courses more relevant for the prospective teacher if they had firsthand awareness of the secondary school curriculum and the problems of secondary school teachers. They might gain such an awareness by visiting secondary schools, participating in methods seminars, or by working directly with prospective teachers in the field-experience phase of their program.

The education of secondary school teachers in geometry should not be thought of as residing solely in the sequence of geometry courses. The undergraduate program in mathematics should have an overall integrity that makes the whole greater than the sum of its parts. Other courses should carry part of the burden for education in geometry. This is particularly true of courses in the algebra sequence in view of current recommendations for secondary school geometry. The topics of vector spaces, linear transformations, linear systems, and vector geometry are commonly treated in linear algebra courses. These topics are clearly related to major curriculum proposals for secondary school geometry, but unfortunately, there is little evidence that these topics have been taught with the view toward influencing the prospective teacher of geometry. Seldom is their relation to evolving programs in the secondary school made clear to prospective teachers. Similarly, the course in modern algebra provides an opportunity to treat the group structure in the context of geometric transformations. Other courses, particularly history of mathematics and foundations of mathematics, provide unique opportunities to enlarge the basis of geometric education for teachers.

Geometry should permeate the school curriculum—it should become a way of thinking. Teachers can make this happen, but only if geometry has not been treated as an isolated area in their own mathematical preparation. Making use of other areas of mathematics in the study of geometry is a step in the right direction, and likewise it is equally important that the concepts and techniques learned in geometry be applied in other mathematical contexts. Teachers need to see the payoff of instruction in geometry. They need to appreciate the fact that the theorems and principles of secondary school geometry have wide application in the study of more advanced areas of mathematics. They need to experience geometry as a problem-solving tool. They need to recognize a geometric basis or motivation for certain concepts in algebra and analysis. The undergraduate curriculum in mathematics provides natural opportunities for such experiences. Too often these opportunities are not consciously exploited. If the teacher does not see geometry as related to other areas of mathematics, any move toward an integrated program in the schools will have limited success. Educators must share the guilt of isolating geometry: this is evidenced in the treatment of geometry in methods textbooks as well as in the school curriculum.

Reference has already been made to the need to be concerned about the spirit in which mathematics is presented to prospective teachers. Teaching methods that generate active involvement in the development of new mathematical ideas are important for all students, but they are especially important for prospective teachers. Departures from the stereotyped lecture methods often associated with college instruction should be frequent. The less expository, more discovery-oriented methods commonly recommended to secondary teachers provide better models for them to emulate. Participation in mathematics-laboratory activities in their training may lead them to experiment with such methods in their own teaching. Who would argue that a passive learner is not likely to become a passive teacher? However, the concern with the spirit in which mathematics is presented goes beyond the teaching methods employed. It also involves the spirit or nature of mathematics that is projected and the methods of mathematical investigation that are used.

Too many secondary school teachers make a sharp and artificial distinction between formal and informal geometry. Somehow, when the student reaches that magical plateau of tenth-grade geometry, informal, or intuitive, approaches are seen as beneath his dignity. Could it be that the teacher's own experience at the college level suggests that when one *really* does mathematics, he discards intuitive models and informal techniques? The content and methods of informal geometry should be considered a legitimate part of the prospective teacher's preparation in geometry. Mathematics departments are abdicating their responsibility if they leave this area entirely to education courses. Informal geometry is, after all, taught by more secondary school teachers than formal, axiomatic geometry. In the final analysis, informal geometry should be characterized by its methods rather than by its content. Almost any mathematical content at any level can be approached informally, and the methods of such an approach are a necessary part of the secondary school teacher's arsenal. It is unfortunate that these methods are not more widely demonstrated at the college level.

A teaching career may span thirty to forty years. The ultimate test of a teacher's ability to adapt to new curriculum developments over such a time span may lie in the extent to which he has developed the inclination and ability for continued self-education in mathematics. Experiences with inquiry-oriented approaches to mathematics at the college level should contribute to this goal. The prospective teacher may not need to become a researcher in mathematics in the sense of discovering or inventing new mathematics, but he does need to develop the skill to "seek out and study concepts that are new to him, and then to synthesize written and especially oral expositions of them designed for others for whom those ideas are also new" (1, p. 32). Is that not the challenge facing any teacher as he approaches a new course or curriculum? Are we providing opportunities for such activity in the context of preservice mathematics courses?

One final observation must be made. Programs for the education of secondary school teachers in geometry should be only one phase of an overall professional move to put to rest, once and for all, both of the following recurring myths in teacher education: that anyone who can teach can teach anything effectively and that anyone who knows his subject matter can teach it effectively.

REFERENCES

1. American Association for the Advancement of Science. *Guidelines and Standards for the Education of Secondary School Teachers of Science and Mathematics.* AAAS Miscellaneous Publication 71–9. Washington, D.C.: The Association, 1971.

2. Birkhoff, George David, and Ralph Beatley. *Basic Geometry.* Chicago: Scott, Foresman & Co., 1941.

3. ———. "A New Approach to Elementary Geometry." In *The Teaching of Geometry.* Fifth Yearbook of the National Council of Teachers of Mathematics. New York: Bureau of Publications, Teachers College, Columbia University, 1930.

4. Brown, Kenneth E., and Ellsworth S. Obourn. *Qualifications and Teaching Loads of Mathematics and Science Teachers.* U.S. Office of Education Circular no. 575. Washington, D.C.: Government Printing Office, 1959.

5. Christofferson, Halbert Carl. *Geometry Professionalized for Teachers.* Oxford, Ohio: The Author, 1933.

6. College Entrance Examination Board, Commission on Mathematics. *Program for College Preparatory Mathematics.* New York: The Board, 1959.

7. Commission on Post-War Plans of the National Council of Teachers of Mathematics. "Second Report of the Commission on Post-War Plans." *Mathematics Teacher* 38 (May 1945): 195–221.

8. Commission on the Training and Utilization of Advanced Students in Mathematics. "Report on the Training of Teachers of Mathematics." *American Mathematical Monthly* 42 (March 1935): 263–77.

9. Committee on the Undergraduate Program in Mathematics of the Mathematical Association of America, Panel on Teacher Training. *Recommendations for the Training of Teachers of Mathematics.* Berkeley, Calif.: The Committee, 1961. Rev. ed., 1966. This also appeared in *Mathematics Teacher* 53 (December 1960): 632–38, 643.

10. ———. *Recommendations on Course Content for the Training of Teachers of Mathematics.* Berkeley, Calif.: The Committee, 1971.

11. Dubisch, Roy. "Teacher Education." In *Mathematics Education,* pp. 285–310. Sixty-ninth Yearbook of the National Society for the Study of Education, pt. 1. Chicago: University of Chicago Press, 1970.

12. Fehr, Howard F. "How Much Mathematics Should Teachers Know?" *Mathematics Teacher* 52 (April 1959): 299–300.

13. Ford, Patrick L. "Mathematics Included in Programs for the Education of Teachers in the Southern Association." Doctoral dissertation, University of Missouri, 1962. Abstract in *Dissertation Abstracts* 23: 543.

14. Joint Commission of the Mathematical Association of America and the National Council of Teachers of Mathematics. *The Place of Mathematics in Secondary Education.* Fifteenth Yearbook of the National Council of Teachers of Mathematics. New York: Bureau of Publications, Teachers College, Columbia University, 1940.

15. Lindquist, Clarence. *Mathematics in Colleges and Universities.* U.S. Office of Education Circular no. 765 (DE-56018). Washington, D.C.: Government Printing Office, 1965.

16. Longley, W. R. "What Shall We Teach in Geometry?" In *The Teaching of Geometry.* Fifth Yearbook of the National Council of Teachers of Mathematics. New York: Bureau of Publications, Teachers College, Columbia University, 1930.

17. Meserve, Bruce E. "Modern Geometry for Teachers." *School Science and Mathematics* 58 (June 1958): 437–41.

18. Moise, Edwin E. *Elementary Geometry from an Advanced Standpoint.* Reading, Mass.: Addison-Wesley Publishing Co., 1963.

19. National Council of Teachers of Mathematics. *A History of Mathematics Education in the United States and Canada.* Thirty-second Yearbook. Washington, D.C.: The Council, 1970.

20. ———. *The Teaching of Geometry.* Fifth Yearbook. New York: Bureau of Publications, Teachers College, Columbia University, 1930.

21. Schumaker, John A. "Trends in the Education of Secondary-School Mathematics Teachers." *Mathematics Teacher* 54 (October 1961): 413–22.

22. Secondary-School Curriculum Committee of the National Council of Teachers of Mathematics. "The Secondary Mathematics Curriculum." *Mathematics Teacher* 52 (May 1959): 389–417.

23. Smith, Lehi T. "Curricula for Education of Teachers." *American Mathematical Monthly* 70 (February 1963): 202–3.

24. Swenson, John A. "Graphic Methods of Teaching Congruence in Geometry." In *The Teaching of Geometry.* Fifth Yearbook of the National Council of Teachers of Mathematics. New York: Bureau of Publications, Teachers College, Columbia University, 1930.

25. Wong, Ruth Eiko Murashige. "Status and Direction of Geometry for Teachers." Doctoral dissertation, University of Michigan, 1968. Abstract in *Dissertation Abstracts* 69: 2410.

26. Young, Gail S. "The NASDTEC-AAAS Teacher Education and Certification Study." *American Mathematical Monthly* 67 (October 1960): 792–97.

Index

Active direct experience. *See* Geometry
Acute angle, 281
Affine approach to Euclidean geometry, 201
  advantages of, 216, 230–31
  aids for teachers of, 230–31
  grade placement of, 217
  student prerequisites for, 217
  suitability of, 217
Affine function, 217
  properties of, 218
  relation of, to affinities, 220–21
Affine geometry. *See* Geometry
Affine line, 218, 326, 327. *See also* Line
  axioms for, 218
  equation of, 225
  example of, 219
  coordinate systems on, 220
Affine plane, 201, 222, 326, 327. *See also* Plane
  axioms for, 222, 224
  coordinate systems in, 206–7, 222, 226
  finite example of, 202, 203
    axioms for, 202
  infinite example of, 205–10
  limitations of, 210, 226
  relation of, to Euclidean plane, 210, 212, 227, 228
  relation of, to projective plane, 201–3
Affine spaces, 326–27
Affine theorems, 210, 212, 226
Affine transformation, 391, 392
Affinity, 220
  image of ray under an, 221
  image of segment under an, 221
  properties of, 221–22
  relation of, to affine function, 220, 221
  relation of, to parallel projection, 224
Albert, Adrian A., 382
Alexander, James W., 418
Algebra
  linear, 379, 456, 459
  structures, 373, 374
Analytic geometry. *See* Geometry
Analytical thinking, 33
Anderson, Richard D., 424
Angle, 120, 253–54, 312, 313
  acute, 281

congruent, 120
corresponding, 313
dihedral, 121
interior postulate, 120
measure of, 120, 313
  additivity of, 122, 125
obtuse, 281
right, 281
Angle-side-angle. *See* ASA
Approaches. *See specific headings*
Arc, 295
  degree measure of, 122
  length, 408–12
  calculus formula for, 408
  path, 408, 411
  path for the helix, 408–9
Archimedes, axiom of, 110
Area, 319
  in elementary methodology, 443
  postulate, 122
Aristotelian definition, 382, 383
Artin, Emil, 376–77, 385, 391
ASA
  postulate, 121
  theorem, 290
Association of Teachers of Mathematics in New England, 55
Axiomatic
  foundations in school geometry, 427, 428
  locally, 386, 393, 394
Axiomatics, 315, 329–31, 384–85, 386, 389, 393–95, 452, 456
Axioms
  Moise, 111, 114
  ruler and protractor, 372–73

Ball, 401
  closed unit, 402
  in a metric space, 402
  open, 401
  open unit, 402
Basis, 262
  orthonormal, 278
Beberman, Max, 71
Betweenness, 220
Betz, William, 14, 18, 57
Birkhoff, George D., 372

Boundary of a simplex, 417
Bourbaki, 372
Brumfiel, Charles F., 444
Bypass postulate, 243

Calculus, 448, 450–51
  and geometry, 405–12
  and path, 405–12
Cambridge Conference on School Mathematics, 22
Cambridge Conference on Teacher Training, 22, 436, 437, 444
Cambridge report, 59
Cartan, Henri, 377–78
Cartesian coordinate system, 262
Cauchy-Schwarz inequality, 277
Ceva. *See* Theorems
Change
  forces for, in school curricula, 397–98, 430–31
  of parameter, 407, 409
  recommendations for, 361–63
Chord, 295
Circle, 214, 230, 294–95, 318, 328
  arc, length of, 230
  area of, 230, 319
  center of, 294
  chord of, 295
  circumference of, 319
  curvature of, 410
  radius of, 294
  unit, 400
  unknotted, 404, 423
Circular functions, 323, 325
City-block metric, 400
Closed
  ball, 402
  interval [0,1], 402
  path, 403
  set, 424
  subsets of the interval [0,1], 424–25
  unit, 402
Coleman, Robert, Jr., 15
College Entrance Examination Board (CEEB), 58–59
  Commission on Mathematics, 116
College geometry, course in, 450, 453, 454
Collinear, 251
Collinearity, 400
Combinatorial
  geometry, 452
  Schoenflies problem, 418
  topology, 413–18
Combinatorics, 382, 386, 390
Commission on Mathematics, CEEB, 116
Commission on Post-War Plans of the NCTM, 58
Committee of Fifteen, 18, 54–55
Committee of Ten on Secondary School Studies, 54

Committee on College Entrance Requirements, 54
Committee on Geometry, 57
Committee on School Mathematics, 59, 60, 70–71, 80
Committee on the Undergraduate Program in Mathematics (CUPM), 438–39, 444, 452, 454, 455
Commutative group, 244
Compact set, 424
Complex, 414–18
  complete five point, 414, 417
  Klein bottle, 416
  Kuratowski's construction theorem, 414
  1-complex, 413–14
  theorem of Flores and Van Kampen, 416–17
  torus, 416
  2-complex, 415–18
  unsolved problem, 417
Components
  of **a** with respect to **b**, 268
  of a vector, 262
Composition, 162
  composite, 162, 176
  definition of, 141
  product, 141
  of transformation, 185
Computers
  as a force of change, 430
  use of, to portray dynamic aspects of geometry, 430–31
Conchyliometry, 391
Cone, 324
Congruence, 120, 229, 441, 442
  definition of, 318
  methodology of, in elementary education, 442
  properties of, 442
Congruent
  angles, 120, 168–69
  figures, 163–65, 293
    definition of, 163
  paths in R^3, 412
  segments, 119, 168–69
  triangles, 120, 171–80 passim, 306–7
Conics, 328
  in an augmented Euclidean plane, 358
  as loci, 348
  paper-folding constructions of, 358, fig. 10.18
  relation to "line at infinity," 215
  as sections of a cone, 358
  using Pascal's theorem to construct, 359, fig. 10.19
Constant curvature, 410
Constructions, 440
  of conics
    paper folding, 358, fig. 10.18
    ruler and compass, 383
    using Pascal's theorem, 359, fig. 10.19

Contemporary geometry course, 453, 454
Content
  course for elementary teachers, 436, 439–43
  of elementary school curriculum, 23
Continuous
  function, 402
  path, 406
Convex set, 119
Convexity, 452, 455
Coolidge, Julian L., 382
Coordinate geometry. *See* Geometry
Coordinates
  approaches using, 116
  function, 406
  in proofs, 129–31, 347, 351, 352
  in textbooks, 132–34
  and transformations, 352
  as used by Descartes, 347
  and vectors, 354–56
Coordinate systems, 310, 312, 313, 316
  affine, 316
  on a line, 119, 205–6, 210, 223, 316
  in a plane, 206–7, 222, 227, 310, 312, 316, 347–50
  polar, 328
  rectangular, 317
  relation of, to affinities, 220
  in space, 324
  three-dimensional, 358
Coplanar, 251
Cosine, 280
  of an angle, 280
  function of, 214, 229
  law of, 214, 228, 282
Coxeter, H. S. M., 382, 391, 392
Coxford, Arthur, 80
Cube, Hilbert, 424–25
Curricula
  college courses, 450, 453, 454
  forces of change in, 397–98, 430–31
  guidelines, 40–41, 48–50
  implementation of new programs, 361–63
  integrated, 427–31
  proposed elementary course, 439–43
  recommendations for teacher education, 363–65
  recommended program in geometry, 341–61
  role of geometry in future, 426–31
  as secondary school subject, 369
  traditional, 427, 428
Curriculum revision, 20
Curvature, 410–12
  of a circle, 410
  constant, 410
  function, 410
  as a geometric property, 412
  of a helix, 410
  of a line, 409, 410

  of a path, 410–12
  and torsion, 412
  zero, 410
Curve
  curvature of, 410
  as an equivalence class of paths, 407
  Jordan theorem, 412
  oriented, 408
  planar, 411
  regular, 409, 410
  represented by arc-length path, 408–9, 411
  torsion of, 411
  of zero torsion, 412
Curvilinear motion, 348, fig. 10.11
Cylinder, 324

Definition, Aristotelian, 382, 383
Definitions, SMSG-GW, 118–22
Dennis, J. Richard, 72
Desargues's, theorem of, 226, 344, fig. 10.2
Descartes, René, 96, 304, 374
Descriptive geometry, 450
Determinant, 264
  third-order, 265
Dienes, Zoltan P., 387
Dieudonné, Jean, 369
Differentiable
  parameter change, 409
  path, 409
Differential geometry, 409, 450–53, 455
Dihedral angles, 121
Dilatation, 136, 147–48, 391
Dilation, 136, 137, 175, 310, 311, 318, 325
  definition of, 147
  properties of, 148
Dilative reflection, 193–95
  definition of, 194
Dimension, 260
Direction, 252
Direction for future programs, 41
Direction numbers
  of a line, 262
  of a plane, 263
Dissection, 386
Distance, 118–19, 220
  between closed subsets of [0,1], 425
  in a Euclidean space, 399
  formula for, 213
  function, 227, 425. *See also* Metric
    defined, 400
    for Euclidean space, 400
    for Hilbert space, 424
  from P to Q, 273
  ratio of, on same line, 210
Distance-preserving function, 399–400
Dot product, 266
Dubrovnik seminar, 373–74
Dynamic aspects of geometry, 430–31

Education
  elementary textbooks, 23, 444
  in-service teacher training, 21
  geometry for elementary teachers, 435–45
  preservice teacher training, 21
  proposed elementary geometry course, 439–43
  recommendations for immediate change, 361–63
  recommendations for teacher education, 363–65
Elementary teachers' education, 435–45
  content of, 436
  course organizers of, 436
  in-service, 21
  language in, 437–38, 440
  methodology, 435, 437
    networks in, 443
    symbolism in, 437–38
    transformational geometry in, 441, 444
    translation in, 441, 442
  pedagogical considerations of, 435
  preservice, 21
  structure of, 438–39
  textbooks, 23, 444
Elements. See Euclid
Elliott, H. A., 444
Elliptic geometry. See Geometry
Empty set. See Set
Enlargement, 136
Equidecomposable, 387
Equivalence
  of differentiable paths, 409
  of paths, 407
Erlanger Programm, 352, 365, 374
Euclid, 96, 97, 98, 104, 117, 369–70, 398, 427
  Elements, 96, 97, 105, 334
  geometry in the style of, 398, 426
  parallel postulate, 105
  postulates, 104–5, 109
Euclidean geometry, 20, 43, 97, 117, 447, 449–55 passim, 456
  defined, 400
  in future school curricula, 397–98, 426–31
  modern topics in, 343–47
  objectives of a school program in, 338–39
  plane, 342–43
  three-dimensional, 357–58
  traditional, 426
  weaknesses of a school program in, 336–38
Euclidean plane, relation of, to affine plane, 201, 212, 227, 228
  adjunction of ideal points to, 204
Euclidean space, 327, 369, 382, 386, 393, 394, 398–400. See also Sphere

  geometry of, 400
  metric for a subset of, 402
  paths in, 405–6
  topology of, 403
  vector of, 399
Euclidean structure theorem, 228
Eudoxus, 97, 110
Euler's formula, 393
Extension, 419
Extension problem
  defined, 419
  relation to fixed-point property, 420
  solution by algebra, 420–23

Fermat, Pierre de, 304
Field(s), 383
  concerned with geometry, 398, 424, 425–26
Finite geometry, 72–73
Fixed point, 420
  property, 420
Fletcher, T. J., 444
Flores, H. I., 416, 417
Formulaire group, 372
Four-color conjecture, 382
Fox-Artin 2-sphere, 404, 418, 423
Fractional numbers, 35
Freudenthal, Hans, 386
Froebel, Friedrich, 15
Function
  circular, 323, 325
  continuous, 402
  coordinate, 406
  curvature, 410
  distance, 227, 400, 425
  distance-preserving, 399–400
  in geometry, 308
  inclusion, 419
  notation for, 406
  torsion, 411–12
Fundamental group, 418

Geoboards, 32, 33
Geometrization, 388–89, 394
Geometry
  active direct experience in, 39
  activities in, 39, 440–44
  aesthetic value of, 39
  affine, 244, 315–17, 375, 453, 456
  affine axioms, 315–16
  in American education, 15
  analytic, 97, 448–52 passim
  approaches using coordinates, 116
  axiomatic foundations of, 426–27, 428
  and calculus, 505–12
  college courses in, 450, 453, 454
  combinatorial, 452
  constructions in, 440
  coordinate, 27, 378, 452, 454–56
  descriptive, 450

differential, 409, 450–53, 455
dynamic aspects of, 430–31
education for elementary teachers, 435–45
as an elementary school subject, 12
elliptic, 371
Euclidean, 369, 382, 386, 393–94, 398, 400, 426, 427. *See also* Euclidean geometry
of Euclidean space, 369, 382, 393, 394, 398–400
in European education, 14
examples of, in the physical world, 30
finite, 72–73
functions in, 308
in future school curricula, 397–98, 426–31
homeomorphism, 402–5
hyperbolic, 371
informal, 449, 460
mathematical fields concerned with, 398, 424, 425–26
metric, 27, 449, 451–53 passim
motion, 27, 43–46
non-Euclidean, 450–53, 456
nonmetric, 11
perspective of, 24–25, 38
physical, 378
point-line-plane, 26
projective, 382–84, 392, 398, 426, 449–53 passim, 456
proposed elementary course in, 439–43
rational-point, 351
recent developments in, 19
recommendations for school program in, 341–61
recommendations for teacher education, 363–65
as a secondary school subject, 369
solid, 26, 447, 448, 450
space, 323, 324
structure of, 438, 439
synthetic, 97, 378, 449, 451–52, 456
as taught by British, 437, 444
textbooks, 444
thirty-six-point, 72
traditional, 426 28
transformational, 27, 33, 43–46, 441, 444
transformations, 79, 352–53, 378, 391, 393, 452, 454, 455, 456, 459
as a unifying theme of mathematics, 34
vector, 379
Geometry with Coordinates (GWC), 116
Gilbert, Charles H., 54
Glass, James M., 57
Glide reflection, 136, 137
axis, 147
definition of, 144
in elementary methodology, 441–42
Graph, 1-complex, 413–14

Graphing, 37
Group, 150, 151, 178, 317, 330, 382, 383, 387
of transformations, 187–95, 371–72
definition of, 187
of isometries, 189, 317–18
fundamental, 418
homotopy, 421–24
symmetry, 187
Gugle, Marie, 57
Guidelines, geometry program, 40–41, 48–50

Half-line, 312
Half-turn, 140, 159, 176
Hall, Marshall, 382
Hanus, Paul H., 53
Helix, 406–7
arc-length path for, 408–9
curvature of, 409–10
Herbart, Johann, 14
Hilbert, David, 96, 109, 111, 372
cube, 424–25
Foundations of Geometry, 109, 114
postulates for geometry, 104, 109
space, 424–26
Hill, Thomas, 15, 53
Historical background on elementary school geometry, 12
Homeomorphism
defined, 402–3
as a geometric property, 403
role in geometry, 403–5
Homotopy theory, 421–23
Hyperbolic geometry. *See* Geometry

Ideal line, 201, 202, 204, 209
relation of parallel lines to, 201, 204
Ideal points, 204, 206, 209
adjunction of, to Euclidean plane, 204
Identity, 141, 159, 177
matrix, 182
translation, 244
Immerzeel, George, 72
Implementation of new programs, 361–63
Inclusion function, 419
Informal geometry, 449, 460
aesthetic value of, 39
Inner product, 266
In-service teacher training, 21
Integrated curricula, 427–31
Interior of an angle postulate, 120
Intersection of plane with closed surfaces, 39
Interval
closed unit, 402
open unit, 402
space of closed subsets of the closed unit interval [0,1], 424–25

Invariant (invariance), 148, 150, 154, 155, 312, 313, 389–92
Inverse, 177
  definition of, 178
  of matrix, 182
Isometry, 151, 229, 291, 312, 314, 455
  definition of, 138, 163
  group of, 189, 317–18
Isosceles triangle, 313, 318

Jeger, Max, 444
Joint Commission of the MAA and the NCTM, 57–58
Jordan curve theorem, 412
Jordan, Janet M., 444

Kite, 167
Klamkin, M. S., 386, 388, 390, 391
Klee, Victor L., 424
Klein bottle, 416
Klein, Felix, 304–5, 371, 375–76
Knot, 404, 423
Kuratowski, Casimir, 414, 417

Lacroix, S. F., 12
Languages in courses for elementary teachers, 437–38, 440
Lattice points, 310
Law of cosines, 214, 228, 282
Law of sines, 285
Lehner, D. N., 392
Leisenring, Kenneth, 431
Length, 119, 319
  of a path, 408–12
  as preserved by motions, 400
  of segment, 312
  of vector, 399
Line, 118, 251
  affine, 218, 326, 327
    equation of, 225
    ideal, 201, 204
    "at infinity," 201
    slope of, 209, 225, 228
    standard forms of, 225
  coordinate system on a, 119
  curvature of, 409, 410
  ideal, 201, 202, 209
  parallel, 121, 169–71
  parametric equations for, 262
  and plane parallel, 121
  and plane perpendicular, 121
  reflection in a, 311, 312, 313, 314
  vector, 326, 327
"Line at infinity," 201, 202
  relation of conics to, 215
Linear
  algebra, 379, 456, 459
  parametric equations, 126–27
  vector space, 201

Linearly dependent, 250
Linearly independent, 250
Locus, 131

MacLean, James R., 444
Magnitude
  of rotation, 139–40, 141–42
  of size transformation, 175–77
Mapping, 310
  kernel of linear, 328
  linear, 328
  range of linear, 328
Marks, Bernhard, 53
Mass point, 319–23
Mathematical Association of America (MAA), 56, 57, 58
  Committee on the Undergraduate Program in Mathematics (CUPM), 438–39, 444, 452, 454, 455
Mathematics Teacher, 56
Matrices, 180–87, 195, 323, 325
  composition of transformations, 184, 185
  definition of, 181
  identity of, 182
  inverse of, 182
  to represent transformations, 182–87
Measure of an angle, 120, 313
Measurement, 35–36
Median of a triangle, 257, 320
Menelaus, theorem of, 226, 344, fig. 10.1
Methodology
  elementary textbooks, 444
  glide reflections, 441–42
  in-service teacher training, 21
  networks in elementary, 443
  perspectives in classroom use of geometry, 24–25, 38
  reflection in elementary, 440–42
  rotation in elementary, 441
  symbolism in elementary, 437–38
  teaching elementary school geometry, 435, 437
  in transformational geometry, 441, 444
  translations in, 441, 442
  volume in elementary, 443
Metric. See also Distance function
  city-block, 400
  defined, 400
  for a subset of a Euclidean space, 402
  for a subset of a metric space, 418–19
  for the set of closed subsets of the closed unit interval [0,1], 425
Metric geometry, 27, 449, 451–53 passim
Metric space
  defined, 400
  open subset of, 401
Midpoint of a segment, 257
Minkowski, Hermann, 426

Mirror, 138
Moise, Edwin E., 104, 110, 418
  axioms, 111, 114
Moore, E. H., 17
Motion
  defined, 400
  orientation-preserving, 410
  preserving length, 400
  rigid, 400
Motion geometry, 27, 43–46
Multiples
  of a, 246
  scalar, of a vector, 399
Multiplication, scalar, 247

National Committee of Fifteen on Geometry Syllabus, 18, 54–55
National Committee on Mathematical Requirements, 56–57
National Council of Teachers of Mathematics (NCTM), 56, 58
  Commission on Post-War Plans, 58
  Committee on Geometry, 57
  Secondary School Curriculum Committee, 58
National Education Association (NEA), 54
  Committee of Fifteen on Geometry, 18, 54–55
  Committee of Ten on Secondary School Studies, 54
  Committee on College Entrance Requirements, 54
n-dimensional, 260
Networks in elementary methodology, 443
Newman, D. J., 392
Non-Euclidean geometry, 450–53, 456
Nonmetric geometry, 11
Norm of a, 273
Number line, 35
  product of real, 424
Numbers, fractional, 35

Obtuse angle, 281
O'Hara, C. W., 382
1-complex, 413–14
One-dimensional, 260
Open
  ball, 401, 402
  set, 401–5
  unit interval, 402
Ore, Oystein, 382
Orientation, 139, 141, 144, 148, 174
  of a curve, 408
  preserving motion, 410
Oriented
  curve, 408
  regular curve, 410
Origin of a coordinate system on a line, 119

Orthogonal, 270–71
Orthonormal basis, 278
Overpostulation, 122, 123

Pappus's theorem, 383
Parabola, 400–401
Parallel, 121, 169–71
  line and plane, 121
  lines, 222, 253, 312–13, 326
    equality of slopes of, 209, 225
    equidistance of, 213
    relation of, to ideal line, 201, 204
  planes, 253, 324, 326
  postulate, 105, 170, 313
  projection, example of, 224
    relation of, to affinities, 224
    relation of, to distance, 227
Parallelogram, 128, 258–59, 313, 316–17, 318, 319, 321, 326
Parameter change
  defined, 407
  differentiable, 409
Parametric equations, 348
  for a line, 262
  linear, 126–27
  for a plane, 263
Pascal, Blaise, 392
Pascal's theorem
  for circles, 345, figs. 10.5–10.7
  for conics, 358, fig. 10.19
  used to construct conics, 358, fig. 10.19
Pasch, Moritz, 372
Path
  arc-length, 408–12
  and calculus, 405–12
  closed, 403
  congruent in R^3, 412
  of constant curvature, 410
  continuous, 406
  curvature of, 410–12
  differentiable, 409
  equivalence, 407
  equivalence of differentiable paths, 409
  in Euclidean space, 405–6
  for helix, 408–9
  length, 408–12
  simple, 403
  torsion of, 412
Perimeters in elementary methodology, 443
Perpendicular, 121, 128, 313, 314
  foot of, 266
  line and plane, 121
  vectors, 399
Perpendicularity, 212, 227, 270
Perspectives on the classroom use of geometry, 24–25, 38
Pestalozzi, J. H., 14, 15
Phyllotaxis, 391
Piaget, Jean, 390
Pick's theorem, 443

Plane, 118, 251, 324, 326, 327
  intersection with closed surfaces, 39
  and line parallel, 121
  and line perpendicular, 121
  parametric equations for, 263
  projective, 201
  reflection in a, 291
  separation, 312
  transformations of, 309–10, 311, 317–
    18, 324
  vector, 326, 327
Plato, 384
Poincaré, Jules, 417
Point, 118
  of division of a segment, 131, 256
  of Euclidean *n*-space, 399
  fixed, 420
  ideal, 204, 206, 209
  lattice, 310
  mass, 319–23
  reflection in a, 311, 312, 314
Point-line-plane geometry, 26
Polyhedral 2-sphere, 418
Postulates
  area, 122
  ASA, 121
  bypass, 243
  Euclid's, 104–5, 109
  Hilbert, 104, 109
  interior of an angle, 120
  parallel, 105, 170, 313
  proportional segments, 121
  protractor, 120
  reflection, 161–62
  ruler, 119
  SAS, 120
  separation, 119
  SMSG, 118–22
  SSS, 121
Preservice teacher training, 21
Problem solving, 390, 392, 394
Product, 141, 176
  definition of, 141
  inner, 266
  of real number lines, 424
  scalar, 399
  of unit intervals, 424
Professionalized subject matter, 448, 450,
    457
Projection, 391, 392
  of **a** on the direction of *l*, 267
  parallel, 224
  of *P* on *l*, 266
  theorem, 284
Projective geometry, 382–84, 392, 398,
    426, 449–53 passim, 456
Projective plane, 201
  as augmented Euclidean plane, 351
  finite, 350, fig. 10.12
  finite example of, 202
    axioms for, 202
  relation of, to affine plane, 201, 202

Proof, 164–69, 174, 315, 329–31
  using coordinates, 347–48
  using transformations, 352, fig. 10.13
  using vectors, fig. 10.15, fig. 10.16
Proportional segments postulate, 121
Protractor
  axiom, 372–73
  postulate, 120
Psychology of the learner, 14
Pythagoras, theorem of, 213, 228, 273
Pythagorean property, 317

Quadrangle theorem, 345–47, fig. 10.8

Radius, 294
Ratio, 256
Rational-point geometry, 351
Ray, 220, 253
  image of, under an affinity, 221
  zero, 120
Ray-coordinate, 120
  system, 120
Real numbers as operators, 247
Recommendations
  for immediate change, 361–63
  for a school program in geometry, 341–
    61
  for teacher education, 363–65
Reflection, 136, 137, 142, 155, 160, 229,
    387, 390, 391
  definition of, 137–38
  in elementary methodology, 440–42
  glide, 136, 137, 144, 441–42
  in a line, 311, 312, 313, 314
  in plane π, 291
  in a point, 311, 312, 314
  postulate, 161–62
  properties of, 138–39, 440–41
Regions, 35
Regular curve, 409, 410
Report of the Committee of Ten on Sec-
    ondary School Studies, 16
Report of the NEA Committee of Ele-
    mentary Education, 16
Riemann, Bernhard, 371–72
Right angle, 281
Rigid motion, 400. *See also* Motion
Rosenbloom, P. C., 387
Rotation, 136, 137, 311, 325
  definition of, 139–40, 163
  in elementary methodology, 441
  magnitude of, 139–40, 141–42
  properties of, 140–41
Ruler
  axiom, 372–73
  postulate, 119

Sanders, Walter J., 72
Sandler, R., 382

SAS
　postulate, 120
　theorem, 289
Scalar, 247
　multiple of a vector, 399
　multiplication, 247
　product, 399
　product of two vectors, 399
Schoenflies theorem, 403–4
　combinatorial, 418
　as an extension theorem, 419–20
　false generalizations of, 404
School curricula. *See* Curricular
School Mathematics Project (SMP), 437, 440
School Mathematics Study Group (SMSG), 20, 59, 100, 104, 112, 116, 439
　definitions, 118–22
　postulates, 118–22
Schori, Richard, 424, 425
Schuster, Seymour, 386, 388
Scott, Lloyd F., 72
Secondary School Curriculum Committee of the NCTM, 58
Secondary School Mathematics Curriculum Improvement Study (SSMCIS), 309
Segment, 220, 253–54, 312
　image of, under an affinity, 221
　length of, 312
　midpoint of, 257
　point of division of, 131, 256
Separation postulates, 119
Set
　closed, 424–25
　compact, 424
　convex, 119
　empty, 401
　open, 401–5
Shear, 185–86
　definition of, 185
　properties, 186
Shibli, Jabir, 57
Side-angle-side. *See* SAS
Side-side-side. *See* SSS
Similar figures, 178–80
　definition of, 178
　properties, 178–80
Similarity, 136, 137, 147, 175–80, 229, 455
　definition of, 147–48, 178
　group, 193
　properties, 148–49
　spiral, 193–95, 325
Similitude, 229
Simplex
　boundary of, 417
　geometric, 413
　standard, 413
Sine, 285

of an angle, 285
　law of, 285
Size transformation, 175–80
　definition of, 175–76
　magnitude of, 175–77
　properties of, 176–77
Slope of a line, 209, 225, 228
Solid geometry, 26, 447, 448, 450
Space, 118
　affine, 326–27
　of closed subsets of the closed unit interval [0,1], 424–25
　Euclidean, 327, 369, 382, 386, 393, 394, 398–400
　geometry, 323, 324
　Hilbert, 424–26
　linear vector, 201
　metric, 400
　vector, 327, 379, 382, 394
　　over the real numbers, 247
Sphere, 294–95
　center of, 294
　of dimension two, 402, 404–5, 418
　in Euclidean *n*-space, 413
　in Euclidean three-space, 402
　lumpy, 405
　in the plane, 402
　polyhedral, 418
　radius of, 294
　unit, 402
　wiggly, 404
Spiral similarity. *See* Similarity
SSS
　postulate, 121
　theorem, 289
Stretch, 186–87
　one-way, 186
　two-way, 186
Structure of geometry, 438, 439
Subgroup, 318
Surface, 415–16
Survey of textbooks, 23
Swenson, John A., 305–9
Symbolism in elementary methodology, 437–38
Symmetry, 162, 166–69, 391
　definition of, 162, 318
　group, 187
Synthetic geometry, 97, 378, 449, 451–52, 456

Tangent. *See* Vector
Teacher education. *See* Curricula; Education; Elementary teachers' education
Technology, 430
Tetrahedron, 322
Textbooks
　for elementary teachers, 444
　survey of, 23
Theorems
　affine, 210, 212, 226

Theorems (*continued*)
  ASA, 290
  Ceva's, 226, 429, 430
  Desargues's, 226, 344, fig. 10.2
  Euclidean structure, 228
  Flores and Van Kampen, 416–17
  formal proofs of, in elementary meth-
    odology, 438, 439, 442
  Jordan curve, 412
  Kuratowski's construction, 414
  Menelaus's, 226, 344, fig. 10.1
  Pappus's, 383
  Pascal's, 345, 358, figs. 10.5–10.7, fig.
    10.19
  Pick's, 443
  projection, 284
  Pythagoras's, 213, 228, 273
  quadrangle, 345–47, fig. 10.8
  SAS, 289
  Schoenflies, 403–4, 412
  SSS, 289
  two coordinate systems, 124
Thirty-six-point geometry, 72
Three-dimensional, 260
Topology, 353–54, 451, 452
  combinatorial, 413–18
  of Euclidean space, 403
Torsion
  and curvature, 412
  function, 411–12
  as a geometric property, 412
  of paths, 412
  zero, 411
Torus, 416
Traditional geometries, 426–28
Transformation, 79, 136, 137, 149–51,
    152, 154–55, 156, 160, 352–53, 378,
    391, 393, 452, 454, 455, 456, 459
  affine, 391, 392
  definition of, 162
  groups of, 187–92, 195, 371–72
  magnitude of size, 175–77
  matrix representation, 182–85
  of the plane, 309–10, 311, 317–18, 324
  proofs using, 352, fig. 10.13
Transformational geometry, 27, 33, 43–46
  in elementary methodology, 441, 444
Translation, 136, 137, 142, 144, 162, 229,
    237, 310, 313, 324
  definition of, 142, 162
  in elementary methodology, 441, 442
  identity, 244
  properties, 144, 241
  variables for, 241
  vector, 144
Trapezoid, 258, 318, 319
Triangle, 253–54, 319
  congruence of, 306–7
  inequality, 277
  isosceles, 313, 318
  median of, 257, 320

  sum of angle measures of, 313
Two coordinate systems theorem, 124
Two-dimensional, 260
2-Sphere
  Fox-Artin, 404–5, 418, 423
  lumpy, 405, 418
  polyhedral, 418
  unit, 404–5

Unit
  ball, 401
  circle, 400
  interval, 402, 424–25
  sphere, 401, 404
  vector, 268
Unit-pair, 118
Unit-point of a coordinate system on a
    line, 119
University of Illinois Committee on School
    Mathematics, 59, 60, 70–71, 80
Usiskin, Zalman, 80
Uspenskii, V. A., 390

Vance, Irvin E., 444
Van Kampen, Egbertus R., 416, 417
Veblen, Oswald, 376, 383
Vector, 144, 248, 315, 326, 354–57
  approach, 452, 454, 455, 456, 459
  of Euclidean space, 399
  geometry, 379
  length of, 399
  lines, 326, 327
  perpendicularity, 399
  planes, 326, 327
  proofs using, figs. 10.15–10.16
  scalar multiple of, 399
  scalar product of, 399
  spaces, 327, 379, 382, 394
    linear, 201
    over the real numbers, 247
  subspaces, 327
  tangent, 411
  unit, 268
Volume, 319
  in elementary methodology, 443

Ward, D. R. B., 382
Wesleyan Coordinate Geometry Group,
    216, 217, 230
West, James E., 424, 425
Whitehead, Alfred N., 392
Whitehead, J. H. C., 383
Whittaker, E. T., 384
Wiederanders, Don, 72
Wilder, R. L., 384

Yaglom, I. M., 444
Yale, Paul B., 391
Young, Gail, 439

Zero-ray, 120
Zero torsion, 411